Promoting Qualitative Research Methods for Critical Reflection and Change

Viktor Wang
University of Montana Western, USA

A volume in the Advances in Knowledge Acquisition, Transfer, and Management (AKATM) Book Series

IGI Global
PUBLISHER of TIMELY KNOWLEDGE

Published in the United States of America by
IGI Global
Information Science Reference (an imprint of IGI Global)
701 E. Chocolate Avenue
Hershey PA, USA 17033
Tel: 717-533-8845
Fax: 717-533-8661
E-mail: cust@igi-global.com
Web site: http://www.igi-global.com

Library of Congress Cataloging-in-Publication Data

Names: Wang, Viktor, 1966- editor.
Title: Promoting qualitative research methods for critical reflection and change / Viktor Wang, Editor.
Description: Hershey : Information Science Reference, [2021] | Includes bibliographical references and index. | Summary: "This book provide readers with a comprehensive array of qualitative research methods, which can be implemented in a variety of contexts for a variety of purposes and offers tools to uncover and facilitate critical reflection and analysis "-- Provided by publisher.
Identifiers: LCCN 2020047954 (print) | LCCN 2020047955 (ebook) | ISBN 9781799876007 (Hardcover) | ISBN 9781799876014 (eBook)
Subjects: LCSH: Qualitative research--Methodology.
Classification: LCC H62 .H24524 2021 (print) | LCC H62 (ebook) | DDC 001.4/2--dc23
LC record available at https://lccn.loc.gov/2020047954
LC ebook record available at https://lccn.loc.gov/2020047955

This book is published in the IGI Global book series Advances in Knowledge Acquisition, Transfer, and Management (AKATM) (ISSN: 2326-7607; eISSN: 2326-7615)

British Cataloguing in Publication Data
A Cataloguing in Publication record for this book is available from the British Library.

All work contributed to this book is new, previously-unpublished material.
The views expressed in this book are those of the authors, but not necessarily of the publisher.

For electronic access to this publication, please contact: eresources@igi-global.com.

Advances in Knowledge Acquisition, Transfer, and Management (AKATM) Book Series

ISSN:2326-7607
EISSN:2326-7615

Editor-in-Chief: Murray E. Jennex, San Diego State University, USA

MISSION

Organizations and businesses continue to utilize knowledge management practices in order to streamline processes and procedures. The emergence of web technologies has provided new methods of information usage and knowledge sharing.

The **Advances in Knowledge Acquisition, Transfer, and Management (AKATM) Book Series** brings together research on emerging technologies and their effect on information systems as well as the knowledge society. **AKATM** will provide researchers, students, practitioners, and industry leaders with research highlights surrounding the knowledge management discipline, including technology support issues and knowledge representation.

COVERAGE

- Cognitive Theories
- Cultural Impacts
- Information and Communication Systems
- Knowledge Acquisition and Transfer Processes
- Knowledge Management Strategy
- Knowledge Sharing
- Organizational Learning
- Organizational Memory
- Small and Medium Enterprises
- Virtual Communities

IGI Global is currently accepting manuscripts for publication within this series. To submit a proposal for a volume in this series, please contact our Acquisition Editors at Acquisitions@igi-global.com or visit: http://www.igi-global.com/publish/.

The Advances in Knowledge Acquisition, Transfer, and Management (AKATM) Book Series (ISSN 2326-7607) is published by IGI Global, 701 E. Chocolate Avenue, Hershey, PA 17033-1240, USA, www.igi-global.com. This series is composed of titles available for purchase individually; each title is edited to be contextually exclusive from any other title within the series. For pricing and ordering information please visit http://www.igi-global.com/book-series/advances-knowledge-acquisition-transfer-management/37159. Postmaster: Send all address changes to above address.

Titles in this Series

For a list of additional titles in this series, please visit: http://www.igi-global.com/book-series

Enhancing Academic Research and Higher Education With Knowledge Management Principles
Suzanne Zyngier (Zyngier Consulting, Australia)
Information Science Reference • © 2021 • 323pp • H/C (ISBN: 9781799857723) • US $195.00

Role of Information Science in a Complex Society
Elaine da Silva (São Paulo State University (UNESP), Brazil) and Marta Lígia Pomim Valentim (São Paulo State University (UNESP), Brazil)
Information Science Reference • © 2021 • 297pp • H/C (ISBN: 9781799865124) • US $195.00

Theoretical and Practical Approaches to Social Innovation
Chamindika Weerakoon (Swinburne University of Technology, Australia) and Adela McMurray (RMIT University, Australia)
Information Science Reference • © 2021 • 265pp • H/C (ISBN: 9781799845881) • US $195.00

Developing Knowledge Societies for Distinct Country Contexts
Nuno Vasco Lopes (University of Minho, Portugal) and Rehema Baguma (Makerere University, Uganda)
Information Science Reference • © 2021 • 296pp • H/C (ISBN: 9781522588733) • US $185.00

Approaches and Processes of Social Science Research
Icarbord Tshabangu (Leeds Trinity University, UK) Stefano Ba' (Leeds Trinity University, UK) and Silas Memory Madondo (CeDRE International Africa, Zimbabwe)
Information Science Reference • © 2021 • 260pp • H/C (ISBN: 9781799866220) • US $195.00

IGI Global
PUBLISHER of TIMELY KNOWLEDGE

701 East Chocolate Avenue, Hershey, PA 17033, USA
Tel: 717-533-8845 x100 • Fax: 717-533-8661
E-Mail: cust@igi-global.com • www.igi-global.com

Editorial Advisory Board

Table of Contents

Detailed Table of Contents

The disruptive effect of digital technologies is felt across almost all aspects of human endeavour. Research is no exception. The disruptive potential of digital technologies is realised through either doing existing activities differently or through the creation of entirely new activities. In research, digital disruption manifests as an expansion of possibilities for research, and as innovation in the tools and methodologies used to carry out research. In the present chapter, qualitative research is discussed with the aim of providing an overview of the influence of digital technologies on qualitative research for the purposes of stimulating further exploration and discussion.

As autoethnography and other methods for self-as-subject research continue to increase in use across doctoral education, this guide proposes to inform the methods in this area specific to critical autoethnography for doctoral scholars desiring to conduct the many forms of social justice research. This includes indigenous research, contemporary feminist research, and arts-based research, which have also seen rise across dissertation research among many disciplines. While many works exist to describe critical autoethnography within specific contexts, few research guides examine critical autoethnography specific for use by the doctoral scholar and specific examples across research focused on societal change and/or disruption of existing power dynamics, lack of parity, or historical trauma from acculturation and/or dispossession. Thus, this chapter offers a concise research methods resource for doctoral scholars and their research supervisors to facilitate use of critical autoethnography across disciplines and among diverse research problems of inquiry.

The case study is a widely used methodology among qualitative researchers irrespective of their philosophical orientation. While positivist and interpretive philosophies are the two most popular research philosophies across diverse research fields, critical realism offers a third alternative. Critical realism is a research philosophy that assumes the existence of an independent reality but also accepts that there may be varied interpretations of it due to a difference in context. Mechanisms are the theoretical building blocks of critical realism, and presence, absence, or interaction of certain mechanisms may result in the presence or absence of certain events. However, limited guidelines are available on conducting a critical realist case study. This chapter fills this gap by providing some practical guidelines on how a CR-based case study may be planned and executed. Practical guidelines are offered for framing the research question, data collection, writing a narrative, coding, and explanation building aligned with a critical realist philosophy.

Grounded theory, since Glaser and Strauss conceived of it, has been applied to myriad research situations. Variation characterises the use of grounded theory, and the landscape of grounded theory is 'messy' and difficult to navigate, especially for novice researchers, and subsequently, many existing studies lack transparency and detail, employing only loosely grounded theory methodologies. The authors argue for the value of grounded theory research in education, especially for teacher researchers, but this value can only be realised if researchers attend in-depth to articulating clearly their methodologies and justifying how the methodology can be considered grounded theory. In the chapter, the authors aim to provide a primer on grounded theory, and to stimulate discussion around grounded theory as a research method in teaching, especially in higher education, for the purpose of generating knowledge as a basis for improving the quality of university teaching.

and organizational change, support from stakeholders affected by the research problem is essential, creating further complexity. Action research may serve as an alternative to more traditional views of social science. In this chapter, the authors describe action research as envisioned by Kurt Lewin, its originator. They show how two variants of action research—action science and collaborative developmental action inquiry—advance insight into how action research can be used to develop personal capability to address system changes that action research seeks to unveil. They conclude with reflections on criteria for rigor and relevance in action research in today's post-modern, complex world.

The purpose of this chapter is to justify the value of Indigenous research paradigms, specifically in the context of research on language acquisition. This argument has implications not only for research on language acquisition and the practice of language instruction but also for qualitative research, more broadly. Specifically, depending on the context of a given research project, it may be critical for educational researchers to value and respect Indigenous epistemologies and worldviews; otherwise, educational research endeavors may be adding to knowledge at the expense of devaluing research participants and local communities.

This chapter is about introducing critical race design (CRD), a research methodology that centers race and equity at the center of educational opportunities by design. First, the authors define design-based implementation research (DBIR) as an equity-oriented education research methodology where teaching and learning is informed by robust, iterative, evidence-based research conducted by multiple stakeholders. Next, they provide a brief overview of critical race theory in education (CRT) as a theoretical and methodological approach that aims to unpack and disrupt the structural inequities experienced by disenfranchised racial groups. They then describe how both education methodologies inform CRD, the authors' emerging anti-racist critical design methodology. Finally, they provide an example where they used CRD to design an online service-learning course that aims to situate the narratives of underrepresented STEM professionals as a curricular resource for nondominant adolescents.

Foreword

As research in a field progresses, new models emerge as possibilities for future research and practice. On a grand scale, research on our Universe unveils new models for images of solar systems. Until recently it was thought that the basic structure of our own solar system and distribution of mass within it would be replicated in other solar systems as we discovered exosolar planets around stars other than our Sun. But as details of those other solar systems emerged, they looked much different from our own in terms of number of planets, mass distribution of the planets and the size of their star. While this opened doors to new research, it also required a reinvestigation into the assumptions and explanations of the evolution of solar systems including our own. Questions emerged about why some stars were larger or smaller than ours. New explanations are emerging about why some planets were certain distances from their star or why solar systems had the larger planets closer to their star which is the reverse of planetary distribution of our own home solar system. New research tools and methodologies needed to be developed and traditional ones needed to be reconsidered and readapted to answer these questions.

Educational researchers face similar challenges and opportunities. They might have a favorite methodology, but throughout their careers it is important to explore other techniques. This all happens against a background of the world around us. In early 2020, the emergence of a global pandemic forced everyone to expand, reorder, reevaluate and reassess assumptions and methodologies, personal philosophies, and comfortable practices.

This book will assist in addressing this huge challenge. Whether a researcher or practitioner, this volume will nudge new ideas and creative practices to lead us forward to a post-pandemic world that is far enhanced from the one that we closed the door on in spring, 2020. But because of this "great pause" in our formerly hectic lives, our minds are freer to roam with fewer constraints into an open space of imagination and innovation and intensity. A speaker at a recent webcast reminded us of two historical events with happy endings. After the devastating plague of the 1300's came the Renaissance. When Cambridge University closed in the mid 1600's due to a plague, Isaac Newton was forced to retreat to his home which many consider gave him the time to do his best work.

Now is the time to access how we can do our best work. This new world will require applying and adapting traditional methods to address current issues such as social justice, anti-racism, and the empowerment of specific populations in the design and implementation of research and teaching and learning practice. This book explores these issues and provides examples of techniques and traditional and emerging research designs to take into account these current societal issues and prompt new ideas for decision making across varying venues and global populations. Contributions to this volume come from across the United States but also from as geographically near-by locations as Barbados and as distant as Australia, India, and Turkey.

This book will provide us with the foundation to propel us forward. It will be useful for researchers and practitioners alike in the Social Sciences, Humanities Information Technologies, Education and International and Intercultural Studies and anyone ready to move our world into a new educational renaissance.

Judith Parker
Teachers College, Columbia University, USA

Preface

Habermas (1971), in recognition of the human need to understand each other, refers to practical or communicative knowledge. Human beings have always been social creatures, instinctively forming groups, tribes, communities, cultures, and nations in order to satisfy their mutual needs. In order for people to survive together in groups and societies, they must communicate with and understand each other. There are no scientific laws governing these communications; when we communicate with others, we interpret what they say in our own way.

This does not mean that communicative knowledge is entirely individual. All societies share and transmit social knowledge, that is, a code of commonly accepted beliefs and behavior. As a society we come to agree on how things should be and are in reference to standards and values, moral and political issues, educational and social systems, and government actions. Communicative knowledge is derived from shared interpretation and consensus and then often becomes reified.

Habermas criticizes communicative knowledge as being too dependent on subjective understanding. He argues that people may misinterpret the world around them based on distorted assumptions about themselves or society. We want social knowledge to be objective and concrete and therefore stop questioning the systems around us, unaware of the distortions that may exist in our assumptions.

The research paradigm used to acquire communicative knowledge is the interpretive paradigm. Data are mostly qualitative, and methods include case studies, grounded theory, phenomenology, narrative inquiry, appreciative inquiry, and self-study.

Another kind of knowledge, which derives from a questioning of instrumental and communicative knowledge, Habermas calls emancipatory. By nature, people are interested in self-knowledge, growth, development, and freedom. Gaining emancipatory knowledge is dependent on our abilities to be self-determining and self-reflective. Self-determination can be described as the capacity to both be aware and critical of ourselves and of our social and cultural context. Self-reflection involves being aware and critical of our subjective perceptions of knowledge and of the constraints of social knowledge.

Emancipatory knowledge is gained through a process of critically questioning ourselves and the social systems within which we live. The philosophical foundation of emancipatory knowledge lies in critical theory. In this paradigm, instrumental and communicative knowledge are not rejected but are seen as limiting. If we do not question current scientific and social theories and accepted truths, we may never realize how we are constrained by their inevitable distortions and errors (the world is flat, the Aryan race is superior). Without the possibility of critical questioning of ourselves and our beliefs, such constraining knowledge can be accepted by entire cultures.

The research paradigm that is relevant for constructing this kind of knowledge is the critical paradigm. Data are always qualitative, and specific methods include action research, participatory research, and feminist research.

Quantitative research unquestionably has a place and is fundamental to scientific advances, but qualitative research delves into what it is to be human. Through qualitative research, we gain insight in communicative knowledge, its rich nature and the mechanisms by which communicative knowledge is formed and interpreted. Qualitative research enables the necessary exploration and critical analysis social systems. Qualitative research methods uncover and facilitate critical reflection on the inevitable assumptions, which shape social behavior and interaction, thereby stimulating and empowering change.

The aim of this volume is to provide readers with a comprehensive array of qualitative research methods, which can be implemented in a variety of contexts for a variety of purposes.

Viktor Wang
The University of Montana Western, USA

REFERENCES

Habermas, J. (1971). *Knowledge and human interests*. Beacon Press.

Acknowledgment

Special thanks go to the editorial advisory board members and reviewers who have reviewed all the chapters around the clock during the blind review process to ensure the quality of every chapter accepted for publication in this volume. Some of them have contributed chapter(s) to this book. Their insights and collegiality are sincerely appreciated. Last but not least, we thank IGI Global, whose editorial and production team has continuously supported my work in this field and other fields. We thank our users and university librarians for recommending this unique book to scholars and practitioners from all disciplines in K-20 Education.

Chapter 1
Qualitative and Digital

Geraldine Torrisi-Steele
https://orcid.org/0000-0002-6045-4114
Griffith University, Australia

Viktor Wang
University of Montana Western, USA

ABSTRACT

The disruptive effect of digital technologies is felt across almost all aspects of human endeavour. Research is no exception. The disruptive potential of digital technologies is realised through either doing existing activities differently or through the creation of entirely new activities. In research, digital disruption manifests as an expansion of possibilities for research, and as innovation in the tools and methodologies used to carry out research. In the present chapter, qualitative research is discussed with the aim of providing an overview of the influence of digital technologies on qualitative research for the purposes of stimulating further exploration and discussion.

INTRODUCTION

Big data, connectivity and social interaction characterise the digital landscape. Close to 60% of the human population are connected to the internet (DataPortal, 2021). The volume of data generated is staggering. It is estimated that more than 2.5 quintillion bytes of data is generated each day (www.techjury.net), the vast portion of which is generated by social media. All of this made possible by the plethora of digital technologies spawned by the advent of the internet back in the 1990's. The digital landscape provides fertile ground for the evolution of new models of knowledge production. It has been long-recognised digital technologies create a trend towards

DOI: 10.4018/978-1-7998-7600-7.ch001

increasing diversity of research activity locations, greater focus on interdisciplinary, multidisciplinary and trans-disciplinary research, and towards greater emphasis on collaboration and informal communication (Houghten, Steele, & Henty, 2003). The global and convenient connectivity enabled by technologies such as cloud computing enable collaboration in research. The image of the lonely researcher working alone dissipated long ago. The ability to collaborate easily regardless of time and place and the array of new and emerging technologies is disrupting the nature of research.

In the present chapter, the changing landscape of research and in particular qualitative research is discussed with the aim of providing a foundation for further exploration and discussion.

THE CHANGING LANDSCAPE OF RESEARCH

A large number of applications has materialised to support research through collaboration: e-Science, e-Social Science, computational social science, e-humanities, e-Research and 'Digital scholarship' to name a few (Dutton, 2013). A move towards increased research accessibility is also evident (Pilaar Birch, 2013). The trend towards collaboration in research is so important that the term 'social shaping' has been applied to digital research as a way of capturing the increasingly collaborative nature of research endeavours. Originating in the social sciences, the term 'social shaping' is used by authors such as Dutton (2013), bringing it to bear on digital research. Dutton acknowledges that 'social shaping' encompasses a great diversity of perspectives, and needs to be better refined if it is to be used as a framework for digital research, it does bring to the fore that research in the digital age has an increasingly social dimension to which social scientists can contribute to the "co-production of the technologies, policies and practices of digital research" (Dutton, 2013, p. 191). Moreover, Dutton (2013) predicted that connectivity of digital technologies, digital research is leaving the realms of purely academic endeavours and into the digital economy. Though now almost a decade ago, Dutton realised that the effect of the digital age on research is greater interaction "between academic communities, business and industry" (Dutton, 2013, p. 191). The connectivity enabled by digital technologies is having the effect of moving research towards global contexts and the exploitation of big data and the creation of new resources for research is rapidly taking hold (Dutton, 2013). Evidently, the traditional image of research as a "solitary endeavour" (Renaud, 2001, p. 59) is under challenge. Collaboration and interaction are one defining feature of research in the digital age. Furthermore, the paper based information resources in research have given way to electronic resources including the web, database and e-journals (Ge, 2010).

Anecdotally, most grad students pursue their research studies rarely, if at all setting foot in a physical library.

Evidently, the advent of digital information and communication technologies are changing how research is conducted. But digital technologies are also changing what research is conducted. There are new sources of data, and new environments in which to conduct research, and new ways of disseminating research. Social media provides a multitude of opportunities for investigating human interaction and behaviour in ways that have never been available before. Researchers in fields such as sociology and ethnography are increasingly engaging in social media contexts to conduct their research.

Rising Popularity of Qualitative Research and Change

Historically, qualitative research has fought to establish legitimacy as a research method, often accused of being less rigorous than quantitative research [[INSERT REF]]. However, conditions in the digital age are conducive to raising the popularity of qualitative research.

Unarguably, digital technologies are changing the nature of social interactions whether for the purpose of business, education, entertainment or simply daily living. To gain a true sense of the impact of digital technologies it must be acknowledge that it is not only that the changes are profound, but also that changes are very rapid, frequent and ongoing. For example, a single tweet from an influential actor maybe enough to bring down or raise stock markets in a matter of hours, on a global scale. It is easy to understand why the 21st century is characterised so often in literature as a dynamic and highly complex period of human existence. It is with some urgency that 'wicked problems are placed at the forefront of many research agendas' [[insert ref]]. The complexities and the magnitude of rapid social change spurs ahead and legitimizes qualitative inquiry in the social sciences (Morse, 2020). "Qualitative research is experiencing a surge of popularity" (Morse, 2020, p. 2), and changes.

Although there are well-established methods of qualitative inquiry, change is inevitable, particularly in the present dynamic, complex and digitally 'hyped' environment. Key factors catalysing change in qualitative methods include

- changes in technology which ease and expedite data collection and analysis as well as enable larger, more widely distributed, sample sizes. Mobile digital technologies enable easy and accessible data collection in the field, and multimodal data collection (e.g., use of video).
- advancement in knowledge and theoretical perspectives about the phenomenon studied
- changing societal conditions

- changing agendas of industry requiring research
- ongoing pressure for methodologies that provide greater rigor to qualitative research (Morse, 2020)

Interestingly, it wasn't so much the advent of the internet which propelled change in qualitative methods. Much of the qualitative research on the internet relied on online questionnaires which followed much the same approach as traditional on-paper questionnaires, with the added convenience of distribution and some 'built' in simple descriptive statistics for data analysis. Making much more impact on qualitative research are emerging computer science technologies such as machine learning (Digsite, 2020)

Change is an evident theme in qualitative research methods. It always has been but conceivably the rate and magnitude of change at present is unprecedented. As noted by Morse (2020), change can be positive or negative and hence, the push for change in qualitative methods of inquiry necessitates evaluation and action to support positive change and resist changes which weaken the field of qualitative inquiry. Changes should be scrutinised remembering that qualitative research is usually long term and "involves prolonged immersion in the field and reflective, complex processes of data collection, interpretation and representation" (Kacen & Chaitin, 2006). Such technologies enable qualitative research to become easily iterative and agile – clearly important attributes in the present rapidly changing environment.

MULTIMODALITY AND DIGITAL INTERACTION

Through digital technologies, researchers may avail themselves of a myriad of tools and techniques, many of them multimodal. For example, video can be a powerful tool for both 'seeing and hearing' data (Salmons, 2017). Among types of video research methods are Participatory video research (Whiting, Symon, Roby, & Chamakiotis, 2016). Participatory video research refers to video diaries etc which are generated by the study participants. The aim of participatory video research is to narrow the 'gap' between the mental models of the researchers and that of the actual participants. In some instances, participants receive training in filming and editing to produce collaborative, community videos (Whiting, Symon, Roby, & Chamakiotis, 2016). Other video methods include videography in which researchers observe and film participants 'in the field' while participants go about their daily lives; Video elicitation in which the video is produced either by participant or researcher and is used to provide a planform for reflection; use of existing videos in which videos have been created independent of the researcher and participant and is commonly used to investigate contextualisation of events and the construction of identities and relationships (Whiting, Symon, Roby, & Chamakiotis, 2016).

In the media rich context of digital technologies, digital qualitative research methods can be highly multimodal. Media choice and integration aside, interactivity is an ever-present element. The analysis of digital interaction is an emerging and rapidly expanding field in qualitative research. Just as traditional interaction analysis is aimed at understanding the co-construction of meaning through 'language in use', digital interaction analysis aims to uncover meanings constructed not only through interaction with other humans via the digital medium, but through interaction with the digital mediums themselves. While some of the traditional techniques for analysing interaction hold in the digital space, the nature of the digital medium necessarily creates a need for change. Qualitative research is highly contextualised, and it is inseparable from the platforms in which the phenomenon is occurring. There is thus an obvious need for the development of interaction – focused scholarship in digital environments including online communication in the various contexts of social media (Meredith, Giles, & Stommel, 2021).

THEORETICAL FRAMEWORKS

Frameworks such as activity theory and community of inquiry are not novel ideas arising from the digital age, but in digital contexts, the attributes of the two frameworks align with the capacities of technology such that researchers are devoting increasing attention to them. We consider four frameworks which are beginning to emerge in very recent literature related to research in the digital age, namely activity theory, actor-network theory, community of practice and technology acceptance framework.

Activity Theory

Activity theory emerges from the work of Vygotsky, Loent'ev and others back in the 1920's. As the name suggests, activity theory is concerned with what people do, with their motivations for doing things, the rules and tools that mediate, or constrain, their activities, as well as the objects or artefacts at which the activity is directed. The fundamental unit of analysis is activity. Tools, whether they be Physical objects or systems (such as language) are used by people to accomplish the activity. What people do (activities) are goal orientated and conscious actions of the person. The activities are mediated by a community and there are divisions of labour and rules. Emphasis is very much on the social factors and interactions that mediate the activity, on the importance of the context within which the activity is carried out, and on the idea that activity is evolving and complex. Thus, activity theory focuses very much on the relationships among person, motive, community, rules, division of labour and outcomes. Coles and Engestrom (1993) view the interacting factors as an activity system or a kind of 'distributed cognition'. Activities are dynamic and change and may be internal and external.

Activity theory is useful as a research framework in the digital age because it offers perspectives on how technology is used in the context of social practices. The system of activity is a useful unit for analysis in applying activity theory to research concerned with digital technologies. For example, Spasser (2002) applies activity theory framework with a case study methodology to investigate technologically mediated social practices to inform digital library assemblage and use. Murphy & Rodriguez-Mnzanares (2014) apply activity theory in depth to consider the use of technology in higher education teaching. Cox (2013) uses activity theory as a framework for a study to gain insights into some of the reasons why some academics are reluctant to add content to open content repositories. Cox maintains that activity theory proved to be a "useful lens" to help identify the factors inhibiting contribution (2013, p. 156). Related to social media use, Rambe (2012) employs the framework of activity theory to explore the academic and student interactions within Facebook in terms of cognitive scaffolding of learners and democratising "student access to knowledgeable peers" (p. 1352). Rambe used the activity system as the unit of analysis and the value of the approach to understand the system is evident in Rambe's summary of findings:

The Facebook activity system was mediated by technological (Facebook's interactional spaces, applications, textual resources) and conceptual tools (symbols, questions, queries, explanations, and answers). Conceptual tools dominated Facebook interactions essentially because lecturer-student and peer-based interactions were question and answer based. .... At the social interactional level, Facebook interaction unfolded through a set of rules like the creation of Facebook accounts, joining group forums, development of "friendship", recognition of the rule of engagement, and contributing to peer-generated postings. Collectively, these rules were both enabling and constraining. Academically motivated students assumed multiple roles as information seekers, information disseminators, collaborative networkers, resource persons and reflectors. Academics also served as instructors, knowledge brokers, mentors and sympathetic coaches. (2012, p. 1352)

Actor-Network Theory

Actor-Network Theory (ANT) has its origins in sociology of science and technology. The fundamental assumption of ANT is that social practice is a result of various elements working together – a network, hence, like Activity theory ANT employs a holistic unit of analysis. According to Law (1992) some questions of ANT, in relation to organisation and systems, are:

How it that it ever seemed otherwise? How is that, at least for a time they made themselves different from use? By what organizational means did they keep themselves in place and overcome the resistances that would have brought them tumbling down much sooner? How was it we colluded in this? (p. 379)

Thus, ANT is not so much concerned with why networks form but rather with how they form and are sustained. Central to ANT is the 'metaphor of heterogeneous networks' – "a way of suggesting that society, organisations, agents and machines are all effects generated in patterned networks of diverse (not simply human) materials (Law, 1992, p. 380). Within ANT, Actants refer to both human and non-human actors. The qualities of Actants, whether human or not, are assumed due to interactions. Three of the major writers in ANT are Bruno Latour, Michel Callon and John Law. Although ANT has been criticised by some for reasons such as ignoring social factors such as racism and gender (Harding, 1991), and by assuming all actors are equal there is no way to address power imbalances. Despite criticism, ANT continues to evolve, and it is a useful framework for research in digital arenas. It makes it possible to explore how interactions with technology influence actions, and how technology is socially constructed. For example, Shirazi (2014) makes use of ANT to bring to the fore the ideologies underpinning Iran's public cloud.]; Steenbruggen, Nijkamp, and van der Vlist, (2014) conceptualises traffic incident management in the Netherlands, in terms of the ANT. Using the ANT framework, the researchers define an information chain which allows them to analyse the interactions of actors involved in the reduction and prevention of road congestion.

Communities of Practice

Wenger's (1998) Communities of Practice offers another framework useful to qualitative research in the digital age. The digital age is defined by connectivity and the concept of communities of practice is fuelled by digital tools which enable communication independent of time and place. Communities of practice are formed by groups of people who have in common a concern or interest in an area or activity and so interact reguarly together in order learn more. A community of practice has three essential characteristics: The domain (the shared domain of interest and therefore community members have a commitment to it); The community (members engaged in shared activity, discussions and relationships that help them learn from each other); The practice (the shared stories, cases, experience etc are an important source of knowledge that is then impelemented into the practice of the members).

Problem solving, seeking experience, reuse of assets, co-ordination and synergy, knowledge mapping are just some of the activities occuring in communities of practice. (Wenger, 1998).

The usefulness of the framework for research is well summarised by Vidich & Lyman (1994, p. 23)

The framework offers both the language and insights to provide a rich description of human practice and the social interactions surrounding practice, thus supporting analysis and deeper understanding of the, 'patterned conduct and social processes of society, and … the bases in values and attitudes on which individual and collective participation is social life rests

The Communities of Practice framework is not peculiar to the digital age, but the advent of social media and the associated synchronous and asynchronous communication tools, presents new opportunities for the formation of Communities of Practice among groups of people, irrespective on their locations and time-zones. Furthermore, as digital tools enable more easily and frequently the connection among diverse groups of people, many activities undertaken in digital contexts can be explored from a community of practice perspective.

Some or the research related to the Communities of Practice are about their use. For example:

Chiu, Hsu, & Wang (2006): McLure & Faraj (2000) use knowledge management as a framework to explore the reasons shy people choose to help others by participating in electronic communities of practice; Hou, (2015) uses an ethnographic approach to investigate perceptions of communities of practice for teachers as professional learning.

In other research, Communities of Practice are used as a theoretical lens through which some activity or phenomena can be examined or better understood. For example: Kimmerle, Thiel, Gerbing, Bientzle, & Halatchliski (2013) use the communities of practice framework to analyze knowledge construction in a web forum in which health related community members exchange information and experiences; Dennen (2014) Uses Community of Practice framework together with Activity theory to examine the norms and activities in a blogging community; Kietzmann, Plangger, Eaton, Heilgenberg, Pitt, & Berthon (2013) use a variation on communities of practice, which they call mobile communities of practice (MCOPs), as a lens to gain understanding of how workers self-organize. The research results in strategic considerations for the management of mobile work. A case study methodology is used; Nam Mak & Chui (2013) employ the Communities of Practice framework to examine how people interact and manipulate power outside of the workplace via Facebook status updates. The study questions if the framework can be used to appropriately conceptualize workplace talk among colleagues after work. Methods of discourse analysis are used to analyze data.

Technology Acceptance

Using Fishbein and Ajzen's (1975) theory of reasoned action as a basis, Davis (1986) developed the original technology acceptance model (TAM) in an attempt to predict an individual's acceptance of technology. Predictive factors include perceived usefulness, external variables, perceived ease of use, attitude to use, all of which predict behavioural intention to use, which in turn predicts actual system use.

Since the inception of the original TAM model there have been a number of extensions to the model including TAM 2 (Venkatesh & Davis, 2000), TAM 3 Venkatesh & Bala, 2008) and the Unified theory of acceptance and use of technology (Venkatesh, Morris, Davis & Davis, 2003). However, the basic concept underlying technology acceptance models (Davis, 1986; Venkatesh & Bala, 2008; Venkatesh & Davis, 2000; Venkatesh, Morris, Davis, & Davis, 2003) is that individuals' reaction to using information technology and their actual use of technology influence each other. Furthermore, individuals' reactions to using technology influence future intentions, which in turn influence the actual use of technology.

TAM has been used extensively across a wide cross-section of disciplines to examining people's attitudes towards acceptance and using of various digital tools and environments. Researchers have often used TAM as a base framework and proposed extended models or variations to it. It is not surprising that with the advent of social media, TAM is being used as a framework for research aimed at better understanding the factors that shape attitudes and behaviours in digital contexts. For example: Lee, Xiong and Hu (2012) seek to highlight the emotional aspects of social media event marketing by adopting and extending the TAM framework. They found that user emotions as exhibited on Facebook event page influenced perceived usefulness, perceived ease of use and perceived enjoyment of the event, but only perceived enjoyment significantly influences the intention toward the event. As a result of the research Lee, Xiong and Hu provide some practical guidelines to event practitioners; Ingham, Cadieux and Berrada (2015) develop a model of e-shopping by integrating the TAM framework in order to achieve a better understanding of marketing with technology; Eagleman (2013) investigated the role of social media within niche sporting organisations in the United States in terms of employees motivation and acceptance to use social media for the achievement of organisational goals;

Other Frameworks

Evidently, Activity theory, ANT, Community of Inquiry, and TAM represent only a very small portion of the theories used as conceptual frameworks for research in the digital age. As researchers engage with research in the digital age, their research spans a large cross-section of research questions and contexts, and thus there are a large number of theories and models used. Ngai, Tao and Moon (2015) conduct an analysis of 45 research paper and identify the theories and models used in each paper. They find 35 different theories used under broad categories of mass communication theories, social behaviour theories and personal behaviour theories – an unsurprising finding given that social media dominates the digital age.

Although the sample of frameworks reviewed is small, it is clear that researchers are repurposing existing frameworks to relatively novel scenarios. Oftentimes, the result of using the framework as a lens to examine new phenomena not only leads to insights about the phenomena, but also to insights about the frameworks themselves. Applying the frameworks in new contexts provides a means of evaluating the usefulness of the framework, and in some cases lead to evolution of frameworks themselves (for example, Nam Mak & Chui (2013)).

One observable trend across the frameworks commonly adopted for digital social research is emphasis on participatory approaches in which the power distance between researcher and participants is significantly decreased. Seemingly, qualitative research is increasingly influenced by the notion of 'collective intelligence', crowd sourcing and reflective of the trend towards flatter organizational structures in which the contribution of individuals is valued. For example, the rise in citizen research, flourishing due to the rising living standards, increased access to knowledge about social research and the relative freedom of the online world (Corti & Fielding, 2016). Social media, of course, plays a key role because it supports the collaboration of very large numbers of people grouping around common interests (Corti & Fielding, 2016).

RESEARCH METHODS IN THE DIGITAL AGE

Many of the methods used in digital research are well established qualitative research methods. For example, case research continues to be widely used for the in-depth investigation of problems in real settings over extended time periods. Similarly focus group research remains a valued approach to gain understanding of the perceptions, attitudes and beliefs of selected populations who have common interest. A moderated discussion, usually of 6-10 participants brought together to discuss a topic of interest for a short period of time. Discussion is moderated and guided by an agenda and an initial set of questions for participants. Given lack of control, findings are not generalizable and so it used for exploratory rather than explanatory research.

The rising complexity of the digital era and the emphasis on wicked problems has increased the appeal of action research and participatory action research. The main steps of Action Research are: Diagnose a problem, Plan an action, Take action, Evaluate action and specify learning. It is once again the connectivity and media rich tools of the digital age that greatly enable action research. Digital technologies have the potential to be inclusive, and accessible. Participants in action research are no longer constrained by geographic location and there is the possibility of action research by globalized communities of practice.

Ethnographic methods for qualitative research are similarly bolstered by digital technologies. According to Coleman (2010) examination of even a single digital genre can be "ethnographically rich" and touch on a variety of analytic frames. That terms such as 'Netography' have emerged to refer to ethnographic approaches to analysing an individual's behaviour on the internet, attest to the popularity of the approach in digital contexts. The impact of the digital era on ethnography is so great that it has warranted the development of 'digital ethnography'.

Data collection tools that are the mainstay of qualitative methods such as direct observation and surveys are being implemented in digital contexts. For example, direct observation (one of the first methods of data collection in virtual worlds) (Ross, Catronova, & Wagner, 2012), surveys and interviews continue to feature prominently in both traditional research contexts and digital contexts.

Direct observation in the virtual world may consist of observing players and or data generated. Data may be gathered either by collaboration with the virtual world developers or using web services (although data from the latter is limited) (Ross, Catronova, & Wagner, 2012).

Surveys can be easily utilised in a range of online environments from email, through to forums and splash screens to fully immersive virtual worlds. As in the real world surveys are a seemingly important tool for digital research because it can provide information about demographics, beliefs, attitudes, habits etc. It is generally agreed the survey will remain an important research too, but Savage and Burrows (2007) encourages researchers to look beyond the historical 'pre-eminence' of surveys towards and consider how data from recent technologies enables "more descriptive data to be deployed in new and powerful ways"(p. 892). One way of following Savage and Burrows advice is to consider the technique of dataveillance in conjunction with surveys (Ross, Catronova, & Wagner, 2012). Dataveillance (Clarke, 1988) refers to the "systematic monitoring of people's actions or communications through the application of information technology."

In immersive virtual worlds, interviews may be conducted between avatars of the researcher and participant. Advantages of conducting interviews with digital media include access to large and diverse populations, very specific access to specific population profiles, and efficient and cost-effective data collection. However, as Dipko, Billington and Brick (2012) remind us, it is important to consider the effect of the mode on participant responses in the virtual world. Dipko, Billington and Brick (2012) compared interview responses from avatars in Second Life, an immersive virtual world, and a real-life interview taken by people behind the avatars. They found that the respondents answered in similar ways in both real-life and in the virtual world. However, Dipko, Billington and Brick (2012) do point out that while their research shows little difference in respondents answer, more exploratory research is needed to confirm. Researchers cannot ignore the possibility that mode-specific adjustments may need to be made.

Given that data collection direct observation, surveys and interviews are useful in digital research, the methodologies which use them such as case research (For example Merrick, Gu and Wang, (2011), Beal and Flynn (2015), Turan and Bayram (2013)), focus groups, action research and ethnographic research also prevails in digital research.

Virtual focus groups are becoming quite popular because they provide fast and cost-effective methods to gather qualified participants from across the globe. Furthermore, since many virtual communications tools use text, the focus group data is often accurately recorded in text form. Authors such as Kruegar (1994), Morgan (1988), Turney and Pocknee (2004).

Innovation in Methods

Whilst the majority of methodologies and data gathering tools in digital research are not new and being used in their original form or being slightly re-purposed, there are innovative data collection methods emerging. One such method is Iconographic tracking (Gries, 2013).

Iconographic tracking enables investigation of the "circulation, transformation, and consequentiality of images across genres, mediums and contexts" (Gries, 2013, p. 332). Iconographic tracking begins by collecting a large data set with image search capabilities. It is essential to acquire a large data set in order to find patterns and trends for later comparison. Gries (2013) calls this the data hoarding stage. According to Gries the second stage entails assembling data into a collection – "a process of sorting through massive amounts of data to locate patterns, trends, and relationships" (p.339); this is a data mining exercise. In the third stage, the data search continues but as a recursive process of refinement and narrowing of the collection categories. Gries (2013) maintains that change in the circulating image is analyzed in terms of "design, form, medium, materiality, genre, function as it enters new associations" (p.343). Insights into image transformation provide information about communication, rhetoric and composition, in a participatory culture that has arisen due to social media, in which 'remixing' is a vehicle for communication (Gries, 2013, p. 343). It useful as a study of transformation to track how the original image designers' message may have changed and what unintended or unexpected consequences this may precipitate (Gries, 2013).

Iconographic tracking is a method aligned with not simply transformation of images, but rather a method relying on vast amounts of data and tools which make that data accessible. Much of the innovation that is beginning to occur in relation to the digital age is anchored in the large amounts of data being generated (Big data) and in the tools being developed that help access and process the data. Traditional methods of dealing with data such as databases often fail in the face of monumental amounts of data and there is a need to develop new technologies and analytical approaches. Text and data mining aims to automate the copying of existing data from a variety of sources and conduct analysis on the basis of patterns, trends and other information. The relevance and importance of data mining in the digital age is made particularly evident by the volume of literature surrounding it and also the establishment of entities such as the Text and Data Mining Working Group in the European Union whose primary objective is the promotion of efficient use of Text and Data Mining (TDM) for research purposes. The relevance of data mining from the perspective of libraries and librarianship is obvious. The flow of information into libraries from digital technologies is accelerating, and in addition, as the world becomes more interconnected and integrated, information transfer is becoming more complicated (Uppal & Chindwani, 2013). Uppal & Chindwandi (2013) undertake a study of an application of data mining to library services. In their study they show how data mining can be used to identify departmental needs, pattern of identification for inventory loss, analysis of book circulation, and gaining information about reader interest – all useful information to help libraries better serve their patrons.

ISSUES AND CONSIDERATIONS

Evidently, and inevitably, qualitative research is changing in the digital era. But however exciting the changes might be, it is important not to lose sight of the fact that digital qualitative research, like any other research, must be carefully considered.

Within digital environments ethical issues of privacy, power and social justice are as important (if not more prominent) in digitally enabled research as they are in traditional research. Although virtual world research shares many of the considerations of real-world research, there are some significant concerns, as identified by Dipko, Billington and Brick (2012, p. 5624):

- A greater risk to individual privacy and confidentiality.
- Challenges in obtaining informed consent.
- Virtual worlds such as Second Life and World of Warcraft allow users to create online identities that can differ significantly from the subjects' real identity.
- Determining an ethically correct approach based on the online venue being used.
- Determining an ethically correct approach based on the global reach of virtual worlds and the multitude of different cultures involved.

Although, on the surface at least, qualitative research is seemingly employing more participatory and empowering approaches, it is dangerous to ignore the issue of the digital divide. The increased ease of access to participants on a global level may easily mask that some participants are still being excluded and perhaps disempowered by digital technologies. It is imperative to ask who is being excluded? Or who's voice is not being heard? Exclusion may occur because of economic reasons (cannot afford the technology), knowledge reasons (do not have understanding or knowledge to be able to use technology) or reasons of infrastructure availability.

From a methodological perspective, we need to consider how the location of the research (real-world or virtual) and the modes of research (social media, virtual worlds etc) may affect our data. Issues of quality and trustworthiness of data gained from relatively anonymous social networks and 'unknown' participants' need to be considered. In digital environment there may be greater opportunity for participants to manipulate surveys or data. We also need to consider and evaluate the extent to which research in the digital world generalises to the real-world and vice versa.

CONCLUSION

Qualitative research is necessarily situated in the social context of the time and all types of research will (and must) evolve in resonance with changes the surrounding environment. Changes of all kinds of environments, social, political, physical and economic will resonate in research activity, influencing what is being researched, and how research is undertaken. The digital era, characterised by social connectivity is raising the profile of qualitative research. The rigor of qualitative research, which has always been a 'sticking point' is positively impacted by digital technologies which enable more systematic processing of larger volumes of qualitative data. And the increasing complexity and volatility of the digital world may be seen to strongly legitimise qualitative research. In the present chapter, the authors have sought to provide a broad sense of the influence of 'digital' on qualitative research. Traditional methods and tools of qualitative research persist and are evidently useful and, in many cases, stand well aside newer approaches. The environment of change is perhaps the opportune time to take stock of 'where we are at' in qualitative research, and consider 'where we want to go', all the while examining assumptions and looking through the ethical lens of social good, and social justice.

REFERENCES

Beal, C. D., & Flynn, J. (2015). Toward the digital water age: Survey and case studies of Australian water utility smart-metering programs. *Utilities Policy*, *32*, 29–37. doi:10.1016/j.jup.2014.12.006

Chiu, C.-M., Hsu, M.-H., & Wang, E. T. (2006). Understanding knowledge sharing in virtual communities: An integration of cosial capital and social cognitive theories. *Decision Support Systems*, *42*(3), 1872–1888. doi:10.1016/j.dss.2006.04.001

Clarke, R. A. (1988, May). Information technology and dataveillance. *Communications of the ACM*, *31*(5), 498–512. doi:10.1145/42411.42413

Cole, M., & Engestrom, Y. (1993). A cultural-historical approach to distributed congition. In G. Salomon (Eds.), Distributed cognitions; Psychological and educational considerations (pp. 1-46). Cambridge University Press.

Coleman, E. G. (2010). Ethnographic Approaches to Digital Media. *Annual Review of Anthropology*, *39*(1), 1–16. doi:10.1146/annurev.anthro.012809.104945

Corti, L., & Fielding, N. (2016, October-December). Opportunities From the Digital Revolution: Implications for Researching, Publishing, and Consuming Qualitative Research. *Digital Representations: Opportunities for re-using and publishing digital qualitative data*, 1-13.

Cox, G. (2013). Researching resistance to open education resource contribution: An activity theory approac. *E-Learning and Digital Media*, *10*(2), 148–160. doi:10.2304/elea.2013.10.2.148

DataPortal. (2021). *Digital Around the World.* https://datareportal.com/global-digital-overview

Davis, F. (1986). *A technology acceptance model for empirically testing new end-user information systems: Theory and results* (Doctoral dissertation). Sloan School of Management, Massachusetts Institute of Technology. http://hdl.handle.net/1721.1/15192

Dennen, V. P. (2014). Become a blogger: Trajectories, norms, and activites in a community of practice. *Computers in Human Behavior*, *36*, 350–358. doi:10.1016/j.chb.2014.03.028

Digsite. (2020). *Qualitative Research is changing*. https://www.digsite.com/qualitative-research-is-changing

Dipko, S., Billington, C., & Brick, P. (2012). The effect of mode on participant responses to qualitative research in virtual worlds. *The American Association for Public Opinion Research (AAPOR) 67th Annual Conference, 2012*. http://www.websm.org/db/12/15291/Web%20Survey%20Bibliography/The_Effect_of_Mode_on_Participant_Responses_to_Qualitative_Research_in_Virtual_Worlds/

Dutton, W. H. (2013). The social shaping of digital research. *International Journal of Social Research Methodology*, *16*(3), 177–195. doi:10.1080/13645579.2013.774171

Eagleman, A. N. (2013). Acceptance, motivations, and usage of social media as a marketing communications tool amongst employees of sport national governing bodies. *Sport Management Review*, *16*(4), 488–497. doi:10.1016/j.smr.2013.03.004

Fishbein, M., & Ajzen, I. (1975). *Belief, Attitude, Intention and Belavior: An Introduiction to Theory and Research*. Addison-Wesley.

Ge, X. (2010, September). Information-seeking behavior in the Digital Age: A multidisciplinary study of academic researchers. *College & Research Libraries*, *71*(5), 435–455. doi:10.5860/crl-34r2

Gries, L. E. (2013). Iconographic tracking: A digital research method for visual rhetoric and circulation studies. *Computers and Composition*, *30*(4), 332–348. doi:10.1016/j.compcom.2013.10.006

Harding, S. (1991). *Whose science? Whose knowledge? Thinking from women's lives*. Cornell University Press.

Hou, H. (2015). What makes an online community of practice work? A situated study of Chinese student teachers' perceptions of online professional learning. *Teaching and Teacher Education*, *46*, 6–16. doi:10.1016/j.tate.2014.10.005

Houghten, J., Steele, C., & Henty, M. (2003). Changing Research Practices in the Digital Information and Communication Environment. Department of Education, Science and Training. Canberra: Commonwealth of Australia.

Ingham, J., Cadieux, J., & Berrada, A. (2015). e-shopping acceptance: A qualitative and meta-analystic review. *Information & Management*, *52*(1), 44–60. doi:10.1016/j.im.2014.10.002

Kacen, L., & Chaitin, J. (2006). "The Times They are a Changing": Undertaking Qualitative Research in Ambiguous, Conflictual, and Changing Contexts. *Qualitative Report*, *11*(2), 209–228.

Kietzmann, J., Plangger, K., Eaton, B., Heilgenberg, K., Pitt, L., & Berthon, P. (2013). Mobility at work: A typology of mobile communities of practice and contextual ambidexterity. *The Journal of Strategic Information Systems*, *22*(4), 282–297. doi:10.1016/j.jsis.2013.03.003

Kimmerle, J., Thiel, A., Gerbing, K., Bientzle, M., & Halatchliski, I. (2013). Knowledge construction in an outsider community: Extending the communities of practice concept. *Computers in Human Behavior*, *29*(3), 1078–1090. doi:10.1016/j.chb.2012.09.010

Law, J. (1992). Notes on the theory of the Actor-Network: Ordering, strategy, and heterogenity. *Systems Practice*, *5*(4), 379–393. doi:10.1007/BF01059830

Lee, O. (2002). An action research report on the Korean national digital library. *Information & Management*, *39*(4), 255–260. doi:10.1016/S0378-7206(01)00094-5

Lee, W., Xiong, L., & Hu, C. (2012). The effect of Facebook users' arousal and valence on intention to go to the festival: Applying an extension of the technology acceptance model. *International Journal of Hospitality Management*, *31*(3), 819–827. doi:10.1016/j.ijhm.2011.09.018

McLure, M., & Faraj, S. (2000). "it is what one does": Why people particpate and help others in electronic communities of practice. *The Journal of Strategic Information Systems*, *9*(2-3), 9. doi:10.1016/S0963-8687(00)00045-7

Meredith, J., Giles, D., & Stommel, W. (2021). *Analysing Digital Interaction*. Palgrave Macmilan.

Merrick, K., Gu, N., & Wang, X. (2011). Case studies using multiuser virtual worlds as an innovative platform for collaborative design. *Journal of Information Technology in Construction*, *16*, 165–188.

Morse, J. (2020). The Changing Face of Qualitative Inquiry. *International Journal of Qualitative Methods*, *19*, 1–7. doi:10.1177/1609406920909938

Murphy, E., & Rodriguez-Mnzanares, M. A. (2014). *Activity theory perspectives on technology in higher education*. IGI Global. doi:10.4018/978-1-4666-4590-5

Nam Mak, B., & Chui, H. (2013). Colleagues' talk and power after work hours: A community of practice in Facebook Status Updates? *Discourse. Context and Media*, *2*(2), 94–102. doi:10.1016/j.dcm.2013.04.002

Ngai, E., Tao, S. C., & Moon, K. L. (2015). Social media research: Theories, constructs, and conceptual frameworks. *International Journal of Information Management, 35*(1), 33–44. doi:10.1016/j.ijinfomgt.2014.09.004

Pilaar Birch, S. P. (2013). Using social media for research dissemination: The digital research video project. *Internet Archaeology, 35*(35). Advance online publication. doi:10.11141/ia.35.4

Rambe, P. (2012). Activity theory and technology mediated interaction: Cognitive scaffolding using question-based consultation on Facebook. *Australasian Journal of Educational Technology, 28*(8), 1333–1361. doi:10.14742/ajet.775

Renaud, M. (2001). Serendipity and policy relevance. In *Social Science for Knowledge and Decision Making*. OECD Publications.

Ross, T., Catronova, E., & Wagner, G. (2012). Empirical research methods in virtual worlds. In Methods in Urban and Planning Studies: Design and Outcomes (pp. 299-311). Information Science Reference. doi:10.4018/978-1-4666-0074-4.ch018

Salmons, J. (2017). *Methodspace.* Sage Publishing. https://www.methodspace.com/seeing-hearing-problem-using-video-qualitative-research/

Sam, C. (2012). Activity theory and qualitative research in digital domains. *Theory into Practice, 51*(2), 83–90. doi:10.1080/00405841.2012.662856

Savage, M., & Burrows, R. (2007). The coming crisis of empirical sociology. *Sociology, 41*(5), 885–899. doi:10.1177/0038038507080443

Shcroder, K. C. (2012, November). Methodological pluralism as a vehicle of qualitative generalization. *Participations, 9*(2), 798–825.

Shirazi, F. (2014). *Interrogating Iran's restricted public cloud: An Actor Network Theory perspective.* Academic Press.

Spasser, M. A. (2002). Realist activity theory for digital library evaluation: Conceptual framework and case study. *Computer Supported Cooperative Work, 11*(1-2), 81–110. doi:10.1023/A:1015288305397

Steenbruggen, J., Nijkamp, P., & van der Vlist, M. (2014). Urban traffic incident management in a digital society: An actor-network approach in information technology use in urban Europe. *Technological Forecasting and Social Change, 89*, 245–261. doi:10.1016/j.techfore.2013.08.038

Turan, F., & Bayram, O. (2013, February). Information access and digital library use in University students' education: The case of Ankara University. *Procedia: Social and Behavioral Sciences*, *73*, 736–743. doi:10.1016/j.sbspro.2013.02.112

Turney, L., & Pocknee, C. (2004). *Virtual focus groups: New technologies, new opportunities, new learning environments*. www.ascilite.org.au/conferences/perth04/procs/pdf/turney.pdf

Uppal, V., & Chindwani, G. (2013, July). An Empirical Study of Application of Data Mining Techniques in Library System. *International Journal of Computers and Applications*, *74*(11), 42–46. doi:10.5120/12933-0008

Vidich, A. J., & Lyman, S. M. (1994). Qualitative methods: their history in Sociology and Anthropology. In Handbook of Qualitative Research (pp. 23-59). Sage Publications.

Wenger, E. (1998). *Communities of practice: learning, meaning, and identity*. Cambridge Press. doi:10.1017/CBO9780511803932

Whiting, R., Symon, G., Roby, H., & Chamakiotis, P. (2016). Who's Behind the Lens? A Reflexive Analysis of Roles in Participatory Video Research. *Organizational Research Methods*, *21*(2), 316–340. doi:10.1177/1094428116669818

Chapter 2
Critical Autoethnography for Social Justice Research in Doctoral Education

Robin Throne
https://orcid.org/0000-0002-3015-9587
University of the Cumberlands, USA

ABSTRACT

As autoethnography and other methods for self-as-subject research continue to increase in use across doctoral education, this guide proposes to inform the methods in this area specific to critical autoethnography for doctoral scholars desiring to conduct the many forms of social justice research. This includes indigenous research, contemporary feminist research, and arts-based research, which have also seen rise across dissertation research among many disciplines. While many works exist to describe critical autoethnography within specific contexts, few research guides examine critical autoethnography specific for use by the doctoral scholar and specific examples across research focused on societal change and/or disruption of existing power dynamics, lack of parity, or historical trauma from acculturation and/or dispossession. Thus, this chapter offers a concise research methods resource for doctoral scholars and their research supervisors to facilitate use of critical autoethnography across disciplines and among diverse research problems of inquiry.

DOI: 10.4018/978-1-7998-7600-7.ch002

INTRODUCTION

In past works, the author with others have reported on the many challenges faced by doctoral scholars who choose autoethnography or other self-as-subject research designs for dissertation research (Lewis & Throne, 2021). Yet, numerous doctoral scholars have reported the means to overcome these impediments to complete rigorous and valuable critical autoethnography of a multitude of phenomena of inquiry for social justice aims. Specifically, for critical autoethnography, Adams (2017, 2020) noted how and why critical autoethnographies can be used as bridges between the individual experience and the larger sociocultural, sociohistorical, sociopolitical, and/or socioeconomic experience. Even more so, critical autoethnography can be used as a mechanism to propose and discover remedy to unequal power dynamics, inequities, injustices, and other personal, cultural, and social destructive dynamics (Adams, 2017).

While challenges may exist within doctoral programs as to the use of critical autoethnographic dissertation research design, transformative research methods have historically gained interest and increased use due to popularities or trends within doctoral research. As Candy (2011) reported, doctoral research has served to formalize research methodologies especially due to their use within dissertation research. Critical autoethnography, specifically, involves blurred research genres and disciplines while addressing institutionalized power and privilege whether emotional, evocative, performative, or arts-based (Marx et al., 2019; Tilley-Lubbs, 2016). Yet, this blurring has also raised concern by some doctoral research supervisors as a research design too nebulous for doctoral research where systematic methods are expected within conventional dissertation research (Forber-Pratt, 2015) or rejected for educational research (Bhattacharya, 2020). Yet, these views diminish the capacity of autoethnography to examine the experiences of marginalized or oppressed voices who may bring forward alternate views to multiplicity of identities necessary to best understand power dynamics and oppression while simultaneously highlighting the interconnectedness of a global human experience (Boylorn & Orbe, 2016). Unlike objectivists or positivist views, this form of research does not separate lived experience from research (Boylorn & Orbe, 2016).

Critical inquiry was generally defined previously by the author within "qualitative research as research undertaken beyond the theoretical to intentionally engage the political discourse to advance the public good, social justice, power structures, or critical consciousness within a socially-just democratic society" (Throne, 2020, p. 173) and this definition is continued here. Thus, doctoral scholars who aspire to explore a phenomenon of inquiry that involves social justice, or pursuit of a dissertation study with a social change aspect, the conventions of critical inquiry may be used or even organic to the research focus. Further, doctoral scholars who desire to explore the phenomenon of inquiry from the experience of the self may additionally involve a critical aspect to explore a phenomenon that has larger societal implications to effect change. When the method chosen is critical autoethnography, these doctoral scholars must remember that the choice may be continually challenged throughout the doctoral research process; yet, this defense may offer the participant-researcher much reward in so doing (Lewis, & Throne, 2021).

Thus, embarking upon critical autoethnography as method for doctoral research requires a vulnerable doctoral participant-researcher who may take a notable risk in proposing critical autoethnography as research method to the research supervisor. Tilley-Lubbs (2016) further described the vulnerability necessary to conduct critical autoethnography to include the courage and grit necessary to navigate the vulnerability needed for the participant-researcher to see beyond one's own words and actions. Intense reflection and introspection must follow to make meaning and draw implications from the interpretive layers of these lived experiences (Tilley-Lubbs, 2016). Specific to the doctoral researcher, this vulnerability may take the form of self-doubt or other insecurities, very normal occurrences within doctoral education as new investigators embark upon new research methods and research experiences (Kennedy, 2020) and the same can be true for critical autoethnographers.

In addition, the very nature of the generative writing involved in autoethnography may require vulnerability, risk, and truth-telling as a form of resistance necessary for social justice research (Lynch & Kuntz, 2019; Marx et al., 2019). The participant-researcher engages in a "researcher gaze" with an "intense reflexivity and introspection" necessary to the research focus through the respective embodied sociocultural lens, which requires the fortitude necessary to see beyond insidious dominant power and oppressive layers and complexities (Tilley-Lubbs, 2016, p. 1). The objective of the chapter is to offer doctoral scholars with guidance and opportunity for the appropriate use of critical autoethnography for dissertation or other doctoral research across various disciplines and from the lens of social justice research intentions.

BACKGROUND

Koro-Ljungberg and Cannella (2017) summarized four decades of historical critical qualitative inquiry and noted the number of epistemological frames used to view critical inquiry. Regardless of whether the inquiry is feminist, indigenous, poststructuralist, postcolonial, or other, it will shape a social or cultural critique of a particular research problem (Koro-Ljungberg & Cannella, 2017). It is from this view the chapter presents a discussion of how critical autoethnography can be used as a relevant and rigorous social justice research method for doctoral research, especially in the current turbulent sociopolitical climate.

For example, Ladhani and Sitter (2020) noted how critical autoethnography can be used in anti-racism research to challenge traditional research methods and remain cognizant of the power dynamics of the participant-researcher and others. From this approach, social change may result for the researcher from the lens of race when employed (Ladhani & Sitter, 2020). The same claim may be made for other social constructs, theoretical frameworks, or social justice objectives used by doctoral researchers who decide to use critical autoethnography for doctoral research. Likewise, Pennant (2020) used an analytic form of critical autoethnography to explore whiteness to better understand the experience of Black female doctoral researchers. The form allowed the participant-researcher to bring her "whole self" into the work and not compartmentalize research experiences from the lifelong experience of self (Pennant, 2020, p. 2). Ohito (2019)

Traditional doctoral research often leaves little room for doctoral scholars who may desire to use nontraditional methods or seek post-qualitative research spaces that better shape their research interests. Yet, when critical autoethnography is embraced as an appropriate method for contemporary research problems that deserve social justice, research supervisors may welcome the opportunity for doctoral scholars to experience less institutionalized methods and instead accept such activist qualitative methodologies (Lynch & Kuntz, 2019). Further, critical autoethnography allows for new learning of the doctoral scholar who seeks illumination of privilege, power, and marginalization in various research settings (Marx et al., 2019). Gant et al. (2019) described the distinction between autoethnography and narrative reflection as autoethnography situates the experience of self within the broader sociocultural context. Similarly, Olmos-López and Tusting (2020) stressed how autoethnography can be used to enrich theoretical understanding of the phenomenon of inquiry. For doctoral researchers, this can be additionally enhanced by a well designed and complex theoretical or conceptual framework by which to view the phenomenon of inquiry in critical autoethnographic designs for doctoral research. Likewise, this approach may also help to immerse the doctoral research into the scholarship surrounding the phenomenon (Olmos-López & Tusting, 2020).

It is also important for doctoral researchers to situate the participant-researcher from within the insider/outsider dualities of the phenomenon of inquiry. Reed-Danahay (2017) noted the retrospective opportunity the participant-researcher may gain when one is ready to view the lived phenomenon from the outsider/researcher viewpoint. This may assist a doctoral researcher to move beyond the insider/outsider duality to explore the phenomenon from the context of the scholarship and theoretical foundations expected within doctoral education. As a doctoral researcher, Edwards (2019) illustrated how the complexities of identity can be explored through critical autoethnography to gain insight and meaning gleaned from lifelong experiences with the phenomenon as supported by social justice activist Audre Lorde[1]:

I had to define myself for myself. It was not enough for my family and mentors to see me as a scholar. I had to see myself that way for the title to become real to me—I had to believe in myself. I hope that my story encourages women of color, particularly from backgrounds like my own, to also see themselves as scholars (p. 34)

Finally, Rodríguez-Mojica et al. (2019) noted the ability to utilize critical autoethnography to enhance the critical consciousness necessary to view the intersectionalities[2] of power and oppression within race, gender, sexual orientation, language, ethnicity, social class, and immigration status so as to enact social justice research. Critical autoethnography may allow the participant-researcher to illustrate the experience of marginality and gain empowerment through the reflexive narratives of self during the exploration of the phenomenon of inquiry (Rodríguez-Mojica et al., 2019). For example, Navarro et al. (2020) used Freire's[3] concept of *conscientization* as a conceptual framework for autoethnographic narratives to facilitate critical consciousness to engage social justice processes in teacher preparation education.

CRITICAL AUTOETHNOGRAPHY AS SOCIAL JUSTICE RESEARCH METHOD

Various foundational critical inquiry and autoethnography researchers have supported the use of critical autoethnography as a social justice research method. Denzin (2016) stressed the need for critical qualitative inquiry to address pressing social justice aims:

This is a historical present that cries out for emancipatory visions, for visions that inspire transformative inquiries, and for inquiries that can provide the moral authority to move people to struggle and resist oppression (p. 8)

Likewise, Ellis and Bochner (2014) highlighted the need for autoethnographers to interrogate self-perceptions and worldviews to better understand others and society to expand perceptions of inequities and injustices. They noted this expanded view may also identify commonalities (Ellis & Bochner, 2014). Similarly, Adams (2017) highlighted the use of critical autoethnography to challenge assumptions, beliefs, or experiential attitudes surrounding these inequities and injustices to effect change of dangerous views and use the tenets of resistance for transformation and social change. Critical autoethnography may bring remedy to personal and cultural inequities, injustices, and offenses (Adams, 2017). Critical theoretical frameworks may even further strengthen the research lens to justify rigorous and relevant doctoral research.

Thus, it is important for doctoral scholars to understand and utilize the specific conventions of critical autoethnography to confront the dynamics of power and privilege among specific phenomena of inquiry within doctoral research. The first step is to embrace Denzin's (2016) recommendation to apply the transformative paradigm and establish a specific paradigmatic perspective that allows for qualitative critical autoethnography as an appropriate research approach to a given doctoral research problem. From this vantage point, critical inquiry is incorporated with the conventions of autoethnography to challenge

prevailing forms of inequality, poverty, human oppression, and injustice. This paradigm is firmly rooted in a human rights agenda. It requires an ethical framework that is rights and social justice based (p. 8).

Data gathered from self can be a valuable source of personal experiences for doctoral research of social injustice and thereby allow for deeper understanding that may lead to new knowledge and remedy for social conflict and inequities whether wrongdoer or victim within such experiences (Adams, 2017). Autoethnography with this critical lens or standpoint allows for exploration of interpersonal and cultural experiences from the perceptions of self to illuminate greater understanding of others, society, and one's place within it (Ellis & Bochner, 2014). Or as Koro-Ljungberg and Cannella (2017) described, it aims to reduce the reproduction of oppression from current power relations amidst the complexities of power and privilege. When this lens involves the transformative or interpretivist paradigmatic perspective, new knowledge may emerge with valuable societal implications for the research focus (Denzin, 2016; Ellis & Bochner, 2014).

As Holman Jones (2016) reminds, the data gathered within critical autoethnography comprises a living body, ever changing, and it is the responsibility of the critical autoethnographer to continually explore the dynamics between theory, the narrative data, and the paradigmatic perspective or theoretical framework of the doctoral study. From these data, previously disempowered voices of experiences with social injustices or inequities may emerge to empower the doctoral participant-researcher to critically examine these very experiences (Ellis & Calafell, 2020). As the authors purported

Autoethnographers must stop writing about social justice, stop writing about reflexivity, and stop writing about privilege. Merely writing about these things constructs lifeless narratives with little utility for achieving autoethnography's goals. Instead, autoethnographers must centralize the actual material bodies that inhabit the text. Our narratives are alive and breathing, just like the bodies our narratives bring voice to and (often) silence. Without attention to the body, the material stakes of our work are easily forgotten, and Others continue to be erased (p. 206)

For example, doctoral scholar Sah (2019) used critical autoethnography to highlight the social injustices experienced in facing expectations for native scholars to assimilate into the target society for uses of British English. Throughout the autoethnography, the author reflected on the power imbalances and ideologies around the expectations for the target language of the dominant culture and deeply explored the experiences of bias and racialization experienced within the dominant language expectations as well as the marginalization of Nepali English (Sah, 2019). This is a cogent illustration for how critical autoethnography can be used to facilitate voices that may otherwise remain silenced, limiting the construction of new knowledge surrounding such oppression, marginalization, and ongoing dominance of an oppressive culture or system.

Finally, Guishard et al. (2018) stressed the responsibility of social science researchers who intentionally examine social justice issues to disrupt notions of traditional research that involve clear bounds between researcher versus researched. However, boundaries in critical autoethnography are not always easily defined especially for marginalized or oppressed groups who have experienced the particular phenomenon of inquiry under examination (Guishard et al., 2018); thus, it may be important for the doctoral scholar to continually clarify these bounds for an audience that may be comprised of traditional researchers who expect these clear distinctions. For doctoral participant-researchers, the exploration of traumatic phenomenon may require additional support beyond the research supervisor as unprepared supervisors may recommend terminating the study when challenges arise rather than offering the support necessary to continue the introspection necessary to complete the study. Or as Warfield (2019) noted, unboundedness is inherent and essential to subjective approaches and essentially nonstatic. When the participant-researcher is bolstered by a support network that allows for deep introspection, meaningful and insightful critical autoethnographic findings may contribute to the contemporary scholarship to substantially inform or disrupt social inequities to effect social change.

The Critical in Critical Autoethnography

Doctoral scholars may be best served by defining the essential nature of the *critical* aspect of critical autoethnography as critical inquiry early in the research manuscript. More specifically, the critical component of autoethnography conjoins the responsibility of the autoethnographer to the tenets of critical inquiry. When narrative data are generated or gathered from archival artifacts for critical autoethnography, the doctoral participant-researcher uses the everyday language of story for theoretical research, bringing an immediate interplay between the self as researcher, participant, and data analyst (Holman Jones, 2016). Thus, the interrogation of one's own lived experience from this critical lens or framework provides opportunity for illumination of the experience and reclamation of the individual to profit a greater social justice aim—whether inequities are due to some form of similarity or difference, or the many injustices surrounding race, gender, class, ethnicity, age, sexuality, religion, physical or intellectual abilities, or other oppression or disempowerment. Bringing a social justice research lens, framework, or paradigmatic perspective to autoethnography can allow the doctoral autoethnographer with the critical reflexivity necessary to conduct critical autoethnography. Likewise, Koro-Ljungberg and Cannella (2017) noted how critical approaches address power to offer sociocultural contributions to national or global problems and the responsibility to pursue ethical research in so doing.

Critical inquiry can be further enhanced when the social sciences merge with the humanities for forms of arts-based research or creative genre incorporated into critical autoethnography. While some researchers consider critical autoethnography as a creative genre in itself; however, for doctoral research, more likely the conventions of qualitative research must be upheld to ensure rigor and trustworthiness. This may be an important distinction for a doctoral scholar to consider in consultation with the research supervisor early in the research process to ensure the research conforms to institutional requirements and the culminating objectives of the doctoral research. Candy (2011) suggested this can also be accomplished when practice-based research may be allowed for the doctoral study whereby creative forms are used for data representation in the doctoral research manuscript. Further discussion follows for how arts-based research may serve as data collection and data representation for critical autoethnography in a doctoral study. Thus, the shape and form of data representation in critical autoethnography holds many possibilities and yet must conform to some degree to institutional expectations for the research manuscript. Doctoral research supervisors who support self-as-subject research methods can be vital collaborators and mentors to guide new investigators for the appropriate and relevant presentation of research findings using critical autoethnography.

Critical Autoethnography for Contemporary Feminist Research

Critical autoethnography conducted from a framework of contemporary feminist research or theory allows the participant-researcher to offer a gendered lens from the principles of feminism. Gender is a social construct influenced by time, geography, and other perspectives. From this particular construct, the critical autoethnographer offers a rigorous writing practice from the participant-researcher's experience thereby connecting the personal to the larger sociological or cultural context (Leavy & Harris,

2018). Contemporary feminist research encompasses many forms from ecofeminism, intersectionality, cyberfeminism, identity politics, and numerous others. Thus, when conducting critical autoethnography from a specific feminist framework, it is essential that the conceptual or theoretical framework for the doctoral research is well founded and clearly designed to hone the particular lens to view the particular research focus. This is essential for any feminist autoethnographic research to ensure the framework encompasses the sociocultural, sociopolitical, and sociohistorical aspects of feminist theory and research and new power relations that evolve with societal change. As gender also incorporates spatial and temporal influences, space and time may also be aspects that also need and deserve clarification and/or explication as Adams (2020) concluded "the task of repairing the past, of forgiveness, of trying to live with others who have and will continue to espouse harmful discourse in unforeseeable ways, is perpetual, uncertain, and infinite" (p. 909).

In addition, it can be essential for the critical autoethnography to clarify the autoethnographer's own participant-researcher positionality as a precursor to the study design that may further inform the lens to view the study focus. In a study of identity politics of international doctoral student-activists, Zilonka et al. (2019) suggested the critical autoethnography can be comprised of personal, political, emotional, and intellectual perspectives thereby bringing further complexity to the lens that can be used to explore the phenomenon from the viewpoint of the autoethnographer as well as to situate the participant-researcher's positionality within the study framework. While important for all post-qualitative critical inquiry, in doctoral research, this must be well defined as components of the autoethnographic design and the clarity of the critical nature of the viewpoint is essential for reader understanding.

Warfield (2019) stressed how the recognition of the participant-researcher viewpoint and clarity surrounding it can bring another level of transparency to the study as "the embeddedness of values and principles in the design of methods and the production of academics as methodologists something that is too often hidden and disguised" (pp. 169-170). This is an important element, especially for justifying the use of critical autoethnography as method in doctoral feminist research, as committee or external reviewers may define what is and is not an appropriate doctoral research method rather than the paradigmatic tenets of critical autoethnography. For example, Pennant (2020) shared how she formulated a multi-faceted theoretical lens for doctoral study that allowed her to embrace the complexities of Black feminist epistemologies, race, intersectionality, and social class dimensions. Thus, when well designed, critical autoethnography can incorporate the complexities of contemporary feminist theory whilst viewing a doctoral research problem from the participant-researcher's theoretical perspective and social justice aims.

Critical Autoethnography Data Collection, Analysis, and Representation

Like the diversity for data representation as presentation of findings in critical autoethnography, data collection and analysis for doctoral critical autoethnography can take many forms, shapes, and approaches. While critical autoethnography may appear to be open-ended and sometimes critiqued as too imprecise or unformulated for rigorous doctoral research; yet, others argue rigorous and systematic study designs for autoethnography can be accomplished for doctoral research (Forber-Pratt, 2015). What is most important is for the doctoral participant-researcher to understand institutional expectations for research design so as to present systemic and clear procedural steps to meet these requirements. For example, Kennedy (2020) utilized reflexive analytic memos and conversations with an academic advisor to generate data for a critical autoethnography of inherent vulnerabilities of a new researcher in a doctoral practicum. The author used personal memories and self-observation to generate the new data and findings were presented as key excerpts of the data from the memos and advisor conversations. (Kennedy, 2020). Some autoethnographers have referred to such data representation as vignettes and offer guidance to present vignettes of key themes identified from analysis.

When a doctoral researcher is unable to gain permission to utilize critical autoethnography as sole method for doctoral research to present as an additional component combined with another critical qualitive research approach. For example, Pennant (2020) incorporated critical autoethnography as a component of a critical qualitative study that utilized semi-structured interviews for data collection. The participant-researcher was then able to include her own reflexivity of the lived experience with the phenomenon of inquiry into data collection and analysis to ascertain her own Black identity amidst white immersion.

Other doctoral scholars, Ajil and Blount-Hill (2020) brought critical perspectives to a collaborative autoethnography to decolonize Eurocentric ways of knowledge production in criminology. Triangulated data were gathered from generative narratives from prompts, reflections on collaborator narratives, and synchronous conversations as well as WhatsApp messaging. Data analysis employed computer-aided qualitative data analysis using a grounded theory inductive approach and findings were presented as dimensions of experience illustrated by key excerpts from the coded data (Ajil & Blount-Hill, 2020). This type of collection and analysis may be appropriate and accepted for culminating doctoral research as it draws from the accepted conventions of traditional qualitative research and can be described within the institutional structures common to dissertation research.

Arts-based data representation is also common for critical autoethnographers to present findings within a research manuscript. For example, Burkholder and Frawley

(2019) gathered data through capture of conversations and cellphone filming as participatory video and video production. Data representation involved visual images as figures from the cellphilming and conversational excerpts to illustrate the findings reported. The conventions of textual and discourse analysis were used for data analysis that highlight how these methods can be combined for unique approaches to analysis and visually evocative forms of findings presentations. In a critical autoethnography of imposter syndrome, Edwards (2019) used a form of storying of life experience, facilitated by evocative self-questioning, that is sometimes used in narrative research to generate new data and moved on to reconceptualization in real time to capture daily self-talk and social media for emergent data. Data representation involved presentation of narrative essays to synthesize a new definition of scholar. Whether the data findings are represented by images, personal narratives, poetry, other creative genre, or other visual forms, the doctoral scholar must justify how and why the study findings are to be represented in alignment with institutional expectations for the doctoral research manuscript or other presentation of findings.

FUTURE RESEARCH DIRECTIONS

In past work with others, the author has stressed the importance and benefits for doctoral supervisors to encourage and foster doctoral research dissemination (Sweat et al., 2021) and recommended doctoral scholars should be encouraged rather than dissuaded to explore self-inquiry research methods, including critical autoethnography, when paradigmatic perspectives align with interpretivist or transformative qualitative research (Lewis & Throne, 2021). This importance certainly applies to the selection of doctoral critical autoethnography for culminating doctoral research projects or the dissertation research. Likewise, doctoral scholars can be prepared to rigorously defend this method choice and understand the risks in proposing this method may bring self-revelation in addition to sociocultural implications for scholarship contributions (Lewis & Throne, 2021; Sah, 2019). When the critical autoethnographic design is successfully defended, the doctoral researcher may also advance researcher agency as well as bolster the confidence to master autoethnographic data analysis and representation of findings (Sweat et al., 2021; Warfield, 2019). Through ongoing dissemination of these studies, the rationale for the adoption of critical autoethnography as appropriate for doctoral research may in turn also be bolstered.

Other scholars have called for additional research and made recommendations for doctoral education to advance the use of critical autoethnography for doctoral research. Lynch and Kuntz (2019) challenged interpretations of an institutional review board (IRB) that deemed critical autoethnography as not-human-subjects-research and thus not requiring IRB review. While this allowed the doctoral research to move forward, the authors indicated the IRB likely did not understand the ethical assurances necessary to a participant-research embarking upon this level of self-introspection and exploration. In spite of this, it is necessary for the doctoral scholar to articulate the ethical assurances to be applied to the critical autoethnography to ensure participant-researcher self-care, especially when the phenomenon of inquiry may involve experiences of lived trauma. Further, the authors called for reconsideration for what constitutes appropriate research areas of focus, measures of data trustworthiness, and to disrupt traditional methods of interpretation to ensure autoethnographers are given space for truth-telling, forthright exploration, and means of deep introspection necessary to the critical autoethnography (Lynch & Kuntz, 2019).

International doctoral scholars, Zilonka et al. (2019) called for more critical autoethnography in U.S. doctoral education to allow for more open reflections of "the challenges of female international doctoral students who dismantle the discriminatory institutional structures" to foster "a more humane future" for academics who may stay within U.S. academia or return post-doctorate (p. 46). In regard to critical qualitative inquiry in general, Koro-Ljungberg and Cannella (2017) called for the necessity of ongoing critical inquiry as pertinent now as it may have been two decades ago to ensure transformative research continues "to inform a contemporary awareness of human privilege" (p. 327). Marx et al. (2019) noted how critical autoethnography may answer this call to foster critical self-examination of educational inequities to allow educators to use systematic exploration as deep introspection to seek understanding, justice, and equity. As critical autoethnography may allow educators, whether doctoral research supervisors or doctoral faculty, to allow doctoral scholars to use the conventions and principles of conscientization and transformational resistance or other critical inquiry for research (Navarro et al., 2020). Further, may doctoral research supervisors allow these varied, complex, and creative forms of theoretical or conceptual frameworks, and data collection, analysis, and representation, as unique as each doctoral scholar so that systemic, rigorous, and appropriate critical autoethnography may be conducted for doctoral research to effect social change and/or social justice aims.

CONCLUSION

The chapter highlights the vulnerability and risk required of doctoral researchers who desire to employ critical autoethnography for doctoral research while simultaneously illustrating the need for more critical autoethnography. While self-inquiry methods have increased in use across doctoral education, this chapter provides guidance specific to the use of critical autoethnography for doctoral scholars who have intention and strong desire to conduct social justice research or effect social change as a focus or component of doctoral research. While many works exist to describe critical autoethnography within specific contexts, few research guides examine critical autoethnography specific for use by the doctoral scholar and specific examples across research focused on societal change and/or disruption of existing power dynamics, lack of parity, or historical trauma from acculturation and/or dispossession. It is essential for the doctoral scholar to justify these critical aspects of the autoethnography amidst institutional guidelines or expectations and formulate an appropriate theoretical or conceptual lens by which to view the social justice intention of the work. The doctoral scholar will likely need to present a compelling case to defend the research design to gain acceptance of critical autoethnography for culminating doctoral research in many disciplines. However, when sole use of critical autoethnography is deemed not acceptable for doctoral research, it may be combined with other forms of critical qualitative inquiry to supplement the participant-researcher's experience with other participants for social justice contributions. Thus, this chapter may complement the many research guides available to doctoral scholars and their research supervisors to facilitate appropriate critical autoethnography across disciplines and among diverse research problems of inquiry with social justice aims.

REFERENCES

Adams, T. E. (2017). Critical autoethnography, education, and a call for forgiveness. *International Journal of Multicultural Education*, *19*(1), 79–88. doi:10.18251/ijme.v19i1.1387

Adams, T. E. (2020). Critical rhetoric, relationality, and temporality: A case for forgiveness. *International Journal of Communication*, *14*, 899–911.

Ajil, A., & Blount-Hill, K. L. (2020). "Writing the other as other": Exploring the othered lens in academia using collaborative autoethnography. *Decolonization of Criminology and Justice*, *2*(1), 83–108. doi:10.24135/dcj.v2i1.19

Bhattacharya, K. (2020). Autohistoria-teoría: Merging self, culture, community, spirit, and theory. *Journal of Autoethnography*, *1*(2), 198–202. doi:10.1525/joae.2020.1.2.198

Burkholder, C., & Frawley, A. (2019). Revisiting and Disrupting Uncritical Diversity Narratives through Autoethnography and Cellphilming. *LEARNing Landscapes*, *12*(1), 63–76. doi:10.36510/learnland.v12i1.979

Candy, L. (2011). Research and creative practice. In L. Candy & E. A. Edmonds (Eds.), *Interacting: art, research and the creative practitioner* (pp. 33–59). Libri.

Crenshaw, K. W. (1991). Mapping the margins: Intersectionality, identity politics, and violence against women of color. *Stanford Law Review*, *43*(6), 1241–1299.

Crenshaw, K. W. (2017). *On intersectionality: Essential writings*. The New Press.

(2016). Critical autoethnography as method choice. InBoylorn, R. M., & Orbe, M. P. (Eds.), *Critical autoethnography: Intersecting cultural identities in everyday life* (pp. 13–26). Routledge.

Denzin, N. K. (2017). Critical qualitative inquiry. *Qualitative Inquiry*, *23*(1), 8–16. doi:10.1177/1077800416681864

Edwards, C. W. (2019). Overcoming imposter syndrome and stereotype threat: Reconceptualizing the definition of a scholar. *Taboo: The Journal of Culture and Education*, *18*(1), 3. doi:10.31390/taboo.18.1.03

Ellis, C., & Bochner, A. (2014). Merging culture and personal experience in critical autoethnography. In R. M. Boylorn & M. P. Orbe (Eds.), *Critical autoethnography: Intersecting cultural identities in everyday life* (pp. 9–10). Routledge.

Ellis, C., & Bochner, A. P. (2000). Autoethnography, personal narrative, reflexivity: Researcher as subject. In N. K. Denzin & Y. S. Lincoln (Eds.), *Handbook of Qualitative Research* (2nd ed., pp. 733–768). SAGE.

Ellis, C. D., & Calafell, B. M. (2020). Toward praxis: Interrogating social justice within autoethnography. *Journal of Autoethnography*, *1*(2), 203–207. doi:10.1525/joae.2020.1.2.203

Forber-Pratt, A. J. (2015). "You're going to do what?" Challenges of autoethnography in the academy. *Qualitative Inquiry*, *21*(9), 821–835. doi:10.1177/1077800415574908

Freire, P. (1970). Cultural action and conscientization. *Harvard Educational Review*, *40*(3), 452–477. doi:10.17763/haer.40.3.h76250x720j43175

Gant, V., Cheatham, L., Di Vito, H., Offei, E., Williams, G., & Yatosenge, N. (2019). Social work through collaborative autoethnography. *Social Work Education*, *38*(6), 707–720. doi:10.1080/02615479.2019.1570109

Guishard, M. A., Halkovic, A., Galletta, A., & Li, P. (2018). Toward epistemological ethics: Centering communities and social justice in qualitative research. *Forum Qualitative Sozialforschung/Forum: Qualitative. Social Research*, *19*(3). Advance online publication. doi:10.17169/fqs-19.3.3145

Kennedy, L. M. (2020). Confessions of a novice researcher: An autoethnography of inherent vulnerabilities. *Qualitative Report*, *25*(6), 1526–1539. https://nsuworks.nova.edu/tqr/vol25/iss6/7/

Koro-Ljungberg, M., & Cannella, G. S. (2017). Critical qualitative inquiry: Histories, methodologies, and possibilities. *International Review of Qualitative Research*, *10*(4), 327–339. doi:10.1525/irqr.2017.10.4.327

Ladhani, S., & Sitter, K. C. (2020). The revival of anti-racism. *Critical Social Work*, *21*(1), 54–65. doi:10.22329/csw.v21i1.6227

Leavy, P., & Harris, A. (2018). *Contemporary feminist research from theory to practice*.

Lewis, C., & Throne, R. (2021). Autoethnography and other self-inquiry methods for practice-based doctoral research. In R. Throne (Ed.), *Practice-based and practice-led research for dissertation development*. IGI Global. doi:10.4018/978-1-7998-6664-0

Lorde, A. (1984). *Sister outsider: Essays and speeches*. Crossing Press.

Lynch, S., & Kuntz, A. (2019). A critical autoethnography of a doctoral students' research journey: Learning to take risks in the academy. *Curriculum Studies in Health and Physical Education*, *10*(2), 156–171. doi:10.1080/25742981.2019.1588762

Marx, S., Pennington, J. L., & Chang, H. (2017). Critical autoethnography in pursuit of educational equity: Introduction to the IJME special issue. *International Journal of Multicultural Education*, *19*(1), 1–6. doi:10.18251/ijme.v19i1.1393

Navarro, O., Shah, J. K., Valdez, C., Dover, A. G., & Henning, N. (2020). Fighting on all fronts: The push, pull, and persistence of social justice educators and the move to reimagine teacher preparation. *Teacher Education Quarterly*, *47*(3), 9–31.

Ohito, E. O. (2019). Thinking through the flesh: A critical autoethnography of racial body politics in urban teacher education. *Race, Ethnicity and Education*, *22*(2), 250–268. doi:10.1080/13613324.2017.1294568

Olmos-López, P., & Tusting, K. (2020). Autoethnography and the study of academic literacies: Exploring space, team research and mentoring. *Trabalhos em Lingüística Aplicada*, *59*(1), 264–295. doi:10.1590/010318136565715912020

Pennant, A. L. M. (2020). My journey into the 'heart of whiteness' whilst remaining my authentic (Black) self. *Educational Philosophy and Theory*, 1–12. doi:10.1080/00131857.2020.1769602

Reed-Danahay, D. (2017). Bourdieu and critical autoethnography: Implications for research, writing, and teaching. *International Journal of Multicultural Education, 19*(1), 144–154. doi:10.18251/ijme.v19i1.1368

Rodríguez-Mojica, C., Briceño, A., & Muñoz-Muñoz, E. R. (2019). Combating linguistic hegemony: Preparing and sustaining bilingual teacher educators in the United States. *Teacher Education Quarterly*, *46*(3), 57–78.

Sah, P. K. (2019). Academic Discourse Socialization, Scaler Politics of English, and Racialization in Study Abroad: A Critical Autoethnography. *Qualitative Report*, *24*(1), 174–192.

Sweat, K., Yedgarian, V., Rigler, K., Maddox, S., & Throne, R. (2021). Doctoral practitioner researcher agency and the practice-based research agenda. In R. Throne (Ed.), *Practice-based and practice-led research for dissertation development*. IGI Global.

Throne, R. (2020). Dispossession of land cultures: Women and property tenure among Lowcountry heirs in the Gullah Geechee Corridor. In U. T. Onyebadi (Ed.), *Multidisciplinary issues surrounding African Diasporas* (pp. 152–174). IGI Global. doi:10.4018/978-1-5225-5079-2.ch007

Tilley-Lubbs, G. A. (2016). Critical autoethnography and the vulnerable self as researcher. In *Re-telling our stories* (pp. 1–15). Brill Sense. doi:10.1007/978-94-6300-567-8_1

Warfield, K. (2019). Becoming method (ologist): A feminist posthuman autoethnography of the becoming of a posthuman methodology. *Reconceptualizing Educational Research Methodology*, *10*(2-3), 147–172. doi:10.7577/rerm.3674

Zilonka, R., Cai, S., Medina, N. E. C., & Chung, G. Y. (2019). "Where are we from?": A critical community autoethnography of place, space, and belonging by Ph. D. international female student-scholar-activists in the US. *The International Journal of Critical Pedagogy*, *10*(1).

ADDITIONAL READING

Briscoe, F. M., & Khalifa, M. A. (2015). *Becoming critical: The emergence of social justice scholars*. SUNY Press.

Dávila, E. R. (2019). Collaboration, community, and collectives: Research for and by the people. In K. K. Strunk & L. A. Locke (Eds.), *Research methods for social justice and equity in education* (pp. 289–294). Palgrave Macmillan. doi:10.1007/978-3-030-05900-2_25

Fisher-Yoshida, B., & Lopez, J. C. (2021). *Redefining theory and practice to guide social transformation: Emerging research and opportunities*. IGI Global. doi:10.4018/978-1-7998-6627-5

Holman Jones, S., & Pruyn, M. (Eds.). (2017). *Creative selves / Creative cultures: Critical autoethnography, performance, and pedagogy*. Palgrave MacMillan.

Mackinlay, E. (2019). *Critical writing for embodied approaches: Autoethnography, feminism and decoloniality*. Springer. doi:10.1007/978-3-030-04669-9

Orbe, M. P., & Boylorn, R. M. (Eds.). (2016). *Critical autoethnography: Intersecting cultural identities in everyday life*. Taylor & Francis.

Phiona, S. (Ed.). (2020). *Critical autoethnography and intercultural learning: Emerging voices*. Routledge.

Plakhotnik, M. S. (2020). Understanding social identity through autoethography: Building intercultural communication competencies in higher education classroom. In *Multicultural instructional design: Concepts, methodologies, tools, and applications* (pp. 1414–1435). IGI Global. doi:10.4018/978-1-5225-9279-2.ch067

KEY TERMS AND DEFINITIONS

Autoethnography: An often-cited definition of autoethnography was offered by Ellis and Bochner (2000) as "auto" (the self), the "ethno" (the culture), and the "graphy" (the research process).

Data Representation: Presentation of research findings in critical autoethnography may differ from traditional qualitative and quantitative research presentation. Data representation is a term used for the various forms selected by the critical autoethnographer to present the essential research findings. While other critical autoethnography may be somewhat unbounded in form, data representation choices for doctoral research may likely need to conform to some degree with institutional manuscript requirements.

ENDNOTES

1 See Lorde (1984).

2 See Crenshaw (1991, 2017).

3 See Freire (1970).

Chapter 3
Guidelines for Conducting a Critical Realist Case Study

Deepak Saxena
https://orcid.org/0000-0002-9331-3799
Birla Institute of Technology and Science, Pilani, India

ABSTRACT

The case study is a widely used methodology among qualitative researchers irrespective of their philosophical orientation. While positivist and interpretive philosophies are the two most popular research philosophies across diverse research fields, critical realism offers a third alternative. Critical realism is a research philosophy that assumes the existence of an independent reality but also accepts that there may be varied interpretations of it due to a difference in context. Mechanisms are the theoretical building blocks of critical realism, and presence, absence, or interaction of certain mechanisms may result in the presence or absence of certain events. However, limited guidelines are available on conducting a critical realist case study. This chapter fills this gap by providing some practical guidelines on how a CR-based case study may be planned and executed. Practical guidelines are offered for framing the research question, data collection, writing a narrative, coding, and explanation building aligned with a critical realist philosophy.

DOI: 10.4018/978-1-7998-7600-7.ch003

INTRODUCTION

Across research fields, two most prevalent research philosophies are positivist and interpretive (Alharahsheh & Pius, 2020; Mingers & Standing, 2017). Positivist studies mostly use quantitative methods (e.g. surveys or experiments) to collect data on variables of interest and to statistically test the hypotheses. It may also use qualitative methods (e.g. interview), but the central assumption is that reality can be 'objectively' captured in terms of quantitative variables (Guba & Lincoln, 1994). In contrast, interpretive studies assume that one can understand the social world by capturing the first-hand experience of the participants and tries to capture the knowledge of participants during investigation (Burrel & Morgan, 1979). For this reason, interpretive studies often use qualitative methods (e.g. interviews or focus groups) for data collection. If any quantitative data is collected, it is used to capture the context rather than being the central focus of a study (Maxwell, 2018).

In contrast to the positivist and interpretive stance, Critical Realism (CR) has been proposed (Bhaskar, 1975, 1989) as a philosophical alternative. In contrast to the positivist assertion that reality exists and is completely comprehendible, critical realists acknowledge that it is imperfectly comprehendible due to our limited cognition and fundamentally complex nature of reality (Burrell & Morgan, 1979). However, at the same time, CR also admits that our interpretations are always bound by the tolerance of an external reality which exists independent of our cognitive processes (Sayer, 2010). In other words, while CR accepts epistemic relativity – that knowledge is always local and historical, it rejects judgmental relativity – that all viewpoints must be equally valid (Mingers, 2014).

Case study is a widely used methodology among qualitative researchers irrespective of their philosophical orientation. Yin (2017) defines case study as an empirical inquiry that investigates a phenomenon in its real-world context using multiple source of evidence. While Yin (2017) remains a highly reputed sourcebook for cast study research, its focus mainly remains positivist. Alternatively, guidance is also available on employing case study in interpretive research (Andrade, 2009). However, except perhaps for Wynn and Williams (2012), not many detailed guidelines are available on conducting a case study with CR philosophy. This chapter fills this gap by providing some practical suggestions on how a CR-based case study may be planned and executed. The next section provides a brief introduction to the philosophy and the nature of explanation in a critical realism. Subsequent section outlines the role of qualitative method and case study in critical realism. This is followed by guidelines for framing the research question, data collection considerations, writing a narrative, coding and explanation building while following a critical realist philosophy. Future research directions are noted before concluding the chapter.

BACKGROUND

Bhaskar (1975) initially proposed CR for natural sciences arguing for deeper explanation beyond statistical generalization espoused by positivist philosophy. Later, he extended it to social sciences (Bhaskar, 1989) arguing for the explanation to move beyond the constructivist accounts of the phenomenon. The critical realist paradigm is 'realist' in the sense that it assumes that an external reality – both social and physical – exists irrespective of our acknowledgement or interpretation (Johnson & Duberley, 2000). For a critical realist, physical reality (e.g. flow of liquid in a pipeline), or a social reality (e.g. knowledge flows in an organization) exists irrespective of our understanding of it. The qualifier 'critical' denotes that our claims about reality are to be accepted with the assumption that there are significant limitations in terms of the objectivity of our knowledge (Mingers, 2014). At the same time, truth claim in CR is more than the result of social construction. For a critical realist, knowledge generation entails both the engagement with an independent reality and ongoing interpretation of it (Johnson & Duberley, 2000). In other words, social constructions process is always bound by an external reality (Sayer, 2010) which exists independently.

Ackroyd and Fleetwood (2000) observe that what prevents critical realists from assuming that the world is merely socially constructed or is concept dependent, is a commitment to materialism. As Sayer (2010) puts it – "Social beings live neither on bread alone nor on ideas and symbols alone" (p. 34-35). For realists, there is always some material substance beneath the concept-dependent nature of social reality. Mingers (2014) note that CR accepts the existence of different types of objects of knowledge (physical, social, and conceptual) which helps in adequately explaining the phenomena. Social structure and culture are said to be inscribed within the material components of the physical artefacts in a way that may help sustain these structural and cultural relations. For instance, our emotional need for connecting with others are inscribed within the social media apps on our smartphones. Therefore essentially, a CR based explanation presents a complex and dynamic interaction between material and non-material.

As proposed by Bhaskar (Bhaskar, 1975, 1989; Mingers, 2014; Mingers & Standing, 2017), CR stratifies reality into three nested domains (Figure 1). The domain of the *real* exist independent of human cognition. It includes the entities, the structures and the causal powers inherent to them as they independently exist. Structure may be defined as "a set of internally related objects and practices" (Sayer, 2010, p. 92) whose causal power, termed a mechanism in CR, are emergent i.e. "cannot be reduced to those of their constituents" (Sayer, 2010, p.119). While the structure is usually directly observable (e.g. the bureaucratic structure), mechanisms are usually inferred by actual events they produce. The domain of the *actual* is considered a subset of the real. It includes the events that occur due to the enactment of the causal powers of the structures and entities. These events may or may not be observed by humans. Finally, the domain of the *empirical* is understood to be a subset of the actual and it consists of our experiences of the events that we are able to perceive and/ or measure. The picture of the reality is thus one of a complex interaction between dynamic, open, stratified systems (Mingers, 2014) where context and structure give rise to certain mechanisms. The interaction of these mechanisms, where one often counterbalances another, causes the presence or absence of actual events.

Figure 1. Three domains of reality in critical realism

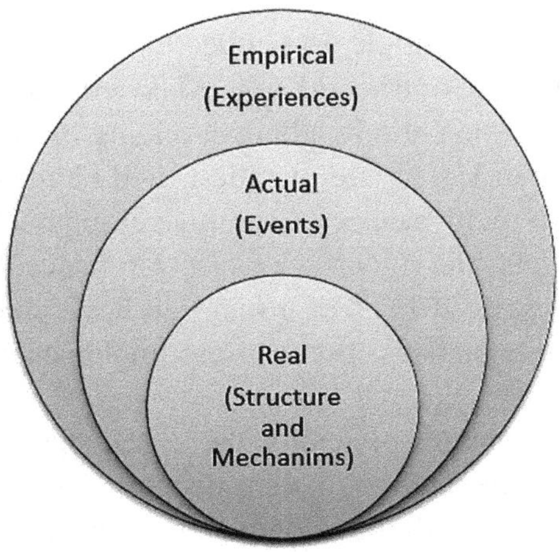

Going beyond statistical regularity (as in positivism) or thick description (as in interpretivism), CR looks for the existence of causal mechanisms (Mingers & Standing, 2017) that result in the phenomena of interest. A CR researcher's goal is to identify the mechanisms that emerge from the components of a physical and social structure to produce the events of interest (Sayer, 2010) or result in a statistical regularity (Tsang, 2014). An event or regularity may be created by presence, absence or interaction among certain mechanisms (Bhaskar 1975, 1989). Therefore, a causal explanation in CR will account for a set of existing and enacted mechanisms, along with the impact of any structural factors and contextual conditions that generated the outcome being studied. It may be noted, however, that a mechanism-based explanation is not like that deterministic laws of natural sciences in which certain inputs lead to certain outputs without an exception (Anderson et al., 2006) of the law. Instead, the outcome of a critical realist study is a 'sometimes-true' theory (Coleman 1964)

by moving beyond the description and providing explanation of unfolding events within a contextual condition, but not predicting anything in a context-free manner.

ROLE OF QUALITATIVE DATA AND CASE STUDY IN CR

Having explained the assumption on reality and the nature of explanatory mechanisms, this section highlights the critical role qualitative data and case study methodology play in a critical realist study.

Qualitative Data in CR

CR recommends the use of multiple data collection methods which could also be understood as elaborate form of triangulation (Maxwell, 2018). Since CR assumes that objectivity can never be entirely attained, relying on many different sources makes it less likely that distorted interpretations will be made (Guba, 1990). It is argued that because of its multi-method and multilevel approaches to causal analysis, CR offers a way to address the rigour–relevance gap in academic research (Wynn & Williams, 2012; Mingers, 2015). For instance, it has a potential to draw clearer links (in the form of mechanisms) between the implementation of interventions and their outcomes.

Depending upon the nature of reality the researcher wants to capture, CR allows for both quantitative and qualitative data collection methods. Therefore, researchers following CR paradigm recommend either mixed (Zachariadis et al., 2013) or qualitative (Wynn & Williams, 2012; Tsang, 2014) data collection methods. However, since CR fundamentally accepts that social systems are inherently open systems and are not amenable to experimentation, qualitative methods are deemed more useful. In other words, relatively less focus is put on quantitative methods for examining statistical regularities, and more focus is put on developing a qualitative understanding of social relations and objects (Sayer, 2010).

Sayer (2010) provides a useful guideline regarding the nature of data and measurements for critical realist studies. For Sayer, the nature of data would depend upon the type of change being studied. Here he differentiates between three types of change: the purely quantitative (e.g. change in income level), that reducible to a qualitative phenomenon (e.g. change in lifestyle), and the changes that is irreducibly qualitative (e.g. change in outlook towards life). For Sayer, the first and second types of change may be mathematized. In the case of the third type of change, however, quantitative measurements would not adequately capture the emergence and enactment of mechanisms, for instance the thought process resulting a change in outlook towards life. Statistical methods usually omit the contextual factors except those codified in the variables selected for measurement (George & Bennett, 2005; Maxwell, 2018). In so doing, quantitative approaches may leave out many contextual and intervening variables that may be crucial for the explanation. Hence, qualitative methods are integral to a critical realist study.

Case Study in CR

Ackroyd (2012) observes that a study based on the tenets of critical realism conceives the social phenomena in terms of underlying processes that comprise them. Thus, research procedures must take account of the processual nature of phenomena. In a processual analysis, the researchers' repetitive questioning about 'how' embodies this constant search for underlying mechanisms (Pettigrew, 2012) which drive the processes. Therefore, the CR methodology is geared towards inquiry in more natural settings, collecting more situational information, reintroducing discovery as an element in inquiry, and soliciting diverse viewpoints (Guba & Lincoln, 1994) to arrive at an explanation that relates closely to the external reality. While Bhaskar did not recommend any specific method of inquiry for conducting a critical realist study, case research strategy is often recommended (Ackroyd, 2012; Easton, 2010; Mingers, 2014; Wynn & Williams, 2012; Tsang, 2014) by the scholars. Easton (2010) argues that CR not only substantiates case research as a research strategy but also provides helpful implications for both theoretical development and research process. Wynn & Williams (2012, 2020) also stress that case study is well-suited to conduct critical realist research since it enables researchers to develop detailed context-sensitive causal explanations of specific phenomena. Finally, Tsang (2014) stresses that the case research strategy augurs well with the critical realist view that closure conditions are rarely achievable in the social sciences.

Different recommendations exist in the literature for the ideal number of cases to study events and processes. Eisenhardt (1989) and Van de Ven (1992) suggest multiple cases for pattern identification and empirical generalization. Pettigrew (2012) prescribes a matched pair of cases such as cases with similar context but dissimilar outcomes or cases with dissimilar context but similar outcomes. However, Dawson (2019) reminds us that the researcher should not equate more cases with higher-level data and instead the focus should be on finding rich details. The aim of a case study is not to make predictions about a sample population but to explore a phenomenon in depth (Yin, 2017; Tsang, 2014). Langley (1999) suggests that the number of cases should be driven by the nature of research question. If the research question aims to provide a holistic understanding of specific events and processes, even one detailed case is deemed enough.

George and Bennett (2005) argue that single, longitudinal case study would be more useful in examining the operation of causal mechanisms in detail. Single case design allows the researcher to inductively observe any unexpected aspects of the operation of a particular causal mechanism, or to identify the contextual conditions that activate the causal mechanism. They note that statistical methods, which omit all contextual factors, except those codified in the variables selected for measurement, essentially leave out many contextual and intervening variables. They also note that using single case, researcher may also use theories on causal mechanisms to give historical explanation of the case. The longitudinal component allows the researcher to study the emergence of causal mechanisms, their interactions, and resultant outcomes with time (Pettigrew, 2012; Dawson, 2014). Next section offers solutions and recommendations on how to plan and execute a critical realist case study.

SOLUTIONS AND RECOMMENDATIONS

This section offers suggestions on each step associated with a critical realist case study – framing a research question, data collection, case writing, analysis, and explanation building. The concept of validity is also discussed in relation to critical realism.

Framing a Research Question

Yin (2017) suggests that case research is suitable for 'what' or 'how' type question. As noted earlier, CR-based research looks to identify mechanisms underpinning events. Consequently, research question for a CR based study may ideally take the form of – 'What are the mechanism that underpin the phenomenon of interest?' (Sayer, 2010) or 'How do certain mechanisms influence the phenomenon of interest?' (Wynn & Williams, 2020). Former is useful when the research is exploratory in nature and the researcher has little or no idea on underlying mechanisms. The latter is useful when there are several candidate mechanisms and the researcher wishes to study the influence of those mechanism on the outcome.

Data Collection Considerations

Scholars (Maxwell, 2018; Sayer, 2010) suggest using the full spectrum of data capturing the social structures, conditions, agency, and events in a critical realist study. It is also recommended to study the mechanisms over time by collecting longitudinal data. Pettigrew (2012) asserts that the only way to reveal the relationship between various levels of analysis is to collect longitudinal data to show how mechanisms at various levels interact to influence the events. Table 1 presents possible sources of evidence (Yin, 2017), along with considerations for a critical realist case study.

Table 1. Data collection considerations in a critical realist case study

Source of Evidence	Data Collection Consideration
Documentation and archival data	• Very useful for identifying the sequence of events, for hypothesizing potential mechanisms, and for identifying interview participants. • Focus on the documentation at multiple level.
Interviews	• Most useful data collection method, especially useful in capturing the operations and outcome of the mechanisms. • Should be flexible enough to allow participants to include what they consider important.
Observations	• Too close to understand the operation of the mechanism, may be used as a triangulation method.
Artifacts	• May be useful in capturing the physical operations and outcome of the mechanisms.

Documentation and Archival Data

Secondary data collection in the form of documentation and archival data may be the starting point for a critical realist case study. Secondary data can be used to identifying the sequence of events, a key requirement for a critical realist analysis, for the case under investigation. Once the sequence of events is identified, it may also help in hypothesizing potential mechanisms that underpin the sequence of events. Moreover, secondary data may also be used to determine key actors as interview participants. Since a diverse set of mechanisms may be active at multiple levels, the secondary data collection accordingly should look for data at multiple levels. For instance, events at the macro level (e.g., an economic recession, a pandemic, or a new government regulation) may influence the events at the organization level (e.g., a low staffing level). Similarly, a technology or social affordance (e.g. ability to share information and being connected with others) may be underpinning a social change in a community (e.g., a social innovation or a movement). Hence, a critical realist study should not only collect data at the level of events, it should also collect data on contextual conditions that may be playing a role in activating the mechanisms under investigation (Maxwell, 2018). Furthermore, since the operation and influence of mechanisms can be observed only over time, secondary data about the past (e.g., economic data, socio-political events, government and newspaper reports) also need to be collected for a critical realist study.

Interviews

Interviews are perhaps the most important data collection method for conducting qualitative research in general, and for a critical realist study in particular (Maxwell,

2018). Interviews are especially useful in capturing the social mechanisms (Avgerou, 2013). The sequence of events identified via secondary data analysis may play a crucial role here. If the researcher has a general idea on the sequence of events, she may probe deeper when interviewing the participants. During the interviews, however, care should be taken in not suggesting any mechanisms per se and capturing the social reality as participants understand it. The interview should be flexible enough to allow participants to include what they consider important. In this regard, the responsive interviewing model developed by Rubin and Rubin (2011) could be a very useful technique. In responsive interviewing, the researcher generally starts with an open-ended question and devises additional questions depending on emerging details. In most cases, it might be useful to spend at least a couple of minutes on the participants' background so that they are at ease with the interviewer and the interviewer also gets a glimpse on their motivations and the mindset.

A typical challenge faced by new researchers is in adhering to the multi-level and the longitudinal element of a critical realist study. In ideal situation, the researcher should try to interview respondents from multiple stakeholder groups that have an influence on or are influenced by the sequence of events. However, it may not be possible due to access and resource limitations. In such situation, the interviewer may ask the participants on their understanding and their view on the events from other levels. For instance, if the policymakers cannot be interviewed on a certain initiative, the implementers and the public make be asked about the policymaking and evaluation (e.g. how should this policy ideally be framed and implemented?). Similar considerations may be adopted regarding longitudinal element. When quantitative researchers use the term longitudinal, they often mean to collect data on same set of variables at regular intervals. However, due to required time and effort, qualitative researchers often do not have this luxury. Hence, a useful strategy for qualitative researchers would be to ask participants about key events from the past and ask them to describe how earlier events influenced subsequent ones. While this strategy runs the risk of post-hoc rationalization (Yin, 2017), it still serves the purpose of capturing the longitudinal element when data is triangulated across participants.

Observations

Observation is a useful mechanism if the events under consideration are still unfolding. The type of observation could be participant observation (in which the researcher also participates in the activities), or non-participant observation (in which the researcher merely notes her observation). While observation is a frequently used method in certain disciplines such as anthropology and sociology, its applicability may be limited in some other disciplines such as management or history. While observation helps in being closer to the field to generate rich descriptions, it may be too close to understand the operation of a mechanism in a critical realist study. Hence, observation may be used as a triangulation device when on the field, and as supporting evidence when presenting the analysis (Maxwell, 2018). During the analysis, coded observations may be linked to emerging understanding of the causal mechanisms involved (Aaltonen & Tempini, 2014).

Artifacts

While physical artefacts are usually ignored as a data collection method by the researchers, sociomaterial thought (Hultin, 2019; Niemimaa, 2014) has resulted in elevating the importance of material artefacts during data collection process. For instance, Fayard and Weeks (2007) exhibit how the floor plan and the location of photocopier / water cooler may influence social interaction in an office setting. Since a critical realist gives equal precedence to the physical and social reality (Johnson & Duberley, 2000; Mingers, 2014), data collection on material artefacts is strongly recommended in a critical realist study. For instance, publicity materials, photographs, and videos may be collected when studying a social movement. As an alternative example, a research in organizational setting may collect the screenshots of the software in use, system logs, or photographs of the office space. With the technological progress, digital technology provides a handy set of physical artefacts for the purpose of research and analysis. As a recent example, Pentland et al (2020) employed digital traces (e.g. computerized event logs and middleware data) to analyze the sequence of events. Similarly, Jong and Dückers (2016) studied the spread of rumors and correction mechanisms based on the content analysis of the tweets posted after an event. A critical realist study, especially the one investigating technological mechanisms, may benefit from collection and analysis of such digital artefacts.

Writing a Narrative

Once the data is collected, a case narrative should be written describing the events, structure, and context (Wynn & Williams, 2012). As suggested by George & Bennett (2005) the narrative should be in the form of process tracing which is highly specific to the case. The narrative should focus on providing pure description (Walcott, 2009) presenting the events as they unfolded *without* any efforts to interpret or analyze so that richness of the case is not compromised. This strategy involves construction of a detailed story from raw data. A critical realist avoids excessive data reduction and tries to present the process as completely as possible (Langley, 1999), including competing narratives from diverse stakeholders (Dawson & Buchanan, 2005; Guba & Lincoln, 1994). Along with the context and experiences, the narrative should also include outcomes (Pettigrew, 2012; Wynn & Williams, 2020) that are observed as a result of the events under investigation. By capturing the context, events, experiences and outcomes, narrative serves as an additional data sources when identifying the mechanisms.

Data Reduction and Coding

The process of qualitative coding and data reduction may begin once the narrative is written. This is suggested to avoid influencing the narrative by ongoing data analysis. Critical realists try to identify mechanisms using retroductive logic (Mingers & Standing, 2017; Tsang, 2014; Wynn & Williams, 2020). This is a combination of both inductive logic (where the researcher starts from empirical data and then arrives at a conclusion, mostly followed in interpretivism) and deductive logic (where the researcher starts from theoretical propositions and then tests with empirical data if the conclusions are correct, mostly followed in positivism). In contrast, in retroduction the researcher moves back and forth between the data and explanation. The researcher needs to work across levels of empirical, actual and real. As Johnson & Duberley (2000) suggest, critical realist explanation moves from a description of a phenomena to a description of a mechanism and structure which either produces the given phenomenon or is a condition for it. Retroduction is iterative in nature as the researcher frequently moves back and forth between the data and explanation (Fletcher, 2017; Saxena, 2019). Through this iterative process, researcher improves the understanding of those mechanisms (Mingers & Standing, 2017; Wynn & Williams, 2020).

Analysis in critical realism often follows a grounded theory type three-stage process (Hoddy, 2019; Miles et al., 2014) for coding and analysis. It starts with the first cycle coding in which codes are assigned directly to the chunk of data. It is recommended that all the interview transcripts and secondary data are coded at this stage. This stage corresponds to the induction phase of the analysis. The second cycle coding takes the first cycle codes and combines them based on their similarity. At this stage, the researcher may also bring codes from the literature. In that case, second cycle coding may correspond to the deduction phase of the analysis. Finally, third cycle coding would try to identify a mechanism by simultaneously combining the second order codes and examining the explanatory power of resultant mechanisms. For interested readers, Saxena (2019) and Fletcher (2017) provide detailed examples of the retroduction process in a critical realist study.

Tsang (2014) notes four types of retroduction which may work as tentative framework for the identification of mechanisms. The four types of retroduction are overcoded, undercoded, creative, and meta-retroduction. In *overcoded retroduction*, the mechanism is drawn from existing mechanism that is shown to have an influence on the events in question. For instance, Affordance is currently a established mechanism in the literature when it comes to theorizing the materiality of artefacts (Volkoff & Strong, 2013, 2018). In *undercoded retroduction*, a number of potential mechanisms are suggested by the current body of knowledge and the researcher selects the most plausible ones that best explains the sequence of events in the specific context. For instance, a researcher may develop a critical realist explanation of change by considering from the four mechanisms – lifecycle, teleological, dialectic, and evolution – that are noted by the researchers in organization change (Van de Ven & Poole, 1995; Van de Ven & Sun, 2011). In *creative retroduction*, the researcher has to invent the mechanism because no suitable ones exist. In this regard, it may also be noted that certain theoretical constructs, although not identified as mechanisms as such, could also be identified as a mechanism in a critical realist sense. For instance, Ackroyd and Fleetwood (2000) note that institutional theory is basically a mechanism-based theory that builds on the mechanism of isomorphism. Similarly, while affordance is currently identified as a mechanism by critical realists (Volkoff & Strong, 2013, 2018), its conceptual origin (Gibson, 1966, 1979) was independent of critical realism. Finally, sometimes observations may not fit our current conceptual schema and will require us to think a completely new set of mechanisms, i.e. *meta-retroduction* resulting in some kind of paradigm shift (Tsang, 2014).

Explanation-Building

A critical realist study uses explanation building (Yin, 2017) as an analytical technique along with the coding exercise. This involves a conversion of the descriptive narrative into an analytical causal explanation (George & Bennett, 2005; Mingers & Standing, 2017; Wynn & Williams, 2020) using hypothesized mechanisms. The goal is to analyze the case study data, codes, and generative mechanisms to build a convincing explanation about the events on 'how' or 'why' something happened. This would also include outlining the inadequacy of rival explanations. Like the coding exercise, explanation-building requires iterations and revisions to achieve greater explanation and theoretical coherence (Yin, 2017). The objective is to identify the most complete and logically compelling explanation of observed events (Wynn & Williams, 2012) using the mechanisms identified. A mechanism-based explanation allows the researcher to move beyond the surface level description (Anderson et al, 2006) by developing a causal reconstruction of events (Avgerou, 2013) and explain the process through which observed outcomes were generated. It may be noted, however, that a mechanism-based explanation is not like the deterministic laws of natural sciences in which certain inputs lead to certain outputs without an exception (Anderson et al, 2006; Mingers & Standing, 2017; Wynn & Williams, 2020). Instead, the theoretical framework developed in a critical realist study should be seen as a 'sometimes-true' theory (Coleman 1964) that provides an explanation of unfolding events within a contextual condition (Avgerou, 2019). As Whetten (1989) notes, sensitivity to context is especially important for theories based on empirical evidence since observations are embedded within their context and should be understood within it.

Ensuring Research Quality

Scholars from different epistemological positions use different terms to denote the criteria to ensure research quality. Yin (2017) discusses four criteria which may be understood as being applicable to positivist case studies – construct validity, internal validity, external validity, and reliability. Lincoln & Guba (1985) propose four criteria to ensure the trustworthiness of interpretive case studies – credibility, transferability, dependability, and confirmability. Both these sets of criteria may be understood as the checks applied on methods of data collection and analysis. However, realist approach to validity pertains not to the method, but to the accounts and conclusions reached by using a particular method in a particular context (Maxwell, 2017). In other words, a realist approach to validity is seen not in relation to the procedures, but in the relationship of an explanation to those things that it is intended to be an account of. Therefore, this study relies on the four criteria proposed by Maxwell (2012) for realist studies.

Descriptive Validity

Descriptive validity relates to factual accuracy of the description of the research when writing the narrative. It involves ensuring that the study reports the activities as seen as physical and behavioral events, rather than solely in terms of meanings expressed by participants or those inferred by the researcher. As Wolcott (2009) notes - "description provides the foundation upon which qualitative research rests" (p. 27).

Interpretive Validity

The issue of interpretive validity arises when the researcher moves from the description to the interpretation of actors' perspectives. Maxwell (2012) notes that since mental phenomena cannot be directly observed, the act of interpretation relies on the meanings expressed by the participants. In this sense, ensuring interpretive validity involves possibly including all the different perspectives expressed by the participants, even if they are divergent (Dawson & Buchanan, 2005; Miles et al., 2014). Here, care should be taken in differentiating between participants' interpretations and researcher's own interpretations.

Theoretical Validity

Theoretical validity goes beyond description/interpretation and explicitly addresses the theoretical perspectives that the researcher brings to the study. The theory may refer to physical events or mental construction, but theoretical validity primarily refers to an account's function as an explanation. Ensuring theoretical validity requires that the theoretical concepts and their relationships expressed in the study bear a consensus within the community concerned with the research. This can be ensured via peer review (Johnson & Duberley, 2000) by discussing the results in conferences and colloquia, and later on in journal submissions.

Generalizability

Generalizability in case research does not take the form of statistical generalization from a sample to a population but takes the form of theoretical propositions that could be used to explain other set of events outside the case (Yin, 2017; Tsang, 2014). In this regard, Maxwell (2017) stresses that compared to external generalizability, internal generalizability is more important for the qualitative researchers in the sense that the findings should be generalizable *within* the case to persons or events that were not directly observed or interviewed.

FUTURE RESEARCH DIRECTIONS

As noted earlier, theoretical outcome of a critical realist case study is arriving at explanatory mechanisms within a contextual condition. However, these mechanisms need to be investigated across diverse context to increase their theoretical validity. At the same time, guidelines need to be developed on more closely researching dynamic and more complex elements of critical realist explanations such as emergence and interaction of mechanisms, and their influence on sociomaterial relations. A related, but slightly tangential research direction would be, on providing similar guidelines on other variants of realism such as agential realism (Shotter, 2014) or dialectical critical realism (Roberts, 2014).

CONCLUSION

The philosophy of critical realism offers an alternative to positivism and interpretivism by simultaneously accepting the existence of an external reality and allowing for diverse interpretations of it. Mechanisms are the basic building block of CR and case study is recommended as a suitable method to develop a mechanism-based explanation. This chapter offers some practical suggestions for planning and executing a critical realist case study which are summarized below:

- The research question may be in the form of 'what' or 'how' and should explicitly look for mechanisms as part of the explanation.
- Multiple data collection methods are to be used to ensure triangulation. Typical data sources in a critical realist case study could be – documentation and archival data, interviews, observation, and artefacts. The data collection should ideally be multi-level and longitudinal.
- Analysis starts from the identification of key events and structuring the narrative around it. There should be no effort to theorize at this stage and the focus should be on providing a comprehensive description.
- Qualitative coding may help in structing the analysis and arriving at the mechanism. The explanatory power of the mechanism(s) would determine the validity of the analysis.
- The analysis process for a critical realist case study is iterative in nature, and continuously moves between the data and explanation to ensure that the theorized mechanisms adequately explain the sequence of events.
- Studies need be done across diverse contexts to determine the generalizability of the explanatory mechanisms.

REFERENCES

Aaltonen, A., & Tempini, N. (2014). Everything counts in large amounts: A critical realist case study on data-based production. *Journal of Information Technology*, *29*(1), 97–110. doi:10.1057/jit.2013.29

Ackroyd, S. (2012). Contemporary Realism and Organisation Change. In D. M. Boje, B. Burnes, & J. Hassard (Eds.), *The routledge companion to organizational change* (pp. 230–244). Routledge.

Ackroyd, S., & Fleetwood, S. (2000). Realism in contemporary organisation and management studies. In S. Ackroyd & S. Fleetwood (Eds.), Realist perspectives on Management and Organisations (pp. 3-25). Psychology Press.

Alharahsheh, H. H., & Pius, A. (2020). A Review of key paradigms: Positivism vs interpretivism. *Global Academic Journal of Humanities and Social Sciences*, *2*(3), 39–43.

Anderson, P. J., Blatt, R., Christianson, M. K., Grant, A. M., Marquis, C., Neuman, E. J., Sonenshein, S., & Sutcliffe, K. M. (2006). Understanding mechanisms in organizational research reflections from a collective journey. *Journal of Management Inquiry*, *15*(2), 102–113. doi:10.1177/1056492605280231

Andrade, A. D. (2009). Interpretive research aiming at theory building: Adopting and adapting the case study design. *Qualitative Report*, *14*(1), 42.

Avgerou, C. (2013). Social mechanisms for causal explanation in social theory based IS research. *Journal of the Association for Information Systems*, *14*(8), 399–419. doi:10.17705/1jais.00341

Avgerou, C. (2019). Contextual explanation: Alternative approaches and persistent challenges. *Management Information Systems Quarterly*, *43*(3), 977–1006. doi:10.25300/MISQ/2019/13990

Bhaskar, R. (1975). *A realist theory of science*. Leeds Books Ltd.

Bhaskar, R. (1989). *The possibility of naturalism: a philosophical critique of the contemporary human sciences*. Harvester Wheatsheaf.

Burrell, G., & Morgan, G. (1979). *Sociological paradigms and organisational analysis* (Vol. 248). Heinemann.

Coleman, J. S. (1964). *Introduction to mathematical sociology*. Free Press.

Dawson, P. (2014). Reflections: On time, temporality and change in organizations. *Journal of Change Management*, *14*(3), 285–308. doi:10.1080/14697017.2014.886870

Dawson, P. (2019). *Reshaping change: A processual perspective* (2nd ed.). Routledge. doi:10.4324/9781351264921

Dawson, P., & Buchanan, D. (2005). The way it really happened: Competing narratives in the political process of technological change. *Human Relations*, *58*(7), 845–865. doi:10.1177/0018726705057807

Easton, G. (2010). Critical realism in case study research. *Industrial Marketing Management*, *39*(1), 118–128. doi:10.1016/j.indmarman.2008.06.004

Eisenhardt, K. M., & Graebner, M. E. (2007). Theory building from cases: Opportunities and challenges. *Academy of Management Journal*, *50*(1), 25–32. doi:10.5465/amj.2007.24160888

Fayard, A. L., & Weeks, J. (2007). Photocopiers and water-coolers: The affordances of informal interaction. *Organization Studies*, *28*(5), 605–634. doi:10.1177/0170840606068310

Fletcher, A. J. (2017). Applying critical realism in qualitative research: Methodology meets method. *International Journal of Social Research Methodology*, *20*(2), 181–194. doi:10.1080/13645579.2016.1144401

George, A. L., & Bennett, A. (2005). *Case studies and theory development in the social sciences*. MIT Press.

Gibson, J. J. (1966). *The senses considered as perceptual systems*. Houghton Mifflin.

Gibson, J. J. (1979). *The ecological approach to visual perception of experimental psychology*. Lawrence Erlbaum.

Guba, E. G. (1990). The paradigm dialog. *Sage (Atlanta, Ga.)*.

Guba, E. G., & Lincoln, Y. S. (1994). Competing paradigms in qualitative research. Handbook of Qualitative Research, 2, 163-194.

Hoddy, E. T. (2019). Critical realism in empirical research: Employing techniques from grounded theory methodology. *International Journal of Social Research Methodology*, *22*(1), 111–124. doi:10.1080/13645579.2018.1503400

Hultin, L. (2019). On becoming a sociomaterial researcher: Exploring epistemological practices grounded in a relational, performative ontology. *Information and Organization*, *29*(2), 91–104. doi:10.1016/j.infoandorg.2019.04.004

Johnson, P., & Duberley, J. (2000). Understanding management research: An introduction to epistemology. *Sage (Atlanta, Ga.)*.

Jong, W., & Dückers, M. L. (2016). Self-correcting mechanisms and echo-effects in social media: An analysis of the "gunman in the newsroom" crisis. *Computers in Human Behavior*, *59*, 334–341. doi:10.1016/j.chb.2016.02.032

Langley, A. (1999). Strategies for theorizing from process data. *Academy of Management Review*, *24*(4), 691–710. doi:10.5465/amr.1999.2553248

Lincon, Y. S., & Guba, E. G. (1985). Naturalistic inquiry. *Sage (Atlanta, Ga.)*.

Maxwell, J. A. (2012). A realist approach for qualitative research. *Sage (Atlanta, Ga.)*.

Maxwell, J. A. (2017). The validity and reliability of research: A realist perspective. In D. Wyse, N. Selwyn, E. Smith, & L. E. Suter (Eds.), *The BERA/SAGE handbook of educational research* (Vol. 1, pp. 116–140). Sage. doi:10.4135/9781473983953.n6

Maxwell, J. A. (2018). Collecting qualitative data: a realist approach. In U. Flick (Ed.), *The SAGE handbook of qualitative data collection* (pp. 19–32). Sage. doi:10.4135/9781526416070.n2

Miles, M. B., Huberman, A. M., & Saldana, J. (2014). Qualitative data analysis: A methods sourcebook. *Sage (Atlanta, Ga.)*.

Mingers, J. (2014). *Systems thinking, critical realism and philosophy: A confluence of ideas*. Routledge. doi:10.4324/9781315774503

Mingers, J. (2015). Helping business schools engage with real problems: The contribution of critical realism and systems thinking. *European Journal of Operational Research, 242*(1), 316–331. doi:10.1016/j.ejor.2014.10.058

Mingers, J., & Standing, C. (2017). Why things happen–Developing the critical realist view of causal mechanisms. *Information and Organization, 27*(3), 171–189. doi:10.1016/j.infoandorg.2017.07.001

Niemimaa, M. (2014). *Sociomaterial ethnography: Taking the Matter seriously. In MCIS 2014 Proceedings*. AIS.

Pentland, B., Recker, J., Ryan Wolf, J., & Wyner, G. (2020). Bringing context inside process research with digital trace data. *Journal of the Association for Information Systems*, 21.

Pettigrew, A. M. (2012). Context and action in the transformation of the firm: A reprise. *Journal of Management Studies, 49*(7), 1304–1328. doi:10.1111/j.1467-6486.2012.01054.x

Roberts, J. M. (2014). Critical realism, dialectics, and qualitative research methods. *Journal for the Theory of Social Behaviour, 44*(1), 1–23. doi:10.1111/jtsb.12056

Rubin, H. J., & Rubin, I. S. (2011). Qualitative interviewing: The art of hearing data. *Sage (Atlanta, Ga.)*.

Saxena, D. (2019). The Search for Mechanisms in Business Research: Reflections on Retroductive Analysis in a Multilevel Critical Realist Case Study. *Electronic Journal of Business Research Methods, 17*(1), 17–27.

Sayer, A. (2010). *Method in social science* (2nd ed.). Routledge. doi:10.4324/9780203850374

Shotter, J. (2014). Agential realism, social constructionism, and our living relations to our surroundings: Sensing similarities rather than seeing patterns. *Theory & Psychology, 24*(3), 305–325. doi:10.1177/0959354313514144

Tsang, E. W. K. (2014). Case studies and generalization in information systems research: A critical realist perspective. *The Journal of Strategic Information Systems*, *23*(2), 174–186. doi:10.1016/j.jsis.2013.09.002

Van de Ven, A. H. (1992). Suggestions for studying strategy process: A research note. *Strategic Management Journal*, *13*(5), 169–188. doi:10.1002mj.4250131013

Van de Ven, A. H., & Poole, M. S. (1995). Explaining development and change inorganizations. *Academy of Management Review*, *20*(3), 510–540. doi:10.5465/amr.1995.9508080329

Van de Ven, A. H., & Sun, K. (2011). Breakdowns in implementing models of organization change. *The Academy of Management Perspectives*, *25*(3), 58–74.

Volkoff, O., & Strong, D. M. (2013). Critical realism and affordances: Theorizing IT-associated organizational change processes. *Management Information Systems Quarterly*, *37*(3), 819–834. doi:10.25300/MISQ/2013/37.3.07

Volkoff, O., & Strong, D. M. (2018). Affordance theory and how to use it in IS research. In R. D. Galliers & M. K. Stein (Eds.), *The Routledge Companion to Management Information System*. Routledge.

Walcott, H. F. (2009). *Writing up qualitative research* (3rd ed.). Sage. doi:10.4135/9781452234878

Whetten, D. A. (1989). What constitutes a theoretical contribution? *Academy of Management Review*, *14*(4), 490–495. doi:10.5465/amr.1989.4308371

Wynn, D. Jr, & Williams, C. K. (2012). Principles for conducting critical realist case study research in information systems. *Management Information Systems Quarterly*, *36*(3), 787–810. doi:10.2307/41703481

Wynn, D. E. Jr, & Williams, C. K. (2020). Recent advances and opportunities for improving critical realism-based case study research in IS. *Journal of the Association for Information Systems*, *21*(1), 50–89. doi:10.17705/1jais.00592

Yin, R. K. (2017). Case study research and applications: Design and methods. *Sage (Atlanta, Ga.)*.

Zachariadis, M., Scott, S., & Barrett, M. (2013). Methodological implications of critical realism for mixed-methods research. *Management Information Systems Quarterly*, *37*(3), 855–879. doi:10.25300/MISQ/2013/37.3.09

Chapter 4
Researching Teaching in Higher Education:
Grounded Theory Concepts

Bahareh Ghodoosi
Griffith University, Australia

Geraldine Torrisi-Steele
https://orcid.org/0000-0002-6045-4114
Griffith University, Australia

Sharmistha Dey
Griffith University, Australia

ABSTRACT

Grounded theory, since Glaser and Strauss conceived of it, has been applied to myriad research situations. Variation characterises the use of grounded theory, and the landscape of grounded theory is 'messy' and difficult to navigate, especially for novice researchers, and subsequently, many existing studies lack transparency and detail, employing only loosely grounded theory methodologies. The authors argue for the value of grounded theory research in education, especially for teacher researchers, but this value can only be realised if researchers attend in-depth to articulating clearly their methodologies and justifying how the methodology can be considered grounded theory. In the chapter, the authors aim to provide a primer on grounded theory, and to stimulate discussion around grounded theory as a research method in teaching, especially in higher education, for the purpose of generating knowledge as a basis for improving the quality of university teaching.

DOI: 10.4018/978-1-7998-7600-7.ch004

INTRODUCTION

The aim of the present chapter is two-fold: To provide a 'primer' on grounded theory, and to stimulate discussion around the usefulness of grounded theory for research to inform the development of teaching frameworks, models and strategies, especially in higher education. Teaching in higher education has been selected as the context of focus for three reasons: Firstly, there is a growing institutional emphasis on improving the quality of learning and teaching in higher education; to remain competitive in a highly globalized, pressured economic environment, learning and teaching quality is pivotal to both the student experience of the institution and student success – both factors of which are crucial to institutions in their struggle to attract students. Secondly, learning and teaching in higher education, although it is gradually transforming, is apparently lagging behind the teaching developments in formal schooling. The reasons for this are numerous but mostly because the teaching context of higher education is uniquely complex with many stakeholders, and faculty who although may be experts in their disciplines, may lack the knowledge and experience in pedagogy. Secondly, education teaching and learning is a particularly complex teaching context, with a diversity of stakeholders. Thirdly, it is proving difficult to overcome the inertia of hundreds of years of 'classical' university lecture models. Academics are oftentimes bombarded with professional development but there may be a disconnect between the tools and approaches presented to them and the actual needs of the discipline context is poorly understood. Furthermore, teaching faculty members are strongly encouraged or even required to take action to improve their teaching based on evidence (of which there are plenty in the form of observations, formal evaluations, informal conversations with students etc.), but these forms of data may not be analysed systematically to understand learning and teaching activity.

The authors of the present chapter posit that grounded theory is a method of research which can be applied by both teaching practitioners and researchers to improve the quality of their teaching and learning in their classrooms. The authors favour the view of grounded theory as 'mindset'. Grounded theory thinking is conducive to the exploration of multiple sources of qualitative data and the approach aligns with principles of constructivist teaching, participatory course design, and teaching for authentic learning. Used iteratively, grounded theory has potential as a basis for in a process of continual improvement grounded theory provides fertile ground for reflection, and insight leading to innovation. However, as we point out in the present chapter, although grounded theory has potential for rich generation of knowledge, the grounded theory landscape is messy and difficult to navigate – a situation exacerbated by existing, poorly detailed empirical grounded theory studies. The authors argue that it is therefore necessary for any researchers adopting the grounded theory mindset to develop an understanding and knowledge of grounded theory, and through that understanding and knowledge, ensure transparency, clarity and detail in presenting their methodologies.

The chapter begins with an overview of grounded theory, its historical origins and its main variants. The integration of mixed methods with grounded theory is also briefly discussed as an emerging trend. Attention is then turned to the use of grounded theory in educational research with some empirical studies summarised with the intent of illustrating typical approaches to educational grounded theory research.

OVERVIEW OF GROUNDED THEORY

The development of grounded theory is attributed to Glaser and Strauss (1967). In working with the experiences of terminally ill patients and those caring for them, Glaser and Strauss questioned the suitability of the deductive and quantitative approaches of the scientific method for what they were doing (Chun Tie, Birks, & Francis, 2019). In their seminal work, *The Discovery of Grounded Theory: Strategies for Qualitative Research,* Glaser and Strauss introduced their ideas and challenged the then commonly held notion that qualitative research was lesser than quantitative research because qualitative research lacked the rigor of quantitative research its deductive methods (Chun Tie, Birks, & Francis, 2019).

The essence of the process of deduction is the identification of a problem, the formulation of a hypothesis based on collected information, and testing the hypothesis to reach conclusions. In deductive methods, the researcher begins with an idea (or has 'deduced') of what is happening ('the theory') and proceeds to test their theory through research. Instead of approaching the situation with a theory or pre-conceived ideas about what is happening, in grounded theory the researcher approaches the situation with an "open mind", and does not have a pre-conceived idea or theory, only formulating the theory after systematic exploration of the situation (Harris, 2015). While deductive methods begin with a theory and the aim is to either confirm or disprove the theory, the aim of grounded theory is to develop the theory or an explanation for what is happening based on data collected through fieldwork in the 'real world' (Harris, 2015). Importantly, grounded theory is an iterative process since early analysis informs further data collection, which when analysed informs further data collection and so on. (Charmz, 2015). In grounded theory, analysis and collection of data therefore occur concurrently throughout the entire research process (Harris, 2015). The iterative approach means that grounded theory process is characteristically comparative (Charmz, 2015) (Harris, 2015). The grounded theorist is constantly "comparing data with data, data with codes of data, codes with codes, codes with [emerging] theoretical categories, and categories with categories", all the while moving the researcher from more concrete "to more abstract and theoretical levels" (Charmz, 2015, p. 402).

Rather than aiming for description of phenomena, the aim of the grounded theory process is to develop understanding with the aim of explaining what is happening. In other words, theory generation is more important than description of the phenomena (Charmz, 2015). Because the task is not to apply theory but rather to look for patterns with a view to gaining new insights and understandings, grounded theory is inherently flexible (Charmz, 2015). The exploration of the phenomena is iterative and adaptative in the presence of emerging evidence. The development of theory necessarily overshadows the importance of seeking to develop "stable structures" (Charmz, 2015, p. 402). Given the exploratory nature of grounded theory and the lack of preconceived ideas to drive the research, it is not difficult to see how grounded theory gains popularity as a qualitative research method in social settings. The nature of the human experience is grounded in personal perspectives and pre-conceived ideas, on the part of the researcher, about what is occurring in social interactions and why it is occurring likely serve only to skew meaning making in the direction of pre-conceptions. While the effect of perspectives, experiences, values and so forth of the researcher can never be completely discounted, grounded theory ameliorates those effects through the systematic exploration of data, pattern seeking, iteration, analysis of categories, constant comparison and an ever-more concrete refinement/reformulation of understandings.

Muller (2014, p. 25) captures beautifully the spirit of grounded theory in stating:

"[grounded theory] makes explicit use of the capabilities that nearly all human share, to be curious about the world, to under- stand the world, and to communicate that understanding to others."

After Glaser and Strauss (1967) published their work, two camps of grounded theory developed as Glaser and Strauss adopted divergent views. Glaser (1978) adopted a positivist philosophy on grounded theory, and Strauss and Corbin (1990) adopting post-positivists perspectives (Chun Tie, Birks, & Francis, 2019). Other philosophical stances have since evolved along a continuum between Strauss and Glaser's standpoints, including Charmaz's (2000).

constructivist grounded theory (Chun Tie, Birks, & Francis, 2019). The end-result of variations in philosophies and methods is that grounded theory methods are apparently complex and can be confusing, especially for the novice researcher.

Following Howard-Payne (2016),The differences between the standpoints of Glaser and, Strauss and Corbin (1990) can be summarised along the dimensions of: ontology and epistemology, role of the researcher, timing of literature review, formulating the research questions, data coding and analysis, and theory verification.

Glaser's epistemology of realism by which findings are "revealed from within data" (Howard-Payne, 2016, p. 53) while for Strauss and Corbin, context is what defines findings, because findings are constructed by inter-subjective understandings. This foundational epistemological difference is the basis of differentiation in methods between the Glaserian and Straussian grounded theory. The researcher embracing Glaser's positivist epistemology will seek to be objective and detached (Howard-Payne, 2016). However, the Straussian researcher will seek to do quite the opposite, striving to be in the 'shoes of the participants so as to best understand the phenomenon as the participants perceive it to be (Howard-Payne, 2016). Unsurprisingly then, Glaser's view is that the literature review is to be conducted only after data analysis (to maintain objectivity and reduce interference) (Howard-Payne, 2016). Consequently, the Glaserian research does not enter the research with pre-formulated research questions, but for the Straussian researcher, having undertaken a literature review, enters with pre-determined research questions (Howard-Payne, 2016). The differences between Glaserian and Straussian schools of thought extend to differences in how data is coded and analysed. Strauss and Corbin (1994) do not always require comparison to show the patterns and trends emerging as categories; even solitary occurrences can be considered (Howard-Payne, 2016).

VARIATION IN GROUNDED THEORY

The works of Glaser and Strauss have ignited and continue to ignite, a great deal of discussion and healthy debate, precipitating much variation in grounded theory schools of thoughts - which is ongoing. The variation in grounded theory and the surrounding debate is testament to the richness of grounded theory because it serves as a fertile ground for new methodological insights (Bryant, 2019). The differences between the schools of thought on grounded theory are for researchers both a source confusion and a source of inspiration and innovation. Awareness and knowledge of differences, forces researchers to carefully consider, and rationalise, what is the most appropriate and productive strategy for the circumstance of their particular research. Mills, Bonner and Francis (2006, p. 25) provide an interesting conception of the various 'types' of grounded theory stating that: "all variations of grounded theory exist on a methodological spiral and reflect their epistemological underpinnings". Mills, Bonner and Francis' (2006) conceptualization of the variations of grounded theory provides an organised view of how the variations arose while keeping in focus the common threads that define the nature of grounded theory methods. Furthermore, Mills, Bonner and Francis' (2006) conceptualization strongly highlights the role of epistemology in research design and how the choice of research design aligns with researchers' understandings and beliefs about the nature of reality. Towards the top end of Mills, Bonner and Francis' spiral is Charmaz's (2000) constructivist grounded theory.

Charmaz's (2000) constructivist grounded theory, as the name suggests, is rooted in constructivism and hence experiences and meanings are co-created or co-constructed with participants (Chun Tie, Birks, & Francis, 2019). From the constructivist grounded theory standpoint, reality is constructed subjectively through the process of interaction between researcher and participant. The values and experiences of both researchers and of participants must be acknowledged as influencing the outcome of their interactions. From this viewpoint, the researcher cannot be objective, and the researcher is as much part of the research as the participants (Mills, Bonner, & Francis, 2006); the outcome of the research is a result of interplay of the values and experiences of both researcher and participants. To acknowledge the subjective nature of the method is to acknowledge the "humanness" of the researcher (Mills, Bonner, & Francis, 2006, p. 26) and also brings attention to the necessity of the researcher examining their assumptions (Charmz, 2015). Charmaz (2000) Challenges the notion that it is possible for a researcher to not be influenced by prior research or even their own experiences and values. The interactions occurring in the research are at the focal point of Charmaz's constructivist grounded theory, and the understanding and meaning of data is the result of "the researchers' interpretative frames as well as what happens during the research, and what they learn about their topics" (Charmz, 2015).

Embedded in post-modern philosophy, Adele Clarke (2003) seeks to contribute to the "regeneration" of grounded theory "after the postmodern turn" (p.554). In the post-modern philosophy, the simplicity, certainty and objectivity of 'reality' found in modernistic philosophy is challenged. Post-modernists hold a much more complex, fluid and situational view of reality, with nothing being beyond questioning. Post-modernism as Clarke (2003) acknowledges the universal influence of post-modernism throughout the academy, and it is diametrically opposed to modernism:

If modernism emphasized universality, generalization, simplification, permanence, stability, wholeness rationality, regularity, homogeneity, and sufficiency, then post postmodernism has shifted emphases to localities partialities, positionalities, complications, tenuousness, instabilities, irregularities, positionalities, contradictions, situatedness and fragmentation – complexities. Postmodernism... is not a unified system of beliefs or assumptions but rather an ongoing array of possibilities (Clarke, 2003, p. 255).

Clarke (2003) maintains that grounded theory must necessarily evolve in consonance with post-modernism. Traditionally, grounded theory methods pivoted on the experience of the individual, which Clarke (2003) agrees is necessary. However, the post-modernist movement brings the importance of situatedness to experience and meaning making. In Clarke's (2003) own words she aims to bring "the full situation of inquiry" to bear on grounded theory. In doing so doing, Clarke (2003) further detaches grounded theory from its positivist heritage, and brings an ecological flavour to grounded theory (by which analysis at levels beyond individual levels are considered, with much more attention to context), and bringing methods which facilitate situational analysis (e.g., mapping situation, variation and differences) and to capture the great discursive diversity of the post-modern era, all the while enabling flexible yet systematic analysis and looking towards 'provocation' rather than substantiating. The creation of situational maps is a cornerstone of Clarke's (2003) methods.

The trend towards a more expansive grounded theory that better captures complexities is also apparent in GoldKuhl and Lind's (2010) multi-grounded theory. Situated the field of information systems (which by nature is multi-disciplinary and therefore requires the frames if various theories) GoldKuhl and Lind's (2010) fundamental tenet is that grounded theory must be grounded more than empirically in data. They argue for a theory development with theoretical, and internal grounding, as well as empirical grounding (Goldkuhl & Lind, 2010). GoldKuhl and Cronholm (2018) explain that 'grounding', following Habermas (1984) refers to analysis and justification of knowledge claims. It is important, GoldKuhl and Conholm (2010) argue, to make clear distinction between theory generation and grounding. Without this distinction there is the danger of neglecting the need to constantly and critically review the grounding of the theory and there must be a continuous process of both theory creation and "checking' (GoldKuhl & Cronholm, 2018). GoldKuhl and Cronholm (2018) maintain that justification can come not from empirical or data sources alone, but can (and must) also come from existing theory and the object of study itself. Hence, from the perspective of multi-grounded theory, the justification of claims must come from the data, comparison with existing theories and from ensuring there is 'congruence' among the elements of the theory itself (GoldKuhl & Cronholm, 2018).

To ignore existing theory and enter the research situation in isolation of existing theory is to risk 're-inventing the wheel', and potentially stagnates the evolution of theory through the process of linking to existing research (GoldKuhl & Cronholm, 2010).

Multi-grounded theory as originally posited by GoldKuhl (1993 as cited in GoldKuhl and Cronholm, 2018) underwent a number of refinements, including bringing to the fore "action knowledge" i.e., any kind of knowledge that could be acted upon and the emphasis on Habermas' (1984) practical rationality was brought to the fore. GoldKuhl later evolved the initial ideas around grounded theory to meld with design research, formulating 'multi-grounded design theory'. Design science is prominent in the field of information systems and GoldKuhl and colleagues were searching for more effective approaches to information systems design. However, design science has an emphasis on hypothesis, which of classical design theory does not. Hence, GoldKuhl and Lind (2010) argue that classical grounded theory in its pure form cannot be used together with design sicence. Grounded theory must be adapted in consonance with the design science frame. In design science drawing on multiple sources of knowledge is key. Furthermore, design research comprises of both design practice (empirical) and meta-design (theory), and knowledge is created through the constant interplay of practice and meta-design. These key characteristics of the design research frame made way for the development of multi-grounded theory.

In their 2018 publication, Goldkuhl and Cronholm clarify the differences and similarities between grounded theory and multi-grounded theory in order to facilitate understanding of the nature of multi-grounded theory. Multi-grounded theory and grounded theory share emphasis on "Taking the word 'grounded' seriously" and "Capturing and explaining context related process and phenomena" (GoldKuhl & Cronholm, 2018, p. 4). Grounding is at the cornerstone of both grounded theory and multi-grounded theory, with multi-grounded theory placing emphasis on empirical, theoretical and internal grounding. In grounded theory, context plays a role in shaping the events, incidents and behaviors. In multi-grounded theory, the importance of context is stressed both for understanding phenomena and on the impact of phenomena. The main differences between multi-grounded theory and grounded theory are in the "pursuit of theory through data" and through theoretical sampling (GoldKuhl & Cronholm, 2018, p. 4). That is, data is at the core of grounded theory and of multi-grounded theory, but in multi-grounded theory, unlike classical grounded theory, there is great emphasis also on the importance of pre-existing theories. Though the usefulness of theory alongside data is acknowledged by Strauss and Corbin (1990, 1998), their caution is that this may occur so long as the presence of theory does not constrain creativity in theory production. GoldKuhl and Cronholm however, advocate for the integration of theory with the research process not because it constrains creativity but because

If one ignores existing theory, there is a risk of reinventing the wheel. As researchers, we often build new knowledge on existing knowledge. An isolated theory development also means that there is a risk for noncumulative theory development. We believe that it is important to relate the evolving theory to established research during the process of theorizing. Existing theory can be used as a building block that supports the empirical data forming the new emergent theory (Goldkuhl & Cronholm, 2010, p. 191).

The aforementioned variations of grounded theory are but some of the most notable variants of grounded theory. There are numerous other variations. The variation in grounded theory standpoints is evidenced by debate about what is and isn't grounded theory and how the numerous variants may (should) be distinguished and classified. Charmaz's (2000) constructivist grounded theory is a commonly considered one of the main three grounded theory variants, the other two being classical or Glaserian, and Straussian. Chun Tie, Birks and Francis (2019) identify three methodological traditions of grounded theory: *traditional GT* associated with Glaser; *evolved* GT associated with Strauss, Corbin and Clarke; and *constructivist GT* associated with Charmaz. A further search of literature around types of grounded theory quickly reveals the magnitude of debate, and a lack of agreement upon classification of grounded theory - Two examples among numerous others: Denzin (2007) identifies seven different types of grounded theory: positivist, post-positivist, constructivist, post-modern, situational and computerized. Frenandez (2012) identifies four: Classic grounded theory, Straussian, Constructivist and Wuest's 1995 Feminist grounded theory. In exploring literature around types of grounded theory, it is evident that "grounded theory" might be now considered more of an umbrella for a host of various methods rather than a term referring to a prescribed singular method. Given the many variations and the ensuing discussion and debate around grounded theory, the landscape of grounded theory can be described as 'messy'. This makes it essential that any researcher, especially novice researchers, embarking on the use of grounded theory are firstly, aware of the history, evolution and debate surrounding grounded theory; and secondly, address explicitly in their work where there methods sit on what can only be described as a continuum of grounded theory approaches. Unfortunately, the latter is difficult to achieve for novice and more seasoned researchers alike. As Dey (1999) observes: There is an irony—perhaps a paradox—here: that a methodology that is based on ''interpretation'' should itself prove so hard to interpret (Dey L., 1999, p. 3).

RIGOR IN GROUNDED THEORY RESEARCH

Grounded theory studies, as interpretative research, can be criticised for lack of rigor or trustworthiness. Strauss and Corbin (1998, pp. 270-272 as cited in Khiat, 2010, p. 1474) provide criteria for evaluating rigor of grounded theory research in the form of the following questions:

1. Are concepts generated?
2. Are concepts systematically related?
3. Are there many conceptual linkages and are the categories well-developed?
4. Do categories have conceptual density?
5. Is variation within the phenomena built into the theory?

6. Are the conditions under which variation can be found built into the study and explained?
7. Has process been taken into account?
8. Do the theoretical findings seem significant and to what extent?
9. Does the theory stand the test of time and become part of the discussions and ideas exchanged among relevant social and professional group?

'DOING' GROUNDED THEORY

The intent of both Glaser and Strauss (1987), and Strauss and Corbin (1998) that researchers explore their own methods of theory generation in a way that is flexible and meets the needs of their investigations, while at the same time adhering to the basic principles of grounded theory (LaRossa, 2005). Considering this intent, it is likely useful to think of grounded theory as a mindset or way of thinking, as descriptive rather than prescriptive. With the aim of assisting researchers, especially novices, various authors have sought to provide more concrete guidance/prescriptive guidance on how to conduct grounded theory and clarify grounded theory methods. There are numerous textbooks written with the aim of guiding researchers to undertake grounded theory, and many papers in which authors seek to provide more prescriptive or concrete guidelines for conducting grounded theory research. Armed with the knowledge of the many variations and continued evolution of grounded theory, researchers will find such resources useful but only if they interpret and make use of the information in accordance with the needs of their research and are self-aware of the epistemological position they take.

In the available literature to help grounded theory novice researchers formulate their research methodology, a simplified but useful framework is provided by ChunTie, Birks and Francis (2019). It is summarised here. The framework captures the essential principles of grounded theory as an inductive and iterative method with emphasis on the necessity of systematic methods for data collection and analysis. The visual representation of the framework as given by Chun tie, Birks and Francis (2019) is shown in Figure 1.

According to Chun Tie, Birks and Francis' (2019) framework, the researcher begins with purposive selection of participants along with the various data sources suited to the research goals. Purposive sampling provides initial data for analysis. Consistent with the basic principles of grounded theory, the framework includes concurrent data collection and analysis wherein collection, coding and analysis of data informs further collection, and analysis. The framework includes theoretical sampling to "identify and follow clues from the analysis, fill gaps, clarify uncertainties, check hunches and test interpretations as the study progresses" (Chun Tie, Birks, &

Francis, 2019, p. 3). Theoretical sampling enables the researcher to "follow leads in the data by sampling new participants or material that provides relevant information (Chun Tie, Birks, & Francis, 2019, p. 5). Theoretical sampling is the mechanism by which the theory is able to evolve, moving from abstract to more concrete and realizes the grounding of the theory in data. Theoretical sampling (as distinct from population sampling) is a process which enables the researcher to follow a 'trail of clues' which eventually lead to the creation of a theory. Theoretical sampling continues throughout the research process with the aim of achieving data saturation for the categories being developed (Chun Tie, Birks, & Francis, 2019). Theoretical sensitivity permeates the entire process. Theoretical sensitivity refers to the researcher having insights into what data is or isn't important to the emerging theory (Chun Tie, Birks, & Francis, 2019).

Figure 1. A grounded theory framework for novices
(Chun Tie, Birks, & Francis, 2019, p. 3).

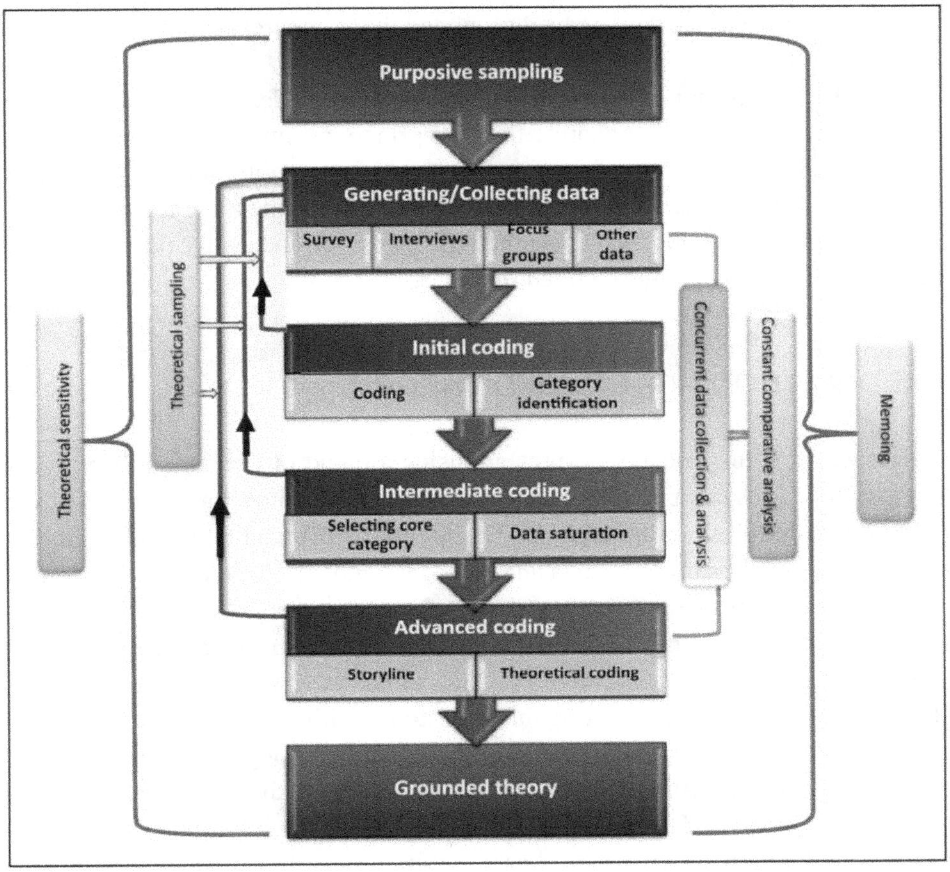

Coding and memoing, are at the core of grounded theory and of the framework. Data analysis in ground theory relies heavily on coding. Memoing is distinct from coding. Memo writing is undertaken by the researcher to capture their ideas and the trail they have followed in their journey through the grounded theory process. Memoing are really 'notes to self' and provide an audit trail or detailed record and documentation to capture the ideas, events and thinking process (Chun Tie, Birks, & Francis, 2019). Memoing is often conducted in other research, but it is of particular importance in grounded theory.

Codes or the labels generated as the researcher interacts with data. Coding as the process by which codes are generated and relationships among them are established. Coding can be open (procedures for developing categories), axial (relationships among categories) and selective (relating the categories to create a cohesive 'story' about the phenomenon) (Chun Tie, Birks, & Francis, 2019). Different coding terms are used by different researchers and authors and the tradition of grounded theory they subscribe to. For example, constructivist may refer to initial coding, focused coding and theoretical coding, instead of using the respective terms, open coding, axial and selective as used in evolved grounded theory genres (Chun Tie, Birks, & Francis, 2019).

Chun Tie, Birks and Francis (2019) framework provides a simplified view of grounded theory method but capture the core principles in which grounded theory is anchored. Conceivably, the framework acts as a foundation for the researcher to develop according to the needs of their research. To implement the framework, the researcher should examine their epistemological position and evolve the framework, in light of other variants of grounded theory.

MIXED METHODS-GROUNDED THEORY

More recently, some research designs are using a blend of Grounded theory with mixed methods research. Mixed methods research integrates both qualitative and quantitative data collection and analysis. The integration of quantitative and qualitative data allows for triangulation of data and provides a more holistic insight than either data type alone (Creswell, 2014). "It [grounded theory] is a research method that fits remarkably well with [mixed methods research]" (Johnson & Turner, 2010)

The integration of mixed methods with grounded theory is a recent trend. Although the benefits of integrating mixed methods with grounded theory methods are often cited by those researchers using the methodology, there is a need for more systematic and consistency in approaches among studies using the technique (Gutterman, Babchuk, Howell Smith, & Stevens, 2019). Gutterman et al. (2019) conducted a literature search of studies integrating mixed methods with grounded theory. Their review of 61 empirical studies, Gutterman et. al (2019) conclude that most published studies provide "very little methodological detail…theoretical development was noticeably absent and majority followed a convergent mixed methods design" (p. 179). Gutterman et. al and others maintain that there are many advantages to bringing together grounded theory and mixed methods research including enhancing substantive theory development, effectively transitioning quantitative trained researchers to systematic methods for qualitative research and assisting with injecting a more systematic approach to handling of qualitative data (Gutterman, Babchuk, Howell Smith, & Stevens, 2019). However, Gutterman et al. is also mindful of the many challenges of bringing together the 'family' of grounded theory methods with mixed methods. The obvious need for further discussion aside, Gutterman et al. recommend that researchers seek for greater transparency in their methodologies by providing details and articulating justification of their methods. Training, mentors and exploring the literature around grounded theory are also important (Gutterman, Babchuk, Howell Smith, & Stevens, 2019). In considering grounded theory and its potentially symbiotic relationship with grounded theory, it is also crucial to insure that grounded theory is not reduced to equate to 'the' qualitative method. Grounded theory, despite its variations and continuing evolution is founded on core principles. To ignore these principles brings 'grounded theory' to be nothing more than a thematic analysis (Carlin & Kim, 2019).

GROUNDED THEORY IN 'CLASSROOM' RESEARCH: SOME EXAMPLES

In the context of education, grounded theory, in various forms has been used primarily used for designing educational interventions and sometimes to improve teaching practice. However, the use of grounded theory methods in education is apparently limited (PrigolI & Behrens, 2019)

As universities strive to be competitive, there is a continuing emphasis of teaching quality in higher education. Teaching faculty members are strongly encouraged or even required to take action to improve their teaching. However, oftentimes attempts to improve teaching are ad hoc, at best driven by the advice of various consultants provided by the institution. Typically, data collected via formal and informal student evaluation and student experience surveys is minimally considered, if at all (especially given the usual challenges of response numbers). Furthermore, teaching faculty are often required to teach new skills or capabilities in response to the changing economic and social environment. Teaching these new skills or capabilities may require the adaptation or development of innovative curriculum and teaching approaches beyond what is already in the faculty members' repertoire of strategies. For example, one of the current challenges is the teaching of data literacy in higher education to close the data literacy gap in industry. Educators and trainers are grappling with how to best teach data literacy and the fact that data literacy is an emerging and dynamic context complicates matters. At present educators are mostly relying on teaching strategies and methods used in other disciplines for teaching of data literacy but these are proving inadequate and there is a need to grow data literacy (Grillenberger & Romeike, 2018). There is therefore a need to better understand data literacy from the perspective of the many stakeholders as a catalyst for developing more effective teaching strategies.

Additionally, in higher education teaching, educators must consider the their teaching not only from the perspective of learning outcomes of their students, but also how the intended learning outcomes and teaching methods satisfy the requirements of the institution and the requirements of the potential employers and even various accrediting bodies outside of the university.

The authors posit that grounded theory thinking can provide a useful foundation for teaching faculty to conduct classroom research as well as for the development of teaching methods and strategies required when a new area of skill or capability is required because

- Grounded theory as a primarily inductive and exploratory method makes room for creative and innovative theory development without the constraint of initial hypothesis
- Grounded theory methods are flexible and are thus adaptable to changing contexts
- Grounded theory, particularly constructivist grounded theory and situated grounded theory is sensitive to stakeholders needs and perspectives and meaning making and experiences resulting from interaction among stakeholders. Grounded theory moves beyond focus on the individual experience to a more holistic understanding of the teaching situation (sometimes the popular student-centred teaching philosophy brings too much focus to the individual and their own construction of knowledge to the detriment of the broader context and the experience created by interactions of the many stakeholders).

Some of the applications of grounded theory to education research aiming to improve teaching. Some representative studies are briefly reviewed below, highlighting their use of grounded theory approaches.

Teräs and Kartoglu (2017) applied grounded theory methodology to improve online professional learning by gaining insights into how professional learning occurs in the online environment. Semi-structured interviews along with observation were used for data collection with seven participants. The interview questions provided an avenue for participants to "reflect on and describe their learning experiences and discuss the impact the course had on their professional learning" (Teräs & Kartoglu, 2017, p. 199). The researchers then followed the coding procedures described in Strauss and Corbin (1988). The initial data was collected through observations and the observations then informed formulation of interview questions and guided the interviews. Interviews which were recorded were listened to again and then transcribed. After transcription the researchers repeatedly read the interviews in order to "immerse in the data, become more familiar with it, and identify points of interest" (Teräs & Kartoglu, 2017, p. 199). Following this, open coding was undertaken to identify and label concepts (the basic unit of analysis) related to "incidents, events and happenings… At the same time, memoing was used to record questions, interpretations, and thoughts that rose from the data, to facilitate a detailed, in-depth analysis that would result in an adequate code" (Teräs & Kartoglu, 2017, p. 199). Concepts were then categorised through the use of concept mapping and by undertaking constant comparison. Axial coding was undertaken following open coding to find relationships among concepts and sub-concepts, reducing the number of codes. "For example, the subcategories of *appreciating collaborative learning, challenges with collaboration,* and *collaborating beyond course requirements* were grouped together under the axial code 'experiencing collaborative learning' (Teräs & Kartoglu, 2017, p. 201). Finally, selective coding was undertaken in which a central category was identified, and the relationship of all other categories was established. This facilitated the integration of categories and the development of theoretical schema (Teräs & Kartoglu, 2017). The results of the study lead to a theoretical model of professional learning.

In another application of grounded theory to education, Khiat (2010) turns to the challenges of teaching mathematics to engineering students. Khiat (2010) explores "engineering students' conceptions of understanding in mathematics learning" (p. 1459). Khiat (2010) as the lecturer for these students, as an insider to the research situation acknowledges the advantages of insider research (such as familiarity and understanding of the context and opportunities to observe students in action), as well as the limitations (such as the power imbalance between lecturer and students as probative of students sharing their views honestly). Touching on the classical view of grounded theory in which the researcher enters might enter the research situation without bias and preconceived ideas, Khiat (2010) stresses the importance of remaining reflexive throughout the research while at the same time acknowledging the value of these experiences – an approach which echoes ideas held in 'progressed' grounded theory methods such as Chamrmaz's constructivist grounded theory, multi-grounded theory and situational grounded theory (although Khiat does not explicitly make these connections). That Khiat (2010) draws on prior experiences is evident in the methodology used. The study involved interviews with 21 students around research questions derived from literature and from the author's own experiences in the classroom. Following Strauss and Corbin, (1990) these concepts were 'provisional' and eliminated if they were deemed theoretically irrelevant. Khiat (2010) used open sampling initially, using semi-structured interviews, and the "initial interview guide changed four times for the purpose of theoretical sampling" (Khiat, 2010, p. 1468), again following Strauss and Corbin (1998) and consistent with the framework for grounded theory research for novices proposed by Chun tie, Birks and Francis (2019). Again, like Teräs and Kartoglu (2017), Khiat (2010) uses open coding, followed by axial and selective coding to arrive at a "substantive theory of engineering students' conception of mathematics learning" (p. 1483).

Similarly concerned with mathematics education, but this time in secondary education, Martinovic and Manizade (2017), follow Charmaz's constructivist grounded theory approach to extend on the existing pedagogical, content, knowledge (PCK) developed by Shulman. Martinovic and Manizade (2017) are motivated to undertake their research using grounded theory because, among other reasons, "Teachers need to be aware of what each and every student is thinking and knowing, to construct meaning and meaningful experiences in light of this knowledge, and have proficient knowledge and understanding of their content to provide meaningful and appropriate feedback . . . ; (Hattie, p 238-239 cited in Martinovic and Manizade, 2017, p.2). In their research, Martinovic and Manizade (2017), note that they do not enter the research as prescribed in classical grounded theory "theoryless", but previously developed theories such as PCK formed the theoretical and conceptual framework and could see the major aspects emerging from [their] data" (Martinovic & Manizade, 2017, p. 7). Principles found in multi-grounded theory (empirical grounding, theoretical grounding and internal grounding) are present in Martinovic and Manizade (2017) application of grounded theory.

Bytheway, (2018) applies grounded theory, not to mathematics, but to understanding work integrated learning. Characteristic of Bytheway's (2018) approach to applying grounded theory methods is awareness of how existing literature and theory should or should not be used in grounded theory approaches. Bytheway (2018) acknowledges the contentious aspect of literature in grounded theory studies, identifying as a principle of grounded theory that literature is not used to generate and initial hypothesis, but that the researcher "must remain as theoretically sensitive as possible " (p. 250). To maintain theoretical sensitivity, the researcher reads widely but a full literature review is only conducted once the theory begins to emerge and the literature is then used as 'more data', and is part of the constant comparative analysis process (Bytheway, 2018).

The understanding of development of various literacies is another area of education in which grounded theory has been applied. McEneaney, Gillete, Farkas and Guziniczack (2016) seek to better understand how digital literacy is developed in e-commerce contexts. Digital literacy has in recent times become an essential capability and because digital literacy requirements are constantly changing as technologies change, there is an ongoing need to understand how these literacies develop, with the aim of improving education in the area. The authors justify their use of grounded theory because grounded theory provides a systematic way of working through 'messy' social situations. McEneaney et. Al. (2016) make explicit that their work is informed by various other theories and perspectives. In this respect, McEneaney et. al.'s (2016) use of grounded theory encapsulates the principles of more expansive views on grounded theory, beyond the constraints of classical grounded theory. The research involved video recording while students, using a think out load protocol, undertook the task of searching online ecommerce sites. Videos were then coded in five phases with an observation phase without coding initially to gain familiarity with participant response, followed by a describing phase to identify 'codeable' properties (McEneaney, Gillette, Farkas, & Guziniczak, 2016). A generalizing and theorizing phase followed. Because multiple people undertook coding of the videos, reliability procedures employed temporal coding (agreement between coders as to the beginning and end of events), and nominal coding (agreement across coders as to the assignment of codes).

Maybee, Carlson, Selbodnik and Chapman (2015) focus on information and data literacy and the need to holistically understand it in order to enable libraries to better support faculty. They address the question of how information and data literacies are addressed in various disciplines. Maybee et. al. (2015) used teams of disciplinary experts consisting of a literacy specialist, data specialists and liaison librarian to analyse discipline syllabi. The procedure involved syllabi familiarization and initial coding, followed by intermediate coding and theme generation. The application of grounded theory in this research is apparently simple, following the principles coding procedures as given in Glaser and Strauss (1967). There is no elaboration on the role of prior knowledge and other literature. With reference to learning within library and information science, Hicks (2018) maintains that within library and information science, grounded theory is needed to more richly understand how people engage and deal with information settings that are constantly changing. The methods of grounded theory, argues Hicks, are in harmony with the needs of exploration in information sciences and therefore grounded theory should be a significant element of the research agenda in information science. Hicks (2018) notes that grounded theory has already a significant history in library and information sciences with techniques such as Clarke's situational analysis already being used to help unravel the complexities of people's engagement with information. However, the emphasis in the use of grounded theory in information literacy to date is learner skill rather than information literacy as a social process to explore how information literacy is constructed in context through human activity (Hicks, 2018). Hick's ideas illustrate well how the value of grounded theory methodologies for evolving knowledge within a discipline.

The applications of grounded theory to educational research briefly described above are just a few examples. Other examples are found in the literature. However, instances of teacher researchers using grounded theory to gain insights to inform their practice are rare. The likely reason is that in classical grounded theory, there is emphasis on the researcher entering the research situation without pre-conceived ideas. Altrighter and Posch (1989) raise questions about the applicability of grounded theory to teacher research and provide some insights into how grounded theory might evolve to serve better teacher research. For example, in classical grounded theory the researcher enters ideally with no prior assumptions or biases. However, reflectivity rather than ignoring theoretical pre-conceptions is crucial for teachers. Altrighter and Posch (1989) point out that Glaser and Strauss would not contradict this view but instead state clearly that it is impossible to remove all 'theory' from the mind before beginning research and the "trick is to line up what one takes as theoretically possible with what one is finding [and] are sources of insight to be cultivated" (Glass & Strauss cited in Altrighter & Posch, 1989, p.24). Another point of possible contradiction between grounded theory and practitioner researchers is that there is an apparent distinction between researcher and practitioner but, Altrighter and Posch (1989) note that the distinction between professional researcher and practitioner only applies to the generation of theory not the application, but, the person who applies theory is effectively a generator of theory also and so the distinction is not great. Altrighter and Posch (1989) argue that while, on the surface, practitioners as researchers is not supported by grounded theory, a deeper look at the theory shows that the application of grounded theory to teacher research is viable. Adding to Altrighter and Posch's (1989) constructivist grounded theory, meaning and experiences are co-constructed in context among participants, researcher and in interaction with the situation. There is an evident synergy here between the act of teaching by which student experiences and learnings are also co-constructed through discourse and interaction with context. It is thus the authors view that grounded theory or rather the grounded theory mindset holds potential to enrich teacher research and the process of learning and teaching. However, the potential is only realised with informed application of grounded theory methodologies, and epistemological stances of the researchers, be they professional researchers or teacher practitioner researchers. Walters (2013) supports this view, given synergies between action research and grounded theory. Walters (2013) argues for the use of grounded theory as a "vehicle for widening teacher involvement in educational research" (p. 72).

CONCLUSION

A theory may be informally thought of as an explanation how or why some-thing works or happens as it does. Theory forms the understandings and are part of the mental models we employ to determine appropriate activity. Broadly, grounded theory is an explanation of some social phenomena that emerges inductively from the systematic analysis of data. Adopting a perspective close to multi-grounded theory, data includes that which is collected through interaction with participants, that which is generated through the progressive analysis of data, and that which is found in existing theories about the phenomena. Glaser and Strauss developed grounded theory when they set out to show that qualitative research can be systematic and rigorous. Since its inception and the subsequent divergence of Glaser and Strauss, grounded theory continues to evolve. Some of the well-recognised variants of grounded theory were briefly discussed with the aim of providing a broad understanding of the nature of grounded theory and to facilitate an appreciation of the 'messy landscape' of grounded theory thinking.

Grounded theory in the present chapter has been discussed from the frame of qualitative research. The authors maintain that it is most important to emphasise that qualitative research does not equate to grounded theory and grounded theory, despite its variants has at the core some principles which identify the methodology as grounded theory. One of the most challenging issues in grounded theory research is the incompleteness of research designs claiming to be grounded theory. Hence, the key message in the present chapter is that to advance grounded theory and to employ it to create needed knowledge of complex social contexts such as education, it is crucial that researchers adopting the grounded theory mindset provide detail and carefully consider their research designs. This alongside ongoing discussion and reflection will likely propel the grounded theory mindset forward.

REFERENCES

Altrighter, H., & Posch, P. (1989). Does the grounded theory approach offer a guiding paradigm for teacher research? *Cambridge Journal of Education, 19*(1), 21–31. doi:10.1080/0305764890190104

Bryant, A. (2007). *The SAGE handbook of grounded theory*. Sage. doi:10.4135/9781848607941

Bryant, A. (2009, September). Grounded theory and pragmatism: The curious case of Anselm Straus. *Forum Qualitative, 10*(3). https://www.qualitative-research.net/index.php/fqs/article/view/1358/2850

Bryant, A. (2019). The varieties of grounded theory. *Sage (Atlanta, Ga.)*.

Bytheway, J. (2018). Using grounded theory to explore learners' perspectives of workplace learning. *International Journal of Work-Integrated Learning, 19*(3), 249–259.

Carlin, A., & Kim, Y. (2019). Teaching qualitative research: Versions of grounded theory. *The Grounded Theory Review, 18*(1). http://groundedtheoryreview.com/2019/12/29/teaching-qualitative-research-versions-of-grounded-theory/

Charmaz, K. (2000). Grounded theory: constructivist and objectivist methods. In N. Denzin, & Y. Lincoln (Eds.), Handbook of qualitative research (pp. 509-535). Sage.

Charmz, K. (2015). Grounded theory: Methodology and theory construction. In *International Encyclopedia of the Social & Behavioral Sciences* (pp. 402–407). Elsevier. doi:10.1016/B978-0-08-097086-8.44029-8

Chun Tie, Y., Birks, M., & Francis, K. (2019). Grounded theory research: A design framework for novice researchers. *SAGE Open Medicine, 7*, 1–8. doi:10.1177/2050312118822927 PMID:30637106

Clarke, A. (2003). Situational analysis: Grounded theory mapping after the postmodern turn. *Symbolic Interaction, 26*(4), 553–576. doi:10.1525i.2003.26.4.553

Denzin, N. (2007). Grounded theory and the politics of interpretation. In A. Bryant & K. Charmaz (Eds.), *The Sage handbook of grounded theory* (pp. 454–471). Sage. doi:10.4135/9781848607941.n21

Dey, I. (2004). Grounded theory. In G. G. C. Seale (Ed.), *Qualitative research practice* (pp. 80–93). Sage. doi:10.4135/9781848608191.d9

Dey, L. (1999). *Grounding grounded theory: Guidelines for qualitative inquiry*. Academic Press.

Frankfurt, H. G. (1958). Peirce's notion of abduction. *The Journal of Philosophy, 55*(14), 593–597. doi:10.2307/2021966

Glaser, B. G. (1978). *Theoretical sensitivity*. Sociology Press.

Goldkuhl, G., & Cronholm, S. (2018). Reflection/commentary on a past article: "Multi-grounded theory: An update. *International Journal of Qualitative Methods, 17*, 1–5.

Goldkuhl, G., & Cronholm, S. (2010). Adding theoretical grounding to grounded theory: Toward multi-grounded theory. *International Journal of Qualitative Methods, 9*(2), 187–205. doi:10.1177/160940691000900205

Goldkuhl, G., & Lind, M. (2010). *A multi-grounded design research process.* Springer. doi:10.1007/978-3-642-13335-0_4

Gutterman, T., Babchuk, W., Howell Smith, M., & Stevens, J. (2019). Contemporary approaches to mixed methods-grounded theory research: A field-based analysis. *Journal of Mixed Methods Research, 13*(2), 179–195. doi:10.1177/1558689817710877

Harris, T. (2015). Grounded theory. *Nursing Standard, 29*(35), 32–39. doi:10.7748/ns.29.35.32.e9568 PMID:25922026

Hicks, A. (2018). Developing the methodological toolbox for information literacy research: Grounded theory and visual research methods. *Library & Information Science Research, 40*(3-4), 40. doi:10.1016/j.lisr.2018.09.001

Howard-Payne, L. (2016). Glasser or Strauss? Considerations for selecting a grounded theory study. *South African Journal of Psychology. Suid-Afrikaanse Tydskrif vir Sielkunde, 46*(1), 50–62. doi:10.1177/0081246315593071

Johnson, R., & Turner, W. (2010). *Grounded theory in practice: Is it inherently a mixed method?* Academic Press.

Khiat, H. (2010). A grounded theory approach: Conceptions of understanding in engineering mathematics learning. *Qualitative Report, 15*(6), 1459–1488.

LaRossa, R. (2005, November). Grounded theory methods and qualitative family research. *Journal of Marriage and Family, 67*(4), 837–857. doi:10.1111/j.1741-3737.2005.00179.x

Martinovic, D., & Manizade, A. (2017). First-page settings open access article using grounded theory to extend existing PCK framework at the secondary level. *Education Science, 7*(60).

Maybee, C., Carlson, J., Selbodnik, M., & Chapman, B. (2015). "It's in the syllabus": Identifying information literacy and data information literacy opportunities using a grounded theory approach. *Libraries Faculty and Staff Scholarship and Research.*

McEneaney, J., Gillette, R., Farkas, W., & Guziniczak, L. (2016). Online commerce as a digital literacy: A grounded theory approach. *The Journal of Literacy and Technology, 17*(3), 219–255.

Mills, J., Bonner, A., & Francis, K. (2006). The development of constructivist grounded theory. *International Journal of Qualitative Methods, 5*(1), 25–35. doi:10.1177/160940690600500103

Muller, M. (2014). Curiosity, creativity, and surprise as analytic tools: Grounded theory method. In J. Olson, & W. Kellogg (Eds.), Ways of knowing in HCI (pp. 49-68). Springer.

Plutynski, A. (2011). Four problems of abduction: A brief history. *HOPOS The Journal of the International Society for the History of Philosophy of Science*, *1*(2), 227–249. doi:10.1086/660746

Prigol, I. E., & Behrens, M. (2019). Grounded theory: Methodology applied in education research. *Educação e Realidade*, *44*(3), 1–20.

Strauss, A., & Corbin, J. (1990). Basics of qualitative research: Grounded theory procedure and techniques. *Sage (Atlanta, Ga.)*.

Teräs, H., & Kartoglu, U. (2017). A grounded theory of professional learning in an authentic online professional development program. *International Review of Research in Open and Distributed Learning*, *18*(7), 191–212. doi:10.19173/irrodl.v18i7.2923

Walters, D. (2013). Grounded practice: Putting the 'self' back into self-evaluation. *Educational Action Research*, *22*(1), 72–92. doi:10.1080/09650792.2013.851612

Chapter 5
Contextualizing Action Research

Emmanuel Adugu
University of the West Indies, Barbados

ABSTRACT

This chapter focuses on designing and conducting action research in diverse settings. Action research is a collaborative approach to problem solving. It involves consultative problem identification, reflects context, encourages reflexive examination, and ultimately encourages and empowers beneficiaries for desirable change. In that regard, it puts all stakeholders at the core of the change process. The process of change from research project conceptualization to analysis and policy implications is thus made more understandable and meaningful to community actors (beneficiaries). The chapter features three empirical models from diverse parts of the world. These are Model 1: Photo-voice as a form action research depicting an underused footbridge in Barbados; Model 2: DANIDA Community Water and Sanitation Project, Ghana; and REACH After School Enrichment Program, USA. All these models show that action research process is people and community-centered, attentive to the views of people as individuals with their own unique needs, resources, and interests.

DOI: 10.4018/978-1-7998-7600-7.ch005

INTRODUCTION

There are different levels that people participate in research that seeks to understand problems facing them in their communities. In that context, the ultimate goal of the research is to generate relevant knowledge to change the dynamics of their livelihood systems from "what is" to "what it ought" to be. Usually, ordinary people participate in research by answering questions posed by "extractive researchers" using questionnaire surveys, or similar approaches to elicit responses. In such cases, the respondents (people who are supposed to be the intended beneficiaries) do not have the opportunity to influence the outcome of the research, as findings are not shared with the respondents, nor is there any opportunity for the people interviewed to check the accuracy of researcher's findings. In other cases, people's participation in research may be interactive. In such contexts, people participate in joint analysis of the research problem—this approach actively seeks multiple perspectives and leads to action plans, empowerment of local people, local control and ownership of research outcomes.

The level of participation of people in any research is shaped by how the researcher(s) perceive the reality of people's problems. Essentially, there are researchers who "erroneously" believe that their perception of the research problem and its diagnosis is an accurate assessment of the situation. In addition to that, such researchers "erroneously" believe that they possess the requisite knowledge and skills to solve the problem in a particular context. For such researchers, their perception of the problem is "more correct" than that of the masses they desire to help. Unfortunately, this attitude of expert "knows-it-all" is carried over from education and training. This calls for change—there is the need for active involvement of people who are conceived to be beneficiaries of any research project. As a result of attitudes of superior knowledge and often bias on the part of researchers, the people they serve with their research do not have the opportunity to participate in the formulation of the research problems and strategies for addressing them. When people participate actively in researches that are conceived to improve the livelihood systems, the chances of undesirable social costs will be lower and hence successful outcomes.

The main objective of this article is to provide the reader with the knowledge and skills for designing and conducting action research in diverse settings. The article is organized into seven distinct but interrelated parts. The first part examines conceptualization of action research as a means of knowledge generation. This is followed by action research cycle. Action research cannot be effectively conducted without a good grasp of community participation. For that reason, the third part of the article links action research to the concepts of participation and community. Subsequently, features of participation; and essential factors for conducting action research are discussed. In order to adequately equip the reader to design and conduct action research, the last two parts of the article present empirical models of conducting action research; and a summary of core issues.

CONCEPTUALIZING ACTION RESEARCH

The ultimate purpose of any social science research is to create new knowledge for assisting people in taking informed decisions with respect to their livelihood systems such as food and environment, human interaction with the built environment, and other factors that critically impinge upon their quality of life. In that context, action research aims at changing the dynamics of people's livelihood systems from "what is" to a desirable one—to improve upon their quality of life. Action research is a systematic investigative process with the ultimate goal of solving an immediate problem or improving practices.

Kurt Lewin is generally credited for coining the concept of action research. Lewin (1946) described action research as proceeding in a spiral of steps, each of which is composed of planning, action, observation and the evaluation of the action. In practice, action research begins with an imperfectly understood felt concern and a desire to take action, suggesting some kind of improvement or change is desirable (Mctaggart, 1994, p. 316). It aims at changing practitioners' practices, their understanding of their practices, and the conditions under which they practice (Kemmis, 2009). In that respect, practical knowledge is created in a localized context. The goal of action research as Toulmin (1996) aptly put it, is to improve practice:

The goal of action research is to improve, not our theories, as in physics or molecular biology, but our practices, as in medicine and engineering. Its interest lies not in abstract conceptual systems, as in mathematics but in local timely knowledge of concrete situations, as in cultural anthropology. (Toulmin, 1996, p. 58)

In essence, practical knowledge is created in a localized context—to change the dynamics of the system from "what is" to a desirable one.

There are many conceptualizations of action research by scholars. For Dickson (2000a; 2000b), it is an inquiry by ordinary people acting as researchers to explore questions in their daily lives, to recognize their own resources, and to produce knowledge and take action to overcome inequities, often in solidarity with external supporters, who are considered as change agents or more precisely, facilitators. Gormley (2001) views action research as a form of investigation where subjugated people work together with research facilitator(s) to gather information and implement solutions to problems in their community. In that regard, action research is about enhancing the capacity of human beings to complement or support each other's learning. It is a systematic enquiry involving all participants. The underlying philosophy is that in the quest for greater effectiveness for addressing social, economic or technical challenges, beneficiary participation is critical.

ACTION RESEARCH CYCLE

Lewin noted that action research is exemplified by the discussion of problems followed by group decisions on how to proceed in finding solutions. Action research must include the active participation of those (normally a group or an individual) who have to carry out the work in the exploration of problems that they identify and anticipate. After investigation of the problems, the group makes decisions, monitoring and keeping note of the consequences. This is followed by regular reviews of progress. The group may decide when a particular plan or strategy has been exhausted and fulfilled or come to nothing. Newly perceived problems are also discussed. This process is referred to as the systems model of action research. The systems model of action research is based on Lewin's notion of antecedents to social change. According to Lewin, the motivation for social change is strongly related to action. Consequently, if people are active in making decisions affecting them, they are more likely to adopt the outcomes. The decision making proceeds in a spiral of steps, each of which is composed of a circle of planning, action and fact-finding about the result of the action as shown in Figure 1 below. The systems model of action research (Figure 1) comprises the following main components: unfreezing, changing, and refreezing.

- Unfreezing: when an individual or group has a puzzlement (difficult situation), the individual or group may become aware of the need to change the dynamics of the system to a desired one.
- Changing: new modes of behavior are explored and tested after substantive examination of the puzzlement.
- Refreezing: new behavior is adopted after series of tests to ensure validity.

As shown in Figure 1, the first state of the action research process begins with identification of steps to achieve a specific goal in a given time period. This is a collaborative process involving the client and change agent. It involves problem diagnosis, where the client becomes aware of the need to change from "what is" to a desirable one. This is then followed by data gathering, feedback of results and consultative plan of action to address the puzzlement (problem identified).

The second stage involves action planning (statement of specific goal achievement over a period of time) and a series of activities that reflect the change objectives. A feedback loop system where the planning and action are linked allows consistency between change objectives and actions.

The results phase is the final stage of the systems model of action research. Evidence is gathered and examined to determine whether desired changes occurred. Measurements are carried out to determine the extent to which planning, action and results are effectively aligned.

Figure 1. Action research cycle

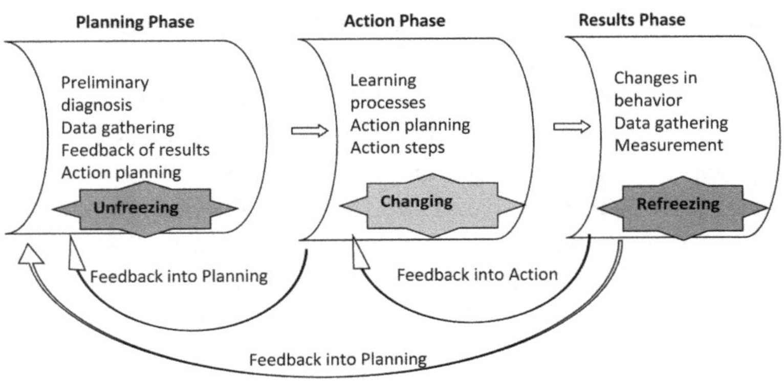

LINKING ACTION RESEARCH TO THE CONCEPTS OF PARTICIPATION AND COMMUNITY

Similar to action research, participation has various conceptualizations. Devas and Grant (2003) view participation as the ways in which citizens exercise influence and control over the decisions that affect them. Inclusion of citizens in decision making on issues that affect them directly or indirectly is another perspective. In that regard, participation is viewed as any type of inclusion of non-state actors in any stage of governmental policy making including implementation (Wesselink, 2011). In terms of development projects, participation is a process through which stakeholders' influence and share control over development initiatives, decisions and resources that affect them (World Bank, 1994a). All stakeholders are involved in the process of inquiry. In addition to that, all aspects of the inquiry, from research focus to data analysis, are undertaken in ways that are understandable and meaningful to participants. The researcher acts as a facilitator, collaborator, and learning resource; participants and facilitator are treated as coequals. The inquiry facilitator recognizes and values participants' perspectives, expertise and simultaneously works to help participants recognize and value their own and each other's expertise.

Action research is also linked to the concept of community. Understanding the concept of community may facilitate better action research designs. There are several conceptualizations of community. Flora and Flora (2004) stated that the term "community" is used in several different ways, all of which focus on groups of people. Community may refer to a place, a location in which members of a group interact with one another (Liepens, 2000, p.7). Liepens views community as having meaning, experience and dialogue— which means reciprocal relationship between members of the community. Colombo and Azzura (2005) supported this conceptualization of community in terms of dialogue. They outlined the idea that community identity is discursively constructed by members in order to lend meaning to experience. In their view, a discursive approach will allow an engagement with an analysis of how community identity is actively constructed by members in order to make sense of the social world. In addition to that, just as the previous scholars, they also shared the geographical or territorial-based and social network relationship conception of a community. Thus irrespective of the varieties of conceptual descriptions of a community, certain common threads that run through are: dialogue and shared meaning; and arena for social action. This has implications for action research design.

Examining a community from the perspective of dialogue and shared meaning has the underpinning approach of symbolic interactionism. It emphasizes interactions of small group and individual processing of the meanings of symbols, rather than on the overall social structure of society and focuses on the symbols within a society, their meanings, and ensuing interactions between individuals and groups based on the meanings of the symbols as well as the subjective nature of the symbols' meaning (Ferrante, 1998). Thus, symbolic interactionism approach is an attempt to make meaning from the actions of members of the social system through interactions in order to construct social reality. As noted by Ferrante (1998), the strength of the symbolic interactionism perspective lies in the fact that it helps in understanding the meanings of social phenomenon drawing from an analysis of actual human interaction. Colombo and Azzura (2005), in discussing the "the discursive construction of community identity," shared in this perspective in order to understand and explain the problematic concept of community.

It is obvious that all the authors are looking at social reality from different "lenses" based on their theoretical perspectives and perhaps ideologies. All these have implications for interventions to change the dynamics of the community from an undesirable state to what is perceived as desirable. The key here is that a clear understanding of the concept of "community" could lead to appropriate research interventions for social change and community development.

FEATURES OF PARTICIPATION IN ACTION RESEARCH

Sherry Arnstein's ladder of participation illustrates the degree of participation from the bottom rungs of non-participation (manipulation and therapy) to the middle rungs of tokenism (informing, consultation and placation) to the highest rungs of citizen power (partnership, delegated power and citizen control). The characteristics of the rungs of Arnstein's ladder are shown in Table 1 below. Low-intensity modes of participation in the form of information-sharing and consultation are associated with the bottom rungs. On the other hand, higher-intensity modes of participation in the form of empowering mechanisms for joint decision and control of beneficiaries over social change initiatives are linked to the highest rungs. These characteristics of participation provide insight into the quest for greater effectiveness in addressing social, economic or technical challenges that critically impinge upon people. It reveals that in order to effectively assist people to take informed decisions about their livelihood systems, preference must be given to more empowering forms of participation over less empowering ones. This claim is consistent with the goal of action research to create practical knowledge in a localized context using a systematic inquiry involving all stakeholders.

Table 1. Arnstein's (adapted) features of citizen participation

Ladder Rung	Underlying Intention
Manipulation	Educating citizens or engineering their support.
Therapy	Curing citizens of their powerlessness.
Informing	One-way information from decision-makers with no channel for feedback or power for negotiation, especially at a late stage in the planning process.
Consultation	Citizens believe that they have participated and decision makers have evidence that they have involved the community.
Placation	Citizens are allowed to advise and participate in planning but decision makers retain the right to judge the legitimacy or feasibility of the advice.
Partnership	Planning and decision making is shared without unilateral change.
Delegated Power	Citizens achieve dominant decision making authority over plans and programs.
Citizen Control	Residents are able to govern a program or institution, be fully in charge and be able to negotiate the conditions of how changes can be made.

It can be argued that in the process engaging in action research, participants are exposed to, and have the opportunity to learn the logic of evidence-based inquiry and reasoning. This is consistent with Lewin's action research model of consultative planning, action and adoption of behavior change. In that context Patton (2002,

p. 184) noted that “through acquisition of inquiry skills and ways of thinking, a collaborative inquiry process can have an impact beyond the findings generated from a particular study.” He continued that “people who participate in creating something tend to feel more ownership of what they have created and make more use of it. Active participants in research therefore are more likely to feel ownership not only of their findings but also of the inquiry process itself.”

Essentially, the following may be considered as important features/guidelines of action research:

- Research and change processes should be negotiated, agreed to, and mutually understood, not imposed, forced, or required.
- Research and change processes should be person centered, attentive to the effects on real people as individuals with their own unique needs and interest.
- The process of how things (action research) are done is as important as the outcome (what is achieved).
- Action and responsibility are shared; unilateral action is avoided.
- Each person or community is unique.
- Each person or community deserves respect.
- Equity, fairness, and mutual respect should be foundations of human interactions.
- The researcher or change agent is nonjudgmental, accepting, and supportive in respecting others’ right to make their own decisions and live as they choose. Source: Adapted from Patton, 2002.

ESSENTIAL FACTORS FOR CONDUCTING ACTION RESEARCH

As discussed earlier, action research is an inquiry by ordinary people acting as researchers to explore questions in their daily lives, to recognize their own resources, and to produce knowledge and take action to overcome inequities, often in solidarity with external supporters or change agents (Dickson, 1999). In a similar vein, action research is a form of investigation where subjugated people work together with research facilitator(s) to gather information and implement solutions to problems (Gormley, 2001). Thus the essential factors for engagement in action research are:

- Problem realization
- Investigation
- Community actors
- Research facilitators
- Information sharing

- Problem solving

A research problem is a statement about a puzzlement or an issue of concern that needs to be investigated and understood in order to generate relevant knowledge for addressing it. This creates a knowledge gap. Usually, there is no need for the conduct of a research if there is nothing new worth finding—it is difficult to justify the research problem if no substantive knowledge gap has been identified. The knowledge gap underscores the significance of the research. The concept of knowledge gap as applied to action research is based on the notion that a research problem has been identified through a consultative approach by all stakeholders.

Once a substantive research problem is identified, the researcher should clearly state the intellectual puzzle or what he or she wishes to explain. Some intellectual puzzles may be causal/predictive or comparative puzzles. Causal/predictive puzzles address issues of causality, involving at least two variables—(a variable is anything that when measured, can produce two or more values or observations). The researcher(s) poses a puzzle about what influences x has on y, or what causes x or y? Comparative puzzles deal with what researcher(s) can learn from the comparing x and y, and how differences and similarities between them can be explained (see Mason, 2002). For example, x and y could be level of education and level of engagement in risky pedestrian behavior respectively.

In order to have a clear conceptualization of the research problem, the following questions are worth considering:

- What is my research question (RQ)?
- Do the RQs express or problematize my intellectual puzzle?
- Are the RQs consistent with each other, and linked to each other?
- Do the RQs add up to a sensible whole?
- Are the RQs coherent and transparent?
- Would anyone but me understand the RQs?
- Do the RQs make possible, and probable, intellectually interesting answers or arguments?
- Are the RQs original and worth asking, as well as grounded in an understanding of the relevant background of the intellectual puzzle? Source: Adapted from Mason, 2002.

A clearly defined research problem allows the researcher(s) to articulate a valid study. The investigator(s) must map the social world. Social mapping is the identification of different social groups using different criteria that fit the local context. Gantz (2006) noted that mapping the social world begins by asking four questions:

- Who are the actors?
- What are their interests?
- What resources do they need to act on those interests?
- How much power do they have to mobilize and deploy these resources?

In order to be effective in action research, it is important to understand the interests of actors and beneficiaries. Actors are persons, groups, of persons—who remember, imagine, choose and reflect on their choices. As shown in Figure 2, people's interests are shaped by their values, needs and resources. For example, a person's value might influence him/her to live an environmentally responsible lifestyle, engage in work ethics, or fulfill religious traditions. Learning to recognize our own interests and that of others we interact with, and the values and resources that shape them is critical to understanding the dynamics that shape our social world (Gantz, 2006, p. 15). This understanding fosters valid contextualization and conduct of action research.

Figure 2. Understanding stakeholder interests in relation to action research *(adapted from Gantz, 2006)*

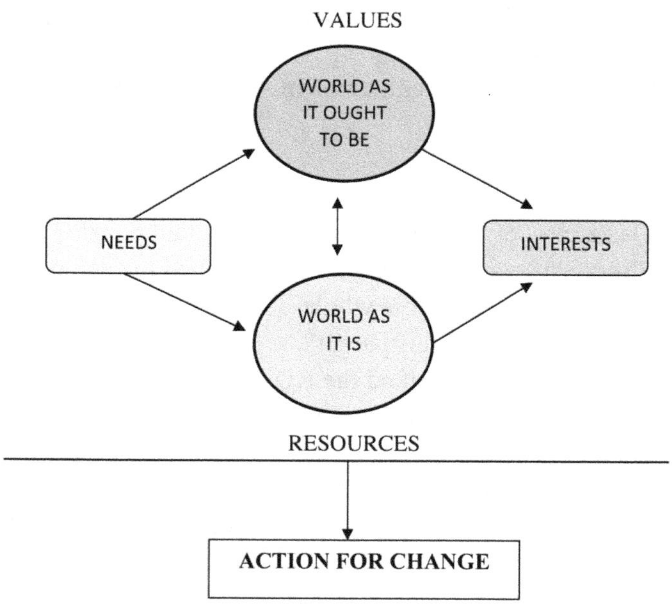

EMPIRICAL MODELS FOR CONDUCTING ACTION RESEARCH

Three empirical models illustrating how to conduct different types of action research are discussed. These are: Photo-voice Project about an underused footbridge in Barbados, in the Caribbean; Danish International Development Agency (DANIDA's) Volta Region Rural Water and Sanitation Agency (VRRWSA), Ghana; and REACH After School Enrichment Program, Hawley, Minnesota, USA.

Model 1: Photo-voice as a Form of Action Research

The discussion so far shows that action research is a collaborative approach to problem solving. It involves consultative problem identification, reflects context, encourages reflexive examination, and ultimately empowers the beneficiaries for desirable change. Photo-voice is an approach to action research where people identify, represent and generate knowledge to address problems in their communities through photographic techniques. The main goals of photo-voice are to: enable people to record and reflect on their community's concerns; promote critical dialogue and knowledge about important issues through discussion of photographs; and develop policy implications (Wang & Burris, 1997). The main stages of photo-voice are:

- Conceptualizing the problem.
- Training
- Research/documentation.
- Discussion & analysis.
- Policy outcome

The conceptualization phase involves the intellectual process of articulating a fitting research problem. In that regard, beneficiaries define their own problems in consultation with the researchers (who are considered as facilitators). This consultative problem identification allows beneficiaries to understand and construct what matters to them in the context of their circumstances rather than having outside researchers define the research problem. Based on the problem identified, the scope, objectives and significance of the study are determined.

At the training stage, community actors (beneficiaries) are instructed on the safe operation of the cameras and how to take pictures. They are also educated on photographic ethics. For example, community actors are made to understand the importance of seeking community members' approval before taking pictures of them or any of their tangible personal properties. They are primarily trained on how to address issues of community cultural values and how that might influence the appropriateness of capturing scenes that are relevant to the problems articulated.

In addition to that, during training, community actors acquire the knowledge and skills to seek out situations that are relevant to their problem coupled with devising themes for taking pictures.

At the research/documentation stage, pictures are taken of scenes that capture the specific subject matter that is being investigated—to highlight research themes. The research themes may include building and environment, community water management and sanitation, road safety, and others. The community photographers then return the cameras to facilitators for the images to be developed.

The discussion and analysis stage comprise series of activities. First, the pictures are displayed to the community. Community actors view each other's picture(s). Each photographer is asked to tell a story behind his/her picture. Normally, questions that may be asked are:

- Why was the picture taken?
- What is the picture depicting?
- Where was the picture taken?
- When was the picture taken?

This questioning process promotes storytelling, critical reflection, dialogue, and contextualization of the pictures. It is important that the facilitator creates an atmosphere that fosters collaborative interpretation of the photographs and also the development of narratives that highlight the research themes. This approach motivates participants to recognize and value their expertise.

Analysis involves organizing the stories behind the pictures (narratives) into conceptual categories or relevant themes. This facilitates the identification of overarching issues affecting the community that need policy intervention for improvement. The research themes that can be shown to emerge convincingly from the pictures are then used to understand the research problem. Further, they are collaboratively used to design appropriate policies for desirable change.

In summary, photo-voice methodology enables community actors to define their own problems, construct what matters to them in consultation with facilitators and policy makers. The photo-vice methodology places all stakeholders at the core of the change process. The process of change from project conceptualization to analysis and policy implications is thus made more understandable and meaningful to community actors (beneficiaries). The photo-voice process is shown in Figure 3.

Figure 4 is a picture of an underused footbridge in Barbados. Barbados is a densely populated island of 431 square kilometers with a population of 284,644. The underused footbridge was constructed in 2012 in a busy residential area. There are sidewalks and side lanes close to the footbridge. Unfortunately, most pedestrians are refusing to use the footbridge. They prefer to cross the road on foot, instead of using the bridge. This poses a risk to motorists and pedestrians. In this situation, residents can use photo-voice to highlight and discourage risky pedestrian behavior and ultimately create change in the community. Table 2 depicts using photo-voice to create awareness on the dangers of not utilizing the footbridge.

Figure 3. Photo-voice research process

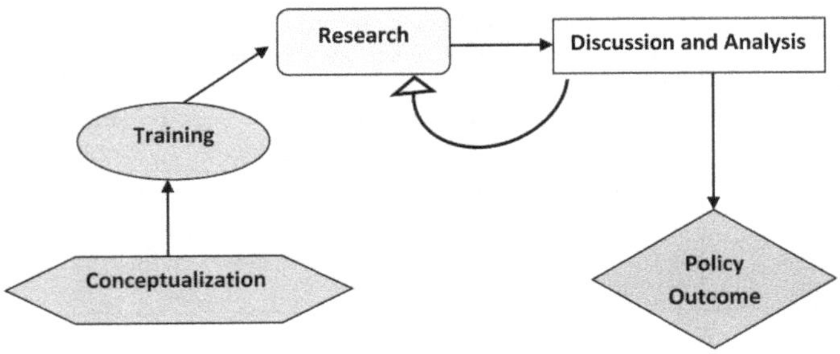

Figure 4. Underused footbridge in Barbados

Table 2. Photo-voice process to create awareness on the dangers of not using footbridge

Stage	Issues
Conceptualization	Road safety, risky pedestrian behavior.
Training	Learn the basics of camera operation, community values and culture, taking ethically appropriate pictures depicting above conceptualizations.
Research/Documentation	Take pictures that capture risky pedestrian behavior, risks to motorists due to pedestrian refusal to use the footbridge.
Discussion and Analysis	Critical discussion and dialogue on: why pedestrians are refusing to use the footbridge, whether the footbridge was a felt need, were there any genuine community participation in the design and construction of the footbridge, to what extent are community values, interests and resources reflected in the design and construction of the footbridge?
Policy Outcome	Recommendations should emphasize why people should care about their risky behavior. Use photos to organize activities such as media campaigns, public rally in a visible location and line up speakers such as road and transport experts, community actors.

Model 2: DANIDA Project, Ghana

Background of the Model

The Danish International Development Agency (DANIDA's) Volta Region Rural Water and Sanitation Agency (VRRWSA) project was initiated in 1991. It emphasizes community ownership and management of rural water schemes funded by DANIDA. The immediate goal was to achieve the provision of adequate and safe water in the areas covered by the project—ultimately resulting in improved health status of project beneficiaries. Unfortunately, there was an elusive quest for effective communication strategy to improve the knowledge of hygienic and sanitary practices. In order to realize this objective, a communication campaign was launched deploying various communication strategies such as using mass media (television and radio) to reach out to beneficiaries.

Figure 5 is a hypothetical model of an action research to study communication channel preferences and accessibility to information based on the Danish International Development Agency Volta Region Rural Water and Sanitation Agency project. This model is hypothetical due to the fact that there is no way to have reliable data (information) on some of its components. In addition to that, the acquisition of such data by direct observation is practically impossible at this time. This model utilizes action research to determine the relationship between community actors' preferred channels, channels used by project (VRRWSA), accessibility to information, and the outcome of the project (desired behavioral change). It comprises the following: need for intervention or action research, community actors' preferred channels for effecting change, different levels of accessibility to innovation information, channels used by project, system factors, and the desired behavioral change.

Boone (1985) defines behavioral change as an alteration in the knowledge, attitude, or skills or any combination of these behaviors in an adult or learner group. This shows the extent to which the gap between "what is" (the situation that needs to be changed) and "what ought to be" has been closed as a result of accessibility to innovation information—in this context, information on community water and sanitation practices. Forbes (1979) defines accessibility as implying the ease of getting to a place. The concept means in general that something is "get-at-able" (Mosley, 1979). In relation to innovation information in this book chapter, accessibility is conceptualized as the ease with which beneficiary of information is able to "get it" from the source (VRRWSA project).

Need for Intervention/Action Research

This is a state of need that demands action research to change the dynamics of the system from "what is" to what it "ought to be" (the desired outcome). This necessitates engagement in action research geared toward generating appropriate knowledge to assist people to take informed decisions that will ultimately lead to the desired outcome. In the context of the VRRWSA project, the need for intervention refers to changes in the water and sanitation practices.

Channels Used by Change Agent (VRRWSA Project)

This deals with tendency among change agents on how to effect change using communication channels normally used in developed nations. It is a reflection of the notion that change agents often have the knowledge to bring about desired changes through the utilization of communication channels that have worked in developed nations. In this case, mass media (television and radio) was used to disseminate information on community water and sanitation practices to beneficiaries.

Community Actors Preferred Channels of Effecting Change

This reflects the peculiar socio-economic circumstances and it is the channels that beneficiaries prefer to be used by the VRRWSA project in the transfer of innovation information on community water and sanitation practices to them.

Relationship Between Community Actors Preferred Channels, Channels Used by the Vrrwsa Project, Different Levels of Accessibility to Innovation Information, and System Factors

The greater the extent to which beneficiaries' preferred channels of effecting change are used, the more likely beneficiaries' accessibility to information and their engagement for knowledge generation, leading to the appropriate desired outcome. This is due to the influence of system factors—which are peculiar circumstances in beneficiary environment that affect interaction between change agents and community actors to create the desired outcome. It is in that context that development research indicates that the socio-economic structure of target audience plays a very important role in any social interventions particularly in communication interventions (Dennis, 1983). It is worthy of note that highest levels of beneficiary accessibility to information will occur when their preferred modes of effecting change are used.

Outcomes

It follows that the desired behavioral change occurs when maximum accessibility to innovation information occurs.

Following the hypothetical model in Figure 5 above, Table 3 shows different types of communication channels (A, B, C & D); number of respondents who indicated channel was used by VRRWAS project (U); number of respondents who preferred a channel but was not used by VRRWSA project (P); and number of respondents who preferred a channel and was used by VRRWSA project (U=P). The higher the number of beneficiaries who indicated that their preferred channels are used by the VRRWSA project (P) relative to number of respondents who indicated channel was used by VRRWAS project (U), higher the chances of accessibility information—since that may be conducive to beneficiary circumstances in terms of occupations, cost, culture, language, etc. The best case scenario is when the number of respondents preferred a particular channel and it was used by VRRWSA project (U=P). This desired ideal state can be achieved engaging in action research where a collaborative inquiry process is used to ascertain the channel preferences of beneficiaries to ensure maximum accessibility to innovation information.

It can be argued based on the hypothetical model in Figure 5 and Table 3 that in order to change community knowledge and practices on water and sanitation, information is best communicated to and with the ultimate users of the information through the use of appropriate channels in order to influence their behavior in the desired direction. Behavior modification can be achieved faster when the communication channels preferred by the target community actors are used in the information exchange process since that ensures maximum accessibility to information. This highlights the importance of dialogue, stakeholder participation, in determining the choices of beneficiaries in order to have the desired outcome. These suggest an active beneficiary that is goal-oriented in channel preference. Further, it reveals the need for active involvement of beneficiaries in determining the channel preferences. Engaging in action research will help create knowledge on channel preference to fit the context of beneficiaries.

Figure 5. A conceptual model of the relationship between communication channels and accessibility to information

Key:
A: Channel Preferred by Beneficiary But Not Used by Source/Sender
B: Channel Preferred by Beneficiary and Used by Source/Sender
C: Channel Used by Source/Sender But Not Preferred by Beneficiary

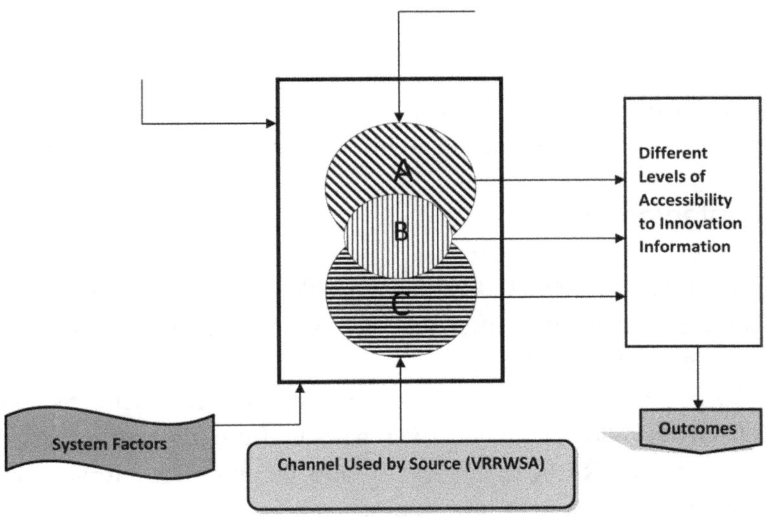

Key:
A: Channel Preferred by Beneficiary But Not Used by Source/Sender
B: Channel Preferred by Beneficiary and Used by Source/Sender
C: Channel Used by Source/Sender But Not Preferred by Beneficiary

Table 3. Communication channels used, preferred and accessibility to information

Type of Communication Channel	Number of Respondents Who Indicated Channel Was Used by Project (U)	Number of Respondents Who Preferred a Channel But Was Not Used by Project (P)	Number of Respondents Who Preferred a Channel and Was Used by Project (U=P)
A			
B			
C			
D			

Model 3: Rural Enrichment and Counseling Headquarters, Inc. (REACH) After School Enrichment Program, Hawley, Minnesota, USA

Community Context

- It is a rural area of Central Minnesota, approximately 4200 people within the community of Hawley, Minnesota.
- Community support for social and human services is provided by REACH.

Need for Intervention

- Community and REACH identified a problem for youth in the area: no areas for youth to gather safely after school. Rise in delinquent behavior, cruising, crime, drug and alcohol use.
- In 1999-2000, a team of social work students were contracted by community members, stakeholders, concerned parents from the school district, and members of the REACH agency to help facilitate an intervention, as the community was unsure of how to proceed.

Preferred Mode for Effecting Change of Community Actors

- The community actors wanted to be very participative, but needed assistance in visioning, brainstorming, and coming up with ideas.
- Preferred role of the change agents was to provide resources and organize events aimed at coming up with solutions, as well as assistance in facilitating outside monetary and material support.

Preferred Mode for Effecting Change of Change Agent

- Change agents (facilitators) were interested mainly in providing tools and other technical assistance that would allow for community members to find their own solutions to their own identified needs.
- Preferred role was as a facilitator for action, rather than decision-making.

Interface of Modes for Effecting Change

- There was a large interface in the preferred modes for change, and the community actors and change agents worked in a manner which allowed for optimal participation of stakeholders.

System Constraints

- Funding and spatial constraints as well as the community power and political system placed some constraints on the actions to be taken.

Different Levels of Involvement for Actors

- There were high levels of involvement for actors who were interested in the participatory process but some stakeholders expected the change agents to enter the situation and make the changes themselves, rather than the community actors in actual fact making the changes.
- Varying levels of participation of change agents based on community needs and personal preferences during different stages of the program.

Outcomes

- Creation of an after school enrichment program which serves many children in the area. The community was able to fund the program through a start-up grant, which the change agents helped facilitate.
- Lessened problems of children not having safe, drug and alcohol free spaces.

Figure 6. A model of the relationship between modes of change between change agents and community actors in the REACH after school enrichment program
Key:
A: Preferred Mode of Effecting Change for Community Actors But Not Used by Change agents
B: Interface Between Modes of Effecting Change Desired by Both Community Actors and Change Agents
C: Change Agent Bias for Effecting Change Not Preferred by Community Actors

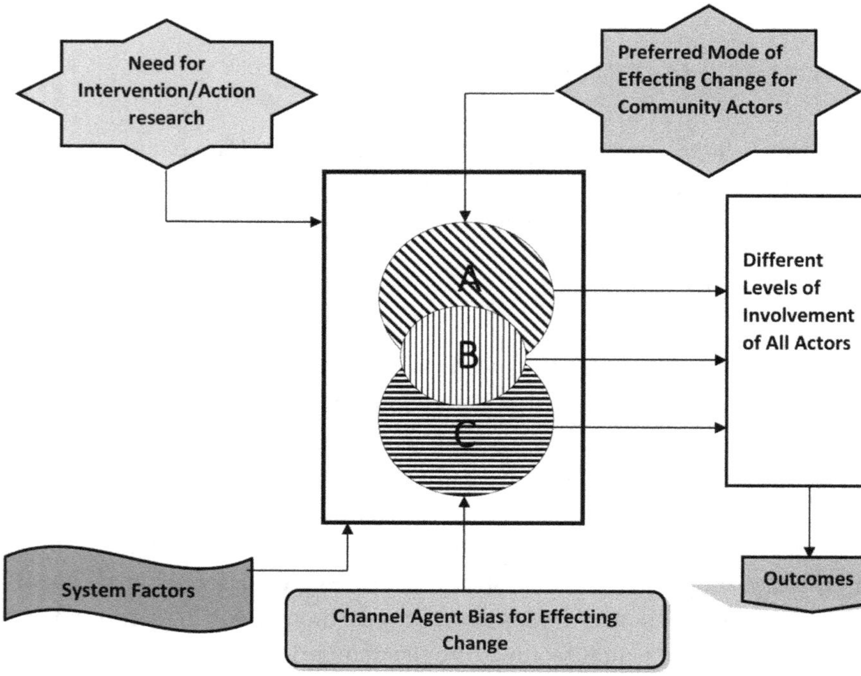

Key:
A: Preferred Mode of Effecting Change for Community Actors But Not Used by Change agents
B: Interface Between Modes of Effecting Change Desired by Both Community Actors and Change Agents
C: Change Agent Bias for Effecting Change Not Preferred by Community Actors

SUMMARY OF CORE ISSUES

Action research is an investigative process where all stakeholders work together to articulate the problem, create the need for change, create intent to change, plan, gather data, analyze and design policy implications that are unique to a particular community. It is an interactive process anchored in equity, fairness, mutual respect where the research problem is not imposed by researcher (who is precisely a facilitator) but rather negotiated, agreed to, and mutually understood. The researcher must create an atmosphere that fosters the active involvement of community actors in the action research process. Further to that, the researcher must be continually aware that his/her intervention occurs in a complex cultural environment. There is therefore the need to have a competent culturally-relevant knowledge, attitude and skills in order to effectively conduct action research in any given community context. Following action research approach may avoid the situation where programs based on expert "knows-it-all" attitudes in the developing world have caused much harm than good because they were not designed in consultation with the intended beneficiaries. Such harm could have at least been avoided or reduced if the people (beneficiaries) are active participants in the researches that form the basis of the programs designed ostensibly to improve their well-being. Fundamentally, action research is localized, values local knowledge, uses a methodology that empowers the people, resulting in ownership of research outcomes.

ACKNOWLEDGMENT

I would like to thank Dr. Shauna Sowga for her insightful comments on model of the relationship between modes of change between change agents and community actors in the Rural Enrichment and Counseling Headquarters, Inc. (REACH), MN, USA after school enrichment program and also for providing valuable information about REACH.

REFERENCES

Boone, E. J. (1985). *Developing programs in adult education*. Prentice-Hall, Inc.

Colombo, M., & Azzurra, S. (2005). The discursive construction of community identity. *Journal of Community & Applied Social Psychology*, *15*(1), 48–62. doi:10.1002/casp.809

Dennis, R. F. (1983). Evaluation models for assessing effects of media-based nutrition education. In Effective communication in primary health care. Food and nutrition bulletin supplement 13, United Nations University.

Devas, N., & Grant, U. (2003). Local government decision-making—citizen participation and local accountability: Some evidence from Kenya and Uganda. *Public Administration and Development*, *23*(4), 307–316. doi:10.1002/pad.281

Dickson, G. (Ed.). (2000a). Participatory action research: Theory and practice. In M.J. Stewart (Ed.), Community nursing: Promoting Canadians' health (2nd ed., pp. 542-563). W.B. Saunders.

Dickson, G. (2000b). Aboriginal grandmothers' experience with health promotion and participatory action research. *Qualitative Health Research, 10*(2), 188–213. doi:10.1177/104973200129118363 PMID:10788283

Ferrante, J. (1998). *Sociology, a global perspective*. Wadsworth.

Flora, C. B., & Flora, J. (2004). *Rural communities: Legacy and change* (2nd ed.). Westview Press.

Forbes, J. (1964). Mapping accessibility. *Scottish Geographical Magazine, 80*(1), 1–12. doi:10.1080/00369226408735915

Ganz, M. (2006). *Organizing: People, power and change*. John F Kennedy School of Government, Harvard University.

Gormley, K. J. (2001). Research as a democratic process: Educational community development in Brazil through participatory research. *Convergence*, 34.

Kemmis, S. (2009). Action research as a practice-based practice. *Educational Action Research, 17*(3), 463–474. doi:10.1080/09650790903093284

Lewin, K. (1946). Action research and minority problems. *The Journal of Social Issues, 2*(1), 34–46. doi:10.1111/j.1540-4560.1946.tb02295.x

Liepins, R. (2000). New energies for an old idea: Reworking approaches to 'community' in contemporary rural studies. *Journal of Rural Studies, 16*(1), 23–35. doi:10.1016/S0743-0167(99)00042-X

Mason, J. (2002). *Qualitative Researching* (2nd ed.). Sage.

McTaggart, R. (1994). Participatory Action Research: Issues in theory and practice. *Educational Action Research, 2*(3), 313–337. doi:10.1080/0965079940020302

Moseley, M. J. (1979). *Accessibility, the rural challenge*. Methuen.

Patton, M. Q. (2002). *Qualitative research & evaluation methods*. Sage Publication, Inc.

Toulmin, S. (1996). Is action research really 'research'? *Concepts and Transformation, 1*(1), 51–62. doi:10.1075/cat.1.1.05tou

Wang, C., & Burris, M. A. (1997). Photovoice: Concept, methodology, and use for participatory needs assessment. *Health Education & Behavior*, *24*(3), 369–387. doi:10.1177/109019819702400309 PMID:9158980

Wesselink, A., Paavola, J., Fritsch, O., & Renn, O. (2011). Rationales for public participation in environmental policy and governance: Practitioners' perspectives. *Environment & Planning*, *43*(11), 2688–2704. doi:10.1068/a44161

World Bank. (1994a). *The World Bank and participation*. Operations Policy Department.

Chapter 6
A Review of Single Case Study and Multiple Case Study Research Designs

Stacey Sneed
https://orcid.org/0000-0001-6317-8939
Texas Tech University, USA

Chau H. P. Nguyen
Texas Tech University, USA

Chrissy L. Eubank
Texas Tech University, USA

Aaron S. Zimmerman
Texas Tech University, USA

ABSTRACT

In the field of education, case study is a widely used qualitative research methodology. While there are a broad range of approaches to case study, the resources available to researchers is limited. Therefore, this chapter aims to shed some light on what defines a case study and some of the possible methodological variations. The goal of this chapter is to provide the reader with a foundational understanding of a case study as a methodology and how this methodology aligns within various research contexts in the field of education.

DOI: 10.4018/978-1-7998-7600-7.ch006

THE CASE STUDY METHOD: DEFINITION AND VARIATIONS

According to Yin (1981), a case study is "(a) a contemporary phenomenon in its real-life context, especially when (b) the boundaries between phenomenon and context are not clearly evident" (p. 59). In other words, a case study focuses on observing a unique situation that occurs (or occurred) in real life; and, the case study method is particularly appropriate when the situation cannot be defined apart from the context in which it occurs (as opposed to a randomized, controlled experiment). The case itself can range from something concrete (e.g., an individual person or organization) to something abstract, e.g., a relationship or idea (Creswell & Poth, 2018; Yin, 2009a).

Epistemological beliefs play an important role in identifying the case, developing research questions, and deciding how the case study will be designed. Researchers must begin the case study research process with an understanding of their epistemological beliefs, because this epistemological perspective will ultimately guide the focus and direction of the research (Bloomberg & Volpe, 2016; Creswell & Poth, 2018; Huff, 2009). In this chapter, we will focus on case studies that align with a positivist or, alternatively, constructivist viewpoint. As the reader will note in the different examples presented in this article, some case studies seek to draw objective, generalizable conclusions based on scientific approaches to data collection and data analysis. This is reflective of the positivistic epistemological approach to case study research (Phillips & Burbules, 2000; Yin, 2009a). The constructivist epistemological approach also seeks to draw conclusions, but this approach is transparent about the role of the researchers' subjectivity in interpreting the data (Yazan, 2015).

Central to the positivistic approach is the idea that "there are social *facts* with an objective reality apart from the beliefs of individuals" (Firestone, 1987, p. 16, italics in original); hence, objectivity, validity, and generalizability are considered the essential tenets of positivism-oriented research (Crotty, 1998). This leads those adopting positivism to believe that construct validity, internal validity, external validity (i.e., generalizability of the findings), and reliability should be emphasized in case study research (Yin, 2012b, 2018).

As opposed to Yin who has more quantitative or positivistic leanings, most contemporary case study researchers hold that constructivism should be the epistemology orienting and informing case study research (e.g., Merriam, 1998; Stake, 1995; Yazan, 2015) because, as Merriam (1998) elucidates, "the primary instrument in qualitative research is human, all observations and analyses are filtered through that human being's worldview, values, and perspective" (p. 22). From the constructivist viewpoint, "knowledge is constructed rather than discovered. The world we know is a particularly human construction" (Stake, 1995, pp. 99-100), and, to use Merriam's (1998) words, "reality is not an objective entity; rather, there are multiple interpretations of reality" (p. 23). Consequently, case study researchers with constructivist tendencies are interested in exploring and grasping the meaning people have constructed through their lived experiences with the world (Merriam, 1998) rather than trying to obtain objective, quantifiable, and generalizable knowledge as those with a positivistic orientation usually do (Crotty, 1998). That is, constructivist case study inquirers do not separate the phenomenon under study from its setting, for it is impossible to study a phenomenon void of its context (Miles & Huberman, 1994).

It is also significant to highlight how case study research tends to begin with a theoretical framework, in contrast to other methodological approaches, such as grounded theory (Merriam, 1985, Yin 2009b). Grounded theory research uses the data and experiences of participants to develop a theory, while case study research begins by identifying a theory to guide choices related to the data sources, data collection, and data analysis.

In this chapter, we will present examples of two commonly used approaches to case study design: single case study and multiple case studies (Bloomberg & Volpe, 2016; Creswell & Poth, 2018). The single case study concentrates on a specific person, group, or organization to provide insight into a larger issue (Creswell & Poth, 2018). This approach is appropriate within a number of different scenarios. For example, single case studies are appropriate in cases that are understood to be rare and unique, where document of the case is essential. Alternatively, representative cases denote cases where there are many examples to choose from but where one case is particularly representative of the phenomenon and, thus, worthy of documentation and exploration. Finally, longitudinal case studies examine a single case over time, thus lending itself to single case study design.

Multiple case studies (or collective case studies) also seek to explore one particular issue or phenomenon, but the design of the study contains more than one well-bounded case. Often, the rationale that supported the single instrumental case study will not apply to the multiple case study; for example, cases that are considered unique will not often provide the opportunity for multiple cases to be formed; however, using multiple case studies is favored by researchers who choose to ascribe to a positivistic epistemology, because the multiple case study design is thought to increase the reliability and validity of the research findings (Yin, 2009b).

DATA SOURCES AND DATA ANALYSIS

A variety of data sources can be used when conducting a case study. The qualitative data sources of documents, interviews, and direct observations align with the constructivist approach to case study design (Yazan, 2015). Yin (2009b), however, suggests the use of both qualitative and quantitative data (such as survey data) to ensure that a high-quality case study is conducted, as multiple data sources help to aid in the process of triangulation (Yin, 2009a). Many case study researchers emphasize the importance of documenting the process of data collection, also referred to as an audit trail (Guba & Lincoln, 1981; Merriam, 1985; Yin, 2009a). Being as transparent and precise as possible about the data sources used in the given case study helps to support the reliability of the study. Yin (2009a) recommends developing a chain of evidence, which helps prevent research from being unreliable, by encouraging the researcher to continuously connect the data in the database to the research questions when drawing conclusions making the link explicit.

In regards to data analysis, a constructivist approach suggests that data collection and data analysis can occur simultaneously (Yazan, 2015), leaving room to develop the study in a flexible way as it progresses. A positivistic approach, in contrast, tends to remain faithful to the pre-established research design, in which case all subsequent data analysis will adhere to the research study's original propositions. Regarding the process of data analysis within the context of multiple case studies, it is worth noting that multiple case studies have multiple cases to analyze within a given study. Hence, the analysis of each individual unit (or case) is referred to as within-case analysis. After each within case analyses is performed, the researcher then conducts a cross-case analysis, which compares the different units within the given study. Analyzing multiple cases, embedded within a given multiple case study, can provide a rich and in-depth analysis of the issue under consideration (Baxter & Jack, 2008; Yin, 2009a).

THE SINGLE CASE STUDY APPROACH

This section begins by defining what is meant by a single case. Miles and Huberman (1994) define a case "as a phenomenon of some sort occurring in a bounded context. The case is, in effect, your unit of analysis. Studies may be of just one case or of several" (p. 25). In much the same way, Clandinin and Johnson (2017) state that "[a] case is defined as a bounded system" (p. 434):

[A] system is a set of interrelated elements that form an organized whole. Using the system metaphor, cases are seen as holistic entities that have parts and that act or operate in their environments. Bounded is added to emphasize that you should identify the outline or boundaries of the system – you must determine what the case is and what it is not. (p. 434, italics in original)

Similarly, Merriam (1998) writes, "I can 'fence in' what I am going to study" (p. 25). Miles and Huberman remind us that "a 'case' always occurs in a specified social and physical *setting*; we cannot study individual cases devoid of their context in the way that a quantitative researcher often does" (p. 27, italics in original). The task of case study researchers is, thus, to "take the reader into the case situation and experience – a person's life, a group's life, or a program's life… allowing the reader to understand the case as a unique, holistic entity" (Patton, 2002, p. 450).

A single-case study is, as its name suggests, a study of a single case or a single bounded system (as opposed to a collective or multiple-case study which has multiple cases or multiple bounded systems) (Kennedy, 1979; Stake, 2005; Yin, 2012a, 2012b). A single case (or a single bounded system) can be an individual, a group of individuals, a classroom of students, a program, an institution, an organization, an event, an activity, a process (e.g., becoming a teacher), or other social phenomenon around which there are boundaries (Clandinin & Johnson, 2017; Creswell, 2007; Merriam, 1985, 1998; Stake, 2005; Yin, 2012a, 2012b).

With respect to the matter of generalizability of single-case studies, Yin (2018) concedes the question, "How can you generalize from a single-case study" (p. 20) is frequently heard. In response to this question, Yin asks the reader/researchers to "consider for the moment that the same question had been asked about an experiment: 'How can you generalize from a single experiment?' In fact, generalizations in the physical and life sciences are rarely based on single experiments" (p. 20). Put another way, "[T]he holistic, single-case study … ha[s] little or no generalizability value" (Yin, 2012a, p. 18). To make this point more clearly, Kennedy's (1979) words are useful:

> *Advocates of single-case methodologies have used a variety of arguments to overcome the problems of sampling limitations, but none has satisfied those who rely on multiple cases to draw generalizable findings from studies. Some authors … have erroneously assumed that by increasing the number of data points on a single case, one has eliminated the problem. But this is not a solution, for these several data points are still based on only one case. (pp. 663-664)*

This chapter now presents four examples demonstrating how single-case studies can be applied to educational research. In these examples, the reader will note that a case can represent an individual, a small group, a program, or an organization.

THE CASE AS AN INDIVIDUAL

The study by Webel, Sheffel, and Conner (2018) is a good example showing how educational researchers can use a single individual as a case in their inquiry. The focus of their investigation was Joni (a pseudonym) and her attempt to implement flipped instruction for mathematics. The bounded system was her fifth-grade class with 14 students in a small, rural elementary school in the Midwestern United States. The authors purposefully selected Joni as their case of interest because she was the only teacher in their sample utilizing flipped instruction in a routine manner. As Webel and colleagues expound, they adopted this methodological design (i.e., an individual as a single case) because the intent of their study was neither to evaluate flipped instruction nor to depict an idealized version of flipped instruction. Rather, their interest was in thoroughly understanding how Joni enacted what she believed to be flipped instruction within her own teaching context. This approach aligns with the epistemological assumption of constructivism. As Yazan (2015) puts it, constructivists "conceive knowledge as being socially constructed and emerging from people's social practices; therefore, [constructivists] conceptualize social reality as being generated and constructed by people and existing largely within people's minds" (p. 138).

The data sources of this study (which were gathered over the course of a school year) included 5 interviews, 6 observations, and 9 instructional videos. The amalgam of these three data collection methods was used as a means of data triangulation. These qualitative data were then analyzed and organized thematically (Merriam, 1985). The authors suggest that their case demonstrated how teachers might adopt flipped instruction to boost the effectiveness of their teaching. The case also highlighted some potential rewards and challenges of this particular instructional practice.

Webel and colleagues do clarify that their findings were drawn from the experience of a single teacher whose teaching setting was, to a certain extent, unique (i.e., a fifth-grade class in a small, rural elementary school), and, thus, their findings are not necessarily generalizable. Webel et al.'s concern is consonant with Stake's (2005) viewpoint: the "epistemological question is…What can be learned about the single case? I will emphasize [that] designing the [case] study [should be] to optimize understanding of the case rather than to generalize beyond it" (p. 443).

THE CASE AS A SMALL GROUP

In the single-case study conducted by Phan and Locke (2015), the phenomenon under scrutiny was Vietnamese EFL (English as a foreign language) teachers' perspectives on sources of self-efficacy for teaching. As the authors point out, "no studies have investigated sources of teachers' self-efficacy in the Vietnamese context" (p. 73), their study thus aimed to explore ways Vietnamese EFL teachers construct their self-efficacy beliefs to advance the development of teacher-self efficacy theory and research. Their aim is in accord with the epistemology of constructivism in that, as Yazan (2015) speculates, "[C]onstructivism puts forward the idea that there are multiple versions of knowledge" (p. 146), and that, to quote Merriam (1998), "[R] eality is constructed by individuals interacting with their social worlds" (p. 6).

The single case was eight Vietnamese EFL teachers and the boundary surrounding the case was a technical university in Vietnam. This case was selected because it was ideally suited to the aim of the research and the researchers' interest (Merriam, 1985). Data were collected by means of individual interviews, teachers' journaling, and observation, which lasted for 7 months. All these qualitative data were inductively coded and analyzed using the thematic analysis method. As Braun and Clarke (2012) illuminate:

An inductive approach to data coding and analysis is a bottom-up approach and is driven by what is in the data. What this means is that the codes and themes derive from the content of the data themselves – so that what is mapped by the researcher during analysis closely matches the content of the data. (p. 58, italics in original)

Phan and Locke (2015) do not directly indicate to what degree their findings can be generalized. Stated more precisely, they do not even mention the term generalization throughout their study because, as Stake (1995) asserts, "single cases are not as strong a base for generalizing to a population of cases as other research designs" (p. 85). However, Phan and Locke do discuss implications of their findings for a larger context – i.e., how to nourish teaching self-efficacy for university-level teachers in higher educational institutions in Vietnam – as their case and such the larger context might share similar features. In other words, the single case study cannot be generalized, yet it still can make contributions (Ellinger & McWhorter, 2016) in the sense that it adds knowledge to existing research literature, it contributes to the development of a theory (Phan & Locke, 2015), and its findings have the potential to be replicated to similar contexts. It should be noted also that, as Kennedy (1979) points out, for studies of single cases, the role of the researcher is to produce the information, but it is the reader that "determine[s] whether it applies to their own situation" (p. 672), i.e., single-case study researchers leave "generalization up to the practitioner" (Kennedy, 1979, p. 672).

THE CASE AS A PROGRAM

The case study by Kostiainen and colleagues (2018) centered on exploring elements of meaningful learning in the context of the ISGB (Interaction Skills in a Group and Network) program. The case was the ISGB program and its bounding was the Department of Teacher Education at the University of Jyvaskyla, Finland. Their participants included 71 undergraduate teacher students who attended the ISGB course. The data collection in this single case study drew merely upon one single source of information: student essays. The data were analyzed employing a step-by-step approach to thematic analysis that included generating initial codes, identifying themes, and naming main categories (see Braun & Clarke, 2012). The researchers included instructors of the program under investigation, and, thus, to minimize potential biases, the data analysis was not done by the responsible teacher but by other members from the team. The act of not letting their personal biases interfere with the reliability and objectivity of the data analysis suggests that the authors advocate the combination of positivist and constructivist orientations towards case study research. The researchers did not try to generalize their case to other contexts. Instead, they encouraged future researchers to investigate what they conceived as meaningful learning in other courses to capture a fuller picture of elements of meaningful learning in the setting of teacher education.

THE CASE AS AN ORGANIZATION

The last example pertaining to single case studies provided in this chapter is an inquiry by Tygret (2017), in which the single case was a school. The purpose of her investigation was to uncover the impact of student teachers on student achievement during their field placement and how cooperating teachers perceived the effectiveness of student teacher's contribution to student growth during that time period. Danielson's (2013) Framework for Teaching was adopted to guide her inquiry. The bounded entity was a local elementary school in Colorado, USA. This school was chosen as her case because, as the author explains, it had hosted student teachers from a local university for years. Hence, several practicing teachers at this school frequently had student teachers in their classrooms from local university for clinical/field experience. In the fall semester, these student teachers were in the classroom only two days a week; however, beginning in January, they took over all the teaching responsibilities.

The data for the inquiry were gathered through two techniques: interviewing and classroom observation. The participants included 7 in-service teachers and the school principal. The interview and observation data were coded and analyzed in cycles. In the first cycle, the author coded and analyzed the data inductively (i.e., codes and themes were derived from the semantic content of the data themselves; see Braun & Clarke, 2012) utilizing *in vivo* coding approach (i.e., using the participants' own language to honor their voice, see Saldaña, 2009). The second cycle involved the author coding and analyzing the data deductively (i.e., the researcher brings theoretical ideas and concepts to their data and uses them to code and interpret their data; see Braun & Clarke, 2012). Hence, in the second cycle, the researcher used Danielson's Framework for Teaching to code her data and identify the emergent themes. In the third cycle, the author constructed pattern themes across the data based on the (first and second cycle) analysis results.

Her findings revealed that all the participating teachers and the principal believed that student teachers exerted a positive influence on student achievement. However, it was not the student teachers alone that produced this influence; but, rather, this influence resulted from the fact that there were two teachers (i.e., student teacher and cooperating teacher) in the classroom working together to teach students. Since this was a single-case investigation, the researcher was, in a constructivist manner, more interested in understanding her case rather than generalizing her findings beyond the present case. As Creswell (2007) recommends, generalizability is "a term that holds little meaning for most qualitative [single-case] researchers" (p. 76). Thus, instead of generalizing her findings, the researcher suggested that further research on the influence of student teachers on student achievement be done in other contexts (such as in middle and high schools) in order to advance the field's understanding of this educational phenomenon.

THE MULTIPLE CASE STUDY APPROACH

The multiple case study approach allows for multiple cases to be analyzed within set boundaries such as time, place, or phenomena, in order to identify common themes and to "generalize to a target population of cases from the results of a purposefully selected sample of cases" (Greene & David, 1984, p. 74; see also Hyett, Kenny, & Dickson-Swift, 2014; Stake, 1995). According to Greene and David (1984), the main qualities of the multiple case approach include "a conceptual framework, a sampling plan, procedures for the conduct of individual cases, and cross-site analysis that tests the limiting conditions of the findings" (p. 75). An influential quality of multiple case study design, therefore, is its ability to use multiple cases within specific contexts to explore and test possible theoretical hypotheses among the cases, allowing for a level of generalizability to similar cases beyond the scope of the study (Hyett et al., 2014; Schwandt, 2001).

A distinct concern within the research community regarding multiple case study design is the inability of the researcher to provide a detailed justification for the specific method utilized in the given study (Thomas, 2011). Several experts have established a framework delineating step-by-step analyses of such case study research (Creswell & Poth, 2018; Merriam, 2009; Stake, 1995), and, in the multiple case studies presented in this article, some aspects of this framework are highlighted to distinguish qualities of the multiple case study approach.

Using a multiple case design, Bray (2011) selected 4 elementary math teachers and how their assumptions about how students learn math affect their teaching methods in responding to student errors. These propositions of teacher beliefs and actions related to instruction contributed to how teachers' knowledge affected their actions related to reform-based mathematics instruction, drawing back to the design of multiple case studies as a way to cross-analyze a specific phenomenon within the context of individual cases. Of importance in this study is the way the author used the individual cases to collectively establish hypothetical justifications for teaching behaviors, including an attempt to hypothesize how teachers' process changed over time as well and how professional development supports this change. Using a comparative analysis of the data, the researcher was able to generalize the research findings beyond the specific sample.

As another example, Daugherty (2009) researched the manner in which teachers may include engineering curriculum in the K-12 classroom curriculum. Specifically, the study focused on how certain types of professional development encouraged teachers to incorporate these methods into their curriculum and instruction. Using the multiple case approach, the researcher was able to compare the professional development programs. A particularly important quality in the design of this comparative study was the researcher's use of discriminate sampling in order to gain access to specific participants, locations, and data to allow for the highest level of quality comparison. The author established specific criteria for participants, and, furthermore, utilized a number of different data sources, including in person interviews with candidates, telephone interviews with leaders of the established professional development programs, an analysis of curriculum documents, and independent observations. In addition to the strategic selection of multiple cases, these multiple data sources for each case were able to strengthen the validity and reliability of the findings (see also Eisenhardt, 1989).

As a third example, Malewski, Sharma, and Phillion (2012) designed a multiple case study and collected multiple sources over 6 years. The researchers were interested in the way that teacher education programs developed preservice teachers' understanding of culturally relevant classroom environments. Data collection included pre-and post-interviews, observations, journals, and audio recordings of informal conversations between the interviewer and the participants. This type of rich data collection allowed researchers to create well developed and well supported themes when comparing the cases.

To consider another example, Orland-Barak and Hasin (2010) selected the multiple case study approach to study the perspectives of expert mentors in five different contexts. Each context distinguished points of view from selected types of mentorship, including supporting teachers as a group in the first years of teaching, mentoring from outside the context of the school, mentoring within the context of the school, mentoring in specific content areas, and individual mentoring. The multiple case study approach was, indeed, the most appropriate approach for this study given that it allowed the researchers to look at the phenomena of mentoring in various contexts to determine, through cross-case analysis, common features of quality mentoring among the cases.

In their study, criteria were developed to identify mentors who exhibited five years of experience with mentoring, a consistent routine of mentoring (twice a week), recommendation from leaders and administrators exemplifying their mentor practices, consideration towards diversity in their role as a mentor, and diversity in the different contexts of mentoring proposed for this study. After narrowing the recommended list of mentors from 15 to 8 with help from experts in general and content-related education, the researchers interviewed leaders and administrators associated professionally with the 8 mentors. Finally, the 8 mentors were approached and asked to consider participation in the study. 5 participants agreed and were selected as ideal examples of mentoring in the various contexts. This rigorous process is exemplary of the strategic sampling approach (Eisenhardt, 1989; Yin, 2018) within the context of a multiple case study. By strategically selecting the participants, the researchers can better justify the extent to which their participants were representative of the specific case under investigation (i.e., outstanding mentorship).

In regards to data collection, the five studies described above included data from semi-structured interviews and observations related to how the mentor described the process of mentoring. Orland-Barak and Hasin (2010) created summaries of this data, identifying major themes of quality mentoring. A within-case analysis was performed to form classifications related to these mentor qualities, and, finally, the cases were analyzed, across cases, to generate universal categories.

As a final example to the multiple case study approach, Dare et al. (2018) comprehensively included a grounded theoretic framework, a bounded context with the nature of the individual cases, and a within and cross-case analysis of the data. The focus of this research was to examine the experiences of middle school science teachers in enacting a multi-disciplinary STEM curriculum into their science classrooms. The study included 9 participants from demographically diverse school districts who participated in a science and math partnership between higher education and K-12 schools. The partnership's aim was focused on increasing the integration of STEM components into elementary and middle school classrooms through professional learning and curriculum support. The program provided teachers with professional learning opportunities, supported by instructional coaching, in developing STEM curriculum units to integrate the multiple disciplines: science, technology, engineering, and math.

The authors justified adopting the multiple case study approach as a viable way to study the experiences by their use of rich descriptive text of the teachers' classroom experiences with the curriculum and interpretive nature the authors employed through cross-case analysis developing common themes between and within cases (Stake, 1995; Yin 2012a). The study was bound by the level of which the participants included the multiple disciplines in their implementation of the curriculum. Stake (1995) alluded to both the interpretive and bounded nature of cases, which suggested that the multiple case study approach used in this study aligned with the participants, context, and nature of the study.

The use of the phenomenological approach to the methods employed in this study provided a clear and descriptive picture of the phenomenon of how individual teachers enact STEM curriculum in middle school science classrooms. This additional approach adds an interesting component to the design of the study in that it enriches not only the collective experiences of the participants within and across cases, but also the individual teacher's encounters with challenges and successes with the STEM curriculum. van Manen (2016) argued the use of phenomenology in research as a way of "creating formative relations between being and acting" (p.13). This study provided rich descriptions of the teacher's individual understanding of the phenomenon as well as their position in the much larger context of the group's experiences.

LIMITATIONS AND GENERALIZATION

Among the above studies (Bray, 2011; Daugherty, 2009; Malewski, Sharma, & Phillion, 2012; Orland-Barak & Hasin, 2010), the authors used various justifications for the context, participants, data collection, and comparative analysis within their cases. How and why researchers choose the context and participants of a study depends upon the intended application of the findings beyond the study.

The presented studies detailed limitations worth review and consideration. Participant selection was identified as a common limitation among the studies. In Daugherty's (2009) research, teacher competency in pre-service programs limited the scope of the study, as how teachers interact with students with a reform-based curriculum for math can depend upon the level of training and capacity the teacher demonstrates. The context of Bray's (2011) setting within a distinct overseas study abroad program to Honduras for pre-service teachers limited the types of observations and interviews performed. Perhaps other contexts such as similar study abroad programs to other countries would have supported a more representative population (Creswell & Poth, 2018). Similar to Daugherty's (2009) study, Malewski and colleagues (2012) acknowledged the limitations of their study in both context and participants, as they recognized how different contexts and participants outside the context of the study could have yielded different results.

Most studies that attempted generalization beyond the scope of the selected participants did so with caution. Stake (1978) claimed that the validity of generalizing beyond studies is a "more effective means of adding to understanding for all readers by approximating, through the words and illustrations of our reports, the natural experience acquired in ordinary personal involvement" (p. 5). Yin (2013) addressed the topic of validity and generalization with case studies cautiously, conceding that often the "analysis remains highly qualitative and may not be very convincing" (p. 322). Yet, he posited that the process of cross-analysis within and between case studies increases the systematic nature of the analysis and improves validity of the data collected. While the multiple case study design allows researchers the freedom to explore the context of purposefully selected participants in detail and to gain a robust knowledge of phenomena occurring, researchers must generalize with care. Every study must be examined for the proper justification of the approach, which, in turns, feeds into the researchers' justifications for generalization (Greene & David, 1984).

CONCLUSION

In this chapter, the authors have presented a number of different applications of the case study methodology. The case study methodology can utilize either a single case or a multiple case approach, each of which has its own unique affordances and limitations. Readers of this chapter should recognize that, before beginning to design a research project that utilizes the methodology of case study, it behooves researchers to consider a number of questions:

- To what extent is the selected case representative or unique?
- Does the author intend to generalize from this case or to explore the uniqueness of the given case?
- What is the unit or size of the case (an individual, a group, or an organization)?
- What selection methods are used when identifying multiples cases to include in the given case study?

- What are the multiple data sources being collected, and how do these multiple data sources contribute to the reliability and validity of the conclusions drawn within the case study?
- Is the researcher able to justify the methodological choices within the design, and how do these methodological choices justify the claims related to validity, reliability, and generalization of the findings from the given study?

The intention of this chapter was to provide readers with a beginning framework that can be used to ask and answer these methodological questions. The authors of this chapter encourage readers to consider how different approaches to the methodology of case study can help researchers to achieve their particular goals within the context of their particular research study. Above all, readers should recognize that there exists more than one approach to the design of case study research.

REFERENCES

Baxter, P., & Jack, S. (2008). Qualitative case study methodology: Study design and implementation for novice researchers. *Qualitative Report, 13*(4), 544–559.

Bloomberg, L. D., & Volpe, M. (2016). *Completing your qualitative dissertation: A road map from beginning to end* (3rd ed.). SAGE Publications.

Braun, V., & Clarke, V. (2012). Thematic analysis. In H. M. Cooper & P. M. Camic (Eds.), APA handbook of research methods in psychology, Vol 2: Research designs: Quantitative, qualitative, neuropsychological, and biological (pp. 57–71). American Psychological Association. doi:10.1037/13620-004

Bray, W. S. (2011). A collective case study of the influence of teachers' beliefs and knowledge on error-handling practices during class discussion of mathematics. *Journal for Research in Mathematics Education, 42*(1), 2–38. doi:10.5951/jresematheduc.42.1.0002

Clandinin, D. J., & Johnson, R. B. (2017). Narrative inquiry and case study research. In R. B. Christensen (Ed.), *Educational research: Qualitative, quantitative, and mixed approaches* (pp. 416–441). Sage.

Creswell, J. W. (2007). Qualitative inquiry and research design: Choosing among five approaches. *Sage (Atlanta, Ga.).*

Creswell, J. W., & Poth, C. N. (2018). *Qualitative inquiry and research design: Choosing among five approaches* (4th ed.). SAGE Publications.

Crotty, M. (1998). The foundations of social research: Meaning and perspective in the research process. *Sage (Atlanta, Ga.).*

Dare, E. A., Ellis, J. A., & Roehrig, G. H. (2018). Understanding science teachers' implementations of integrated STEM curricular units through a phenomenological multiple case study. *International Journal of STEM Education, 5*(1), 1–19. doi:10.118640594-018-0101-z PMID:30631694

Daugherty, J. (2009). Engineering professional development design for secondary school teachers: A multiple case study. *Journal of Technology Education, 21*(1), 10–24. doi:10.21061/jte.v21i1.a.1

Eisenhardt, K. M. (1989). Building theories from case study research. *Academy of Management Review, 14*(4), 532–550. doi:10.5465/amr.1989.4308385

Ellinger, A. D., & McWhorter, R. (2016). Qualitative case study research as empirical study. *International Journal of Adult Vocational Education and Technology, 7*(3), 1–13. doi:10.4018/IJAVET.2016070101

Firestone, W. A. (1987). Meaning in method: The rhetoric of quantitative and qualitative research. *American Educational Research Association, 16*(7), 16–21. doi:10.3102/0013189X016007016

Greene, D., & David, J. L. (1984). A research design for generalizing from multiple case studies. *Evaluation and Program Planning, 7*(1), 73–85. doi:10.1016/0149-7189(84)90027-2

Guba, E. G., & Lincoln, Y. S. (1981). *Effective evaluation*. Jossey-Bass.

Huff, A. S. (2009). Ontology and epistemology. In *Designing research for publication* (pp. 108–126). Sage.

Hyett, N., Kenny, A., & Dickson-Swift, V. (2014). Methodology or method? A critical review of qualitative case study reports. *International Journal of Qualitative Studies on Health and Well-being, 9*(1), 1–12. doi:10.3402/qhw.v9.23606 PMID:24809980

Kennedy, M. M. (1979). Generalizing from single case studies. *Evaluation Quarterly, 3*(4), 661–678. doi:10.1177/0193841X7900300409

Kostiainen, E., Ukskoski, T., Ruohotie-Lyhty, M., Kauppinen, M., Kainulainen, J., & Mäkinen, T. (2018). Meaningful learning in teacher education. *Teaching and Teacher Education, 71*, 66–77. doi:.tate.2017.12.009 doi:10.1016/j

Malewski, E., Sharma, S., & Phillion, J. (2012). How international field experiences promote cross-cultural awareness in preservice teachers through experiential learning: Findings from a six-year collective case study. *Teachers College Record, 114*(8), 1–44. PMID:24013958

Merriam, S. B. (1985). The case study in educational research: A review of selected literature. *The Journal of Educational Thought, 19*(3), 204–217.

Merriam, S. B. (1998). *Qualitative research and case study applications in education*. Jossey-Bass Publishers.

Merriam, S. B. (2009). *Qualitative research: A guide to design and implementation* (3rd ed.). Jossey-Bass.

Miles, M. B., & Huberman, A. M. (1994). Qualitative data analysis: An expanded sourcebook. *Sage (Atlanta, Ga.).*

Orland-Barak, L., & Hasin, R. (2010). Exemplary mentors' perspectives towards mentoring across mentoring contexts: Lessons from collective case studies. *Teaching and Teacher Education, 26*(3), 427–437. doi:.tate.2009.05.009 doi:10.1016/j

Patton, M. Q. (2002). Qualitative research and evaluation methods. *Sage (Atlanta, Ga.).*

Phan, N. T. T., & Locke, T. (2015). Sources of self-efficacy in Vietnamese EFL teachers: A qualitative study. *Teaching and Teacher Education, 52*, 73–82. doi:10.1016/j.tate.2015.09.006

Phillips, D. C., & Burbules, N. C. (2000). *Postpositivism and educational research.* Rowman & Littlefield.

Saldaña, J. (2009). The coding manual for qualitative researchers. *Sage (Atlanta, Ga.).*

Schwandt, T. A. (2001). *Dictionary of qualitative inquiry.* Sage Publications.

Stake, R. E. (1978). The case study method in social inquiry. *Educational Researcher, 7*(2), 5–8. doi:10.3102/0013189X007002005

Stake, R. E. (1995). The art of case study research. *Sage (Atlanta, Ga.).*

Stake, R. E. (2005). Qualitative case studies. In N. K. Denzin & Y. S. Lincoln (Eds.), *Handbook of qualitative research* (pp. 443–466). Sage.

Thomas, G. (2011). A typology for the case study in social science following a review of definition, discourse, and structure. *Qualitative Inquiry, 17*(6), 511–521. doi:10.1177/1077800411409884

Tygret, J. A. (2017). The influence of student teachers on student achievement: A case study of teacher perspectives. *Teaching and Teacher Education, 66*(1), 117–126. doi:10.1016/j.tate.2017.04.005

Webel, C., Sheffel, C., & Conner, K. A. (2018). Flipping instruction in a fifth grade class: A caseof an elementary mathematics specialist. *Teaching and Teacher Education, 71*, 271–282. doi:10.1016/j.tate.2018.01.007

Yazan, B. (2015). Three approaches to case study methods in education: Yin, Merriam, and Stake. *Qualitative Report, 20*(2), 134–152.

Yin, R. (2013). Validity and generalization in future case study evaluations. *Evaluation, 19*(3), 321–332. doi:10.1177/1356389013497081

Yin, R. K. (1981). The case study crisis: Some answers. *Administrative Science Quarterly, 26*(1), 58–65. doi:10.2307/2392599

Yin, R. K. (2009a). *Case study research: Design and methods* (4th ed.). SAGE Publications.

Yin, R. K. (2009b). How to do better case studies. In L. Bickman & D. J. Rog (Eds.), *The Sage handbook of applied social research methods* (pp. 254–282). SAGE Publications. doi:10.4135/9781483348858.n8

Yin, R. K. (2012a). *Applications of case study research* (3rd ed.). Sage.

Yin, R. K. (2012b). Case study methods. In H. M. Cooper & P. M. Camic (Eds.), APA handbook of research methods in psychology, Vol 2: Research designs: Quantitative, qualitative, neuropsychological, and biological (pp. 141–155). Washington, DC: American Psychological Association. doi:10.1037/13620-009

Yin, R. K. (2018). *Case study research and applications: Design and methods* (6th ed.). Sage.

Chapter 7
Action Research Approaches

Karen E. Watkins
University of Georgia, USA

Aliki Nicolaides
University of Georgia, USA

Victoria J. Marsick
Columbia University, USA

ABSTRACT

The contemporary use of action research draws on the exploratory, inductive nature of many qualitative research approaches—no matter the type of data collected—because the type of research problems studied are complex, dynamic, and located in rapidly changing contexts. When action research is undertaken to support social and organizational change, support from stakeholders affected by the research problem is essential, creating further complexity. Action research may serve as an alternative to more traditional views of social science. In this chapter, the authors describe action research as envisioned by Kurt Lewin, its originator. They show how two variants of action research—action science and collaborative developmental action inquiry—advance insight into how action research can be used to develop personal capability to address system changes that action research seeks to unveil. They conclude with reflections on criteria for rigor and relevance in action research in today's post-modern, complex world.

DOI: 10.4018/978-1-7998-7600-7.ch007

INTRODUCTION

This paper explores the initial framing of action research as a much needed alternative research approach in the social sciences within a positivist research tradition, and then offers emerging variations both in organizational practice and in a doctoral program to demonstrate how action research has evolved. With this evolution, new criteria for rigor are needed. This chapter offers a new perspective on validity in action research.

The Lewinian Tradition

Kurt Lewin (1946)— inquiring into complex social issues in his era—concluded that research methods needed were really a "type of action-research, a comparative research on the conditions and effects of various forms of social action, and research leading to social action." (p. 35). He argued that such research was not 'lower' than pure science, but quite the opposite- since it needed to uncover the underlying laws governing social action. Lewin noted that traditional social science research focused on fact-finding data and diagnosis. Yet, given the complexity of social problems such as intergroup relations, diagnosis was not enough. It needed to be "complemented by experimental comparative studies of the effectiveness of various techniques of change." (p. 37) Drawing on the example of the bombing of Germany, he noted that in social action, once a plan was made, reconnaissance had to serve several functions. The proposed action had to be evaluated against actions taken and whether results were above or below expectations to help planners assess what was or was not working. Researchers could then make mid-course corrections to both the initial and overall plan. Lewin's process of change began with collaborative fact-finding with those who sought to understand the social problem more clearly. Together they did the reconnaissance needed to further refine their plans, improve the plan, take actions, and monitor results, leading to an iterative cyclical process of action and evaluation that helped calibrate the best steps to take to resolve the social problem under study. For Sussman and Evered (1978) Lewin's action research "provides a mode of inquiry for evolving criteria by which to articulate and appraise actions taken in organizational contexts " (p. 599).

Lewin discovered the benefit of collaborative reflection and dialogue in the Commission on Community Interrelations (CCI). Lewin and the other interveners held evaluation meetings at the end of each day of training, in which trainers reflected on what went well, what they had observed, and what they could do differently. This reflection clearly led to changes in the pedagogy implemented the next day. As Lewin (1946) commented, "This and similar experiences have convinced me that we should consider action, research and training as a triangle that should be kept together for the sake of any of its corners. It is seldom possible to improve the action pattern without training personnel" (p. 42). On one occasion, Lewin invited a participant who had voiced feedback during the day to join them. When the participant disagreed with the trainers' interpretation of the day's event, Lewin realized the value of this joint reflection- and a new approach ensued (Glassman, Erdem, & Bartholomew, 2013).

Lewin stressed that social scientists need to be aware of the moral obligation of using action research tools for good, and recognized the results of action experiments such as his work to address minority problems had the potential to create a lasting effect on the history of the country. He concluded, "It is equally clear, however, that this job demands from the social scientists an utmost amount of courage. It needs courage as Plato defines it. It needs the best of what the best among us can give, and the help of everybody." (Ibid, p. 46) While acknowledging the level of ambiguity inherent in such important change work, he was also mindful that this work requires moral courage and a faith over fear that is uncommon in pure science. Yet Lewin (1944) said "It is essential that a democratic commonwealth and its educational system apply the rational procedures of scientific investigation also to its own process of group living." (p. 200)

Lewin saw action research as a new research methodology that grew out of his earlier theories of learning and of the social field. He argued that behavior change is a function of the interaction of the person in their environment (B = f(P,E) in the social context (Lewin, 1936). Traditional research of his time seemed inadequate in the face of the dynamic, ever changing context, with multiple intervening variables. Susman and Evered (1978) refer to Lewin's laboratory as a change experiment that heralded a new form of social research "which combined generation of theory with changing the social system through the researcher acting on or in the social system. The act itself is presented as the means of both changing the system and generating critical knowledge about it" (p. 586).

Lewin's (1947) analyses of the force fields surrounding social change led him to conclude that change was advanced by reducing barriers to its implementation. Coghlan and Jacobs (2005) found this to be consistent with Lewin's earlier writing on re-education. In effect, one barrier to change is our own internal conceptions of a situation, our attitudes or resistance toward changing it. Re-education can thus becomes transformative. Coghlan and Jacobs further stressed Lewin's assertions that social action is 'steered by perception,' and that false perceptions are not changed by accurate information (p. 448). Rather, "Lewin's notion of re-education is that change in cognition, values, and behavior is intimately linked to membership and identification with groups." (p. 453). The group helps encourage and reinforce new behavior- and for Lewin, re-education has not occurred until the new cognitions or attitudes are evident in individuals' and group's behavior (Ibid. p. 454)

The emphasis on education is significant. Glassman, et al. (2013) emphasize the adult education roots in the evolution of action research, through Lewin and Lippitt's collaborators, Bradford and Benne, influenced by Lindemann and Dewey. Argyris was deeply influenced by Lewin and always referred to his role as an educator (Bolman, 2017?). Bolman (2017?) likened Argyris' own contributions to action research as those that Argyris attributed to Lewin: "Lewin did three things: he was committed to understanding reality as his participants understood it, he used a combination of so-called 'normal' science with a narrative-integrative approach, and he tested his ideas by trying to change the things he was studying" (Woodell, 2003) (Bolman, 2016, p. 20).

Learning is at the heart of action research. Glassman (2013) make this clear in their comment, "AR attempts to do two things simultaneously: create a more vibrant community capable of nonhierarchical, dynamic, democratic decision-making processes and create a change-based intervention that leads to more efficient, satisfying, and especially lasting problem-solving capabilities". (Glassman et al., 2013, p. 283). Argyris (1968) characterized his form of action research as competence acquisition- and distinguished it from therapy. Watkins and Wilson, 1997 referred to Argyris himself as "the reluctant adult educator" (in Jarvis, 1997, p. 193).

Clearly, this type of research changes the roles of the researcher and the researched. "Action research constitutes a kind of science with a different epistemology that produces a different kind of knowledge, a knowledge which is contingent on the particular situation, and which develops the capacity of members of the organization to solve their own problems." (Sussman & Evered, 1978, p. 601). These authors also noted that enacting action research develops skills that "are interpersonal and problem-defining. Competence is developed in interpretation and judgment, in establishing problem-solving procedures, acting in contingent and uncertain situations, learning from one's errors, generating workable new constructs from one's experiences." (Ibid., p. 599).

Evolving Qualitative Approaches to Action Research

Checkland and Holwell (1998) asserted that "With the increased acceptance that 'social reality' is not a given, but is the changing product of a continual intersubjective discourse, there has been in the last decade an increased interest in qualitative research. AR and its many variants are increasingly being treated as serious and appropriate alternatives to the hypothesis testing which is at the core of research methodology in natural science. But proponents of AR need to recognize the limits to the claims they can make for the validity of their approach" (p. 20).

The legacy of Lewin seen in contemporary approaches to action research is that knowledge is socially constructed. Collaborative, participatory, democratic, group processes dominate emerging approaches. Stakeholders who are tightly tied to the study's social action concerns are involved in multiple ways – as members of an action research team, as sources of "reconnaissance" data, and/or as experimental subjects. Checkland and Holwell (1998) describe action research as developed by Lewin as a cycle that begins with a **framework** of ideas that are embodied in a **methodology** that is **applied** to an area of concern that yields **learning** about both the framework of ideas and the area of concern. This formula (**F + M + A = L**) essentially describes all research methods. Action research is conducted in the social context, however, and must adapt as the situation unfolds. Over time, as complexity of social situations has increased, social scientists have evolved variations of AR that combine research with solutions suited to the changing environment that continue to build human capacity for their implementation.

CONTEMPORARY VARIATIONS OF ACTION RESEARCH

Watkins and Brooks (1994) described the multiple variants of action research as "action technologies" to indicate that they were different approaches or tools to generate changed action. They noted these common characteristics: "formation of groups from among those who have the problem, reflection on problems in groups, collection of data around the problems, group analysis and group feedback; group-designed interventions to attempt to solve the problems" (p. 100). Their emphasis on the role of the group continues the tradition of a collaboratively designed and enacted process. Their framework alters Lewin's focus on social change to one of solving problems, though the involvement of key stakeholders and groups still roots these modifications in the social context. Contemporary variations often combine research with personal and social development. Action Science embraces the re-educative role of action research, and the transformation of mental models that prevent implementing change. Finally, collaborative developmental action inquiry emphasizes the person in their context and their developmental capacity to change and transform themselves and their systems.

Action Science

Action science is Argyris' (1982) unique approach to action research constructed around the ethical concerns of producing valid information, doing no harm, and creating free and informed choice. Described as a science of interpersonal action, action science is an approach to researching problematic social encounters for the purpose of transforming future actions in similar situations. Argyris proposed that there are two dominant patterns of social interaction—one that is characterized by attempts to control others, the situation, and the outcome (Model I); and one that is focused on learning, i.e., creating situations that allow all involved to learn their way through to a new way of seeing that will lead to new ways of acting (Model II) (Argyris, 1982).

In action science, individuals share a personal case using a 2-column format, writing what was said or done in the right hand column, and what was thought or felt but not said in the left hand column. A small group works together on each case to identify the theory-in-use driving the person's behavior in a collaborative learning climate characterized by the Model II principles of valid information, free and informed choice, and internal commitment to the change. Individuals seek to transform the belief system that prevents them from acting effectively in these problematic situations and to enact a Model II dialogue that creates enhanced capacity for individuals to learn and to act differently when these interpersonal problems recur.

Action science shares elements in common with other action research approaches:

- Action science is conducted in groups (collaborative)
- With cycles of taking action and reflecting (iterative)
- Using directly observable data to inform future action (evidence-driven)
- That is publicly shared and tested against an ideal ethical standard [Model II] (democratic transparency)
- Until new behavior is generated consistent with that standard and that transforms the interaction from one dominated by control to one characterized by a learning intent (re-education and change)

Enacting action science requires generating cases of problemtatic interpersonal incidents. Cases are written in two columns—the right hand column contains the recalled incident- in dialogue and script notations and the left hand column contains what was thought and felt but not expressed. Argyris (1995) called what was written in the right column directly observable data—and said this is the kind of data necessary to create actionable knowledge. "Actionable knowledge has been defined as information that actors could use, for example, to craft conversations that communicate the meanings they intend. Actionable knowledge has to specify how to produce meanings but leave actors free to select the specific words" (p. 21). Argyris noted "… re-education and change programmes should produce relatively directly observable data about these clients' reasoning and actions" (p. 23). This data is the person's own words- selected by the individual. In this way, the case work meets the ethical demands of an intervention that requires valid information, free and informed choice, and internal responsibility to implement change (Argyris, 1995). Argyris uses the term action science because "Action science is about understanding and producing action. Action, regardless of how large and complex or small and simple, requires the execution of a solution, production of the invention in the everyday life in which the problem is discovered, and evaluating the effectiveness of the production. A second foundation of action science is its methodology. Learning is said to occur when the diagnosis (insights, etc.) and the inventions are actually produced" (p. 25).

Argyris, Putnam, and Smith (1985) contrasted action research with traditional qualitative methods by pointing to differences in the degree of detachment or objectivity vs engagement and intersubjectivity. This separation of the researcher from the researched disappears in action research; and researchers are expected to reflect on their own biases and deeply held views of the world, particularly their need to control the purposes of the research in such a way that could cause harm to others. Action research is catalytic; it knowingly intends to change or transform. This brings with it significant ethical demands Argyris' (1970). "Argyris hypothesized that there are organizational norms that tend to suppress emotionality and conflict. As a result, people in higher ranks are unable to communicate fully with each other, and a cycle of decreasing interpersonal competence and organizational effectiveness develops" (Alderfer, 1967, p. 443). In action science, the goal is to uncover and transform these dysfunctional norms in order to create organizations and people who learn more effectively. This makes action research considerably less detached and objective with a high need for ethical practice based on new norms of valid information, free and informed choice and internal commitment to the changes.

Like action research as Lewin conceived it, action science requires the courage of Plato in pursuit of change goals. What must change is individuals' and systems' fundamental belief systems and ways of interacting particularly when most stressed—in difficult, conflict situations.

Collaborative Developmental Action Inquiry

Developed by Torbert (2004), Collaborative Developmental Action Inquiry (CDAI) focuses on developing individuals' and organizations' ability to engage in simultaneous action and inquiry for greater awareness and effectiveness in the moment. Torbert (2013) describes CDAI as a "meta-paradigmatic approach of scientific inquiry that integrates first-person, adult spiritual inquiry and consciousness development in the emerging present with second-person, transformational mutuality-seeking social scientific inquiry over a lifetime, and third-person, intergenerational, objectivity-seeking social scientific inquiry and its effects" (p. 265). CDAI seeks to build capacity to triangulate among the subjective, inter-subjective, and objective aspects of real-time action and inquiry. In this way, it extends traditional social science paradigms that privilege and focus on objective or third-person, "after-the-fact" knowledge.

CDAI's formation was influenced by action science and constructive-developmental theory. Specifically, it incorporates and builds on action science's second-person communication practices and parts of speech, and on single-, double-, and triple-loop learning (Foster, 2012). Torbert (2004) extends action science's concepts of advocating and inquiring to include four parts of speech: framing, advocating, illustrating, and inquiring. Framing describes the purpose for the communication and setting the context; advocating states an assertion or position; illustrating provides concrete details that support the advocacy; and inquiring involves questioning and inviting input from others. According to Torbert, using these four parts of speech improves mutuality and receiving better, accurate information from others.

CDAI extends action science's discussion of single- and double-loop learning to include triple-loop learning. Single-loop learning is defined as an attempt to solve problems with minimal variation in methods without questioning underlying assumptions about how work is supposed to be done (Argyris & Schon, 1974). In the context of organization, it is a mere behavioral change that aims for problem fixing. Its interest is in finding out what the problem is and ways in which things could be done better rather than asking why the problem occurs. In contrast, double-loop learning is defined as a process of inquiring into the assumptions or mental models that govern our actions (Argyris & Schon, 1974). It is a total reframing of our cognitive schema, which could lead to fundamental changes in our behaviors. In this sense, behavioral changes derived from double-loop learning are more powerful and transformative than those from single-loop learning because the former entails a deep-level change in our cognitive framework. In triple-loop learning, a person probes their deeper assumptions and ways of knowing, often linked to their context, that further inform strategy and actions. Triple-loop learning is a form of learning that goes beyond single- and double-loop learning. It is a level of learning that has the potential to transform the very deepest in ourselves. It serves as the basis for processes of the most fundamental and profound change, or in other words radical innovation in ourselves (Peschl, 2007).

In action science, Argyris characterizes these assumptions as universal models – oriented toward either control or learning. Torbert, deeply influenced by the work of Gurdjieff which he studied for 25 years (Taylor, 2017), drives more toward Gurdjieff for whom the "central inquiry of this spiritual work is how one can develop an impartial, post-cognitive consciousness amidst the stresses of every- day action" (p. 1327). In addition, Torbert draws on adult developmet theory to explore how one's deeper assumptions and ways of knowing are constrained by one's developmental capacity. While all three levels of learning are necessary, triple-loop learning is considered a prerequisite for true and lasting transformation because it requires a fundamental shift in how one makes meaning, a "post-cognitive attention" (Taylor, 2017, p. 1329).

CDAI is both action-oriented and developmentally focused—the latter by probing for increased levels of awareness and more complex meaning-making capacity. Developmental Action Inquiry considers the three different levels: the first-person (subjective), the second-person (intersubjective), and the third-person (objective) (Torbert, 1976; Torbert, 1999). More specifically, in the midst of action, the questions of how we can be aware of our flow of attention in real-time; how we can find ourselves in mini-communities doing with inquiring in each moment; and how we can timely inquire our psycho/social relations within the third-person dynamics (organizational or social) as well as ourselves and interactions with others and act accordingly become the key interest of Developmental Action Inquiry (Torbert, 1999). Also, its developmental feature that emphasizes the importance of one's developmental capacity in learning and the belief in the potential for cycles of growth and transformation is what explicitly distinguishes Developmental Action Inquiry from Argyris' version of Action Science (Torbert, 1999). Without adequate developmental capacity, one cannot effectively initiate, learn through, complete, and internalize the various loops of feedback given to them (Kwon & Nicolaides, 2017; Nicolaides, 2015). This is especially important in many interpersonal contexts such that the lack of developmental capacity of any individual involved in the same situation would make it difficult for mature reflection and true transformation to happen. The collaborative nature of CDAI enables participants to better engage in critical reflection by exposing and encouraging them to consider others' perspectives, facilitating their transformative learning experiences (Kwon & Nicolaides, 2017; Nicolaides, 2015). Unlike traditional research, collaboration also overcomes the power of the researcher over the participants and creates mutually transforming power among them, which leads to an open and authentic environment where both parties can potentially experience transformative learning and grow (Torbert & Associates, 2004).

Specifically, in CDAI, individuals draw on cases to describe four territories of experience: one's intentions or purposes, one's plans or strategies, one's behaviors or actions, and the third-person outcomes that occur as a result of these (Torbert & Associates, 2004). CDAI seeks effective, simultaneous action and inquiry through two mechanisms:

- Greater mutuality between people and groups that allows for more open and honest information sharing,
- More complex meaning-making capacity and greater personal awareness on multiple levels (Fisher, Rooke, & Torbert; 2003; Foster, 2012; Torbert & Associates, 2004).

CDAI offers an approach and a set of tools that promote first-, second-, and third-person inquiry and practice. CDAI is a form of action research that reflects deeply on prior action using a developmental lens and a process of inquiry that reveals deeper insights that have the potential to transform how an individual thinks and acts in the world.

ACTION RESEARCH IN DOCTORAL STUDIES

The use of action research to solve problems and simultaneously develop capacity through exploratory inquiry can be seen in a doctoral program in Learning, Leadership and Organization Development at the University of Georgia that uses action research as a primary culminating activity. The approach is highly consistent with the program's mission, " to lead change through action research". It challenges students to use theory to implement change at the individual, group, and system levels—while generating new knowledge (Coghlan & Brannick, 2014).

Action research begins with a problem—generally one that requires investigation that, if solved, will improve the social conditions of those affected. The problem as defined by stakeholders is examined—as Lewin said—with discussions, negotiations, immersion in the context, and marshaling evidence. Through this process, the problem is redefined and reframed to fit what has been learned both in the literature and through local reconnaissance. Theory is a lens for going deeper into the nature of the problem and becomes a guide to intervention.

Cycles of action [intervention] and reflection occur next. Each cycle can potentially reframe or even invalidate the problem's original conception. The quality of the data collected to monitor the results of the action experiments taken enables this rethinking, as does the quality of the group process of the action research team as they navigate this process.

Data are collected throughout about the nature of the problem, the effectiveness of interventions, and the dynamics of the action research team process. One assumption in the UGA approach is that the team could be a potential source of error in the process, and a thorough understanding and documentation of what happens in team meetings is one way to identify and interrupt these errors.

Data are collected early on to verify that the problem selected is actually a problem—both locally and beyond the borders of the selected context. Data may be collected through documents within the organization, sometimes from state or national databases, or other research studies on the topic. Students may also design a survey or needs assessment, re-analyze an existing organizational survey, or use a validated instrument to identify the current state relative to the problem of interest. Finally, the action researcher keeps a research journal, noting instances of evolution of the understanding of the problem through the lens of the theory guiding the study, process issues within the team, and possible outcome data as it unfolds.

As with all qualitative research, the rigor of this approach is tied to the credibility of the documents, potential triangulation among the multiple sources of evidence, and particularly any validation the student can find that the problem in the local setting is significant beyond the borders of their context and not readily solved by means discovered elsewhere. Ultimately, validity is primarily found in the training and skill of the researcher interventionist themselves. Five courses on action research and applied research methods, cognates in adult education, organizational development, and facilitating change, and additional doctoral courses support the students' growing skills in the design, conduct and dissemination of the action research project.

Another source of data is transcripts of action research team meetings and of summary interviews with action research team members. Interviews can shed light on what members learned through the action research process and about the research purpose. In some cases, these data are also the only source of evidence on how the interventions worked.

In most instances, the team and key stakeholders collect and review additional data to monitor the chosen interventions. Sources could include additional interviews with stakeholders and clients; questionnaire and survey data; descriptive statistics of changes in participation rates, retention data, or other relevant outcome data; or products developed by the team that are then reviewed by the team and key stakeholders.

With so much data collected, an important step in action research is data reduction. First, data are identified and eliminated that do not fit the research purpose. Then data that best fit each research question are identified and a final dataset created for analysis that might be both inductive and deductive. Inductively, the action researcher reads through data and notes what is significant. He/she uses open coding (Strauss & Corbin, 1990) to look for patterns and themes. Then, a more deductive approach may winnow interesting but irrelevant information (Miles, Huberman, & Saldana, 2013). The process is often iterative to eliminate conversations ancillary to the team's focus.

Data analysis varies with the type of data the researcher is analyzing, i.e. documents, transcripts, or questionnaires. Students at UGA often use critical incident technique using guidelines by Flanagan (1954) and by Ellinger and Watkins (1998).

Writing the research report is a daunting undertaking, given the sheer volume and complexity of the data generated in an action research project that can span several months as well as the challenge of depicting a systemic look at what has occurred. Students often divide their findings into two chapters—a chronologically rendered case study of what they did in the project and a discussion based on the research purpose. The case study gives transparency to the process, and is often a strong practitioner report of what worked and what did not work to address the original problem. The second findings chapter explores what was learned about the research questions, particularly about how well the theory guiding the work illuminated the problem and provided guidance for addressing it. Whether or not action researchers separate these two sets of findings, inclusion of both is essential in the report (Coghlan & Brannick, 2014). Table one gives a checklist that outlines what faculty expect in each stage of the action research dissertation process.

Table 1. Action research dissertation checklist

Key Elements	Optimal Implenetation	Acceptable Adaptations
Client Organization	Work in system in which you have some authority	Create own satellite "organization/ team"; work with multiple units
Researchable Problem	Collaboratively evolve problem or issue as part of working together; verify in the literature, and in collaboratively collected data; collaboratively create action experiments to address the issue	Organization defines key problem/ challenge/opportunity with you; you collect data to better define the problem, use the literature to illuminate possible interventions;
Action Research Team	Work with carefully chosen stakeholders with interest in and knowledge of the issue & capacity to act on it	Work with volunteers interested in the problem
Initial Data Collection on the Problem	Using both existing and new organizational data, verify and expand problem understanding	Work with both internal and external data on the problem as evidenced nationally; in the research literature
Interventions, Action Experiments	Team identifies possible interventions from theory, literature, and data to try; implements them, & assesses results	Team uses an existing or ongoing intervention as a source of data to learn more about the problem, to create more effective adaptations of initial interventions
Data Collection on the Results of Interventions	Team identifies indicators of success, creates pre and post measures of effectiveness	Team uses existing assessment information to assess change
Data Collection on the Action Research Team Process	Team meetings transcribed and evaluated for critical incidents of learning about the problem by team members & Team members interviewed for critical learning about the research purpose and the initial problem through the action research process.	Additional data might include Researcher field notes supplemented with questionnaire data to assess interventions or??
Final Case Study Report	Both description of the action research process, results, & high level conceptual analysis of what occurred within individuals, the team, and the system relative to the purpose & research questions	Thematic analysis of data to supplement the chronological story presented as an additional findings chapter

There are many examples of these projects- a participatory action research project with farmers in the Caribbean, a study of emotion work and compassion fatigue among hospice care workers, a study of transformation of a community theatre board to promote racial integration, a study of talent management in a higher education coordinating board, etc. A hallmark of many of these studies is a change in the system that simultaneously signals to the organization enhanced leadership skills on the part of the student and potentially of the action research team. Many students were promoted as a direct result of this process and some of the changes implemented persist today (Graham-Cagney & Watkins, 2014). Still other dissertations chronicled the failure of the system to change and explored both barriers within the system and within the intervenor[s]. One remarkable study looked at how an action research study designed to explore and enhance workplace civility was completely derailed by a culture of incivility. As Lewin noted, this work also requires courage—an interpersonal courage— that makes action research unique.

Faculty in this program also found that the experience of teaching in this program was challenging. *"But it was like, I don't know, eating stones, you know? Grinding stones, like it's different."* This feeling of doing something countercultural in the academy was offset by the excitement of learning together. ""*We had to read a lot, work really hard to come up with this innovative EdD. So we started thinking about what is signature pedagogy . . . . . . what's going to differentiate this degree from any other degree? What is EdD, what is PhD, and how should we look at them?"* They found power in the collegiality of the experience *"So we had a truly collegial, a sense of we can be different, we can think differently, always keeping this interest of the student at heart, that's the bottom line. You have a different idea, I have a different idea, how do we make this work so it enhances the student's experience."* But the ultimate reward was seeing what happened to the students: "*I think that the other thing they feel empowered about, which makes it transformative, is that they've enacted a change. Something has changed; something in that organisation. And it's a lasting kind of imprint that they leave on that organisation."* (From Graham-Cagney & Watkins, 2014, p. 2).

MOVING PAST MODERNIST IMPOSITIONS OF RIGOR IN ACTION RESEARCH

Rigor and validity need to be differently understood in this context. The inductive, constructivist qualitative paradigm—or even abductive reasoning that seeks the best possible explanation from limited information—are used in action research to open thinking to views of multiple stakeholders and changing circumstances. A core idea is to interrupt habitual practice by exploring and inspiring innovative alternatives

(Zandee, 2013) in action research. How do action researchers escape the impositions of an objective truth to embrace the messy reality of subjective and inter-subjective inquiry in action? Whereas empirical positivism aims at *universalisable, valid certainty in reflection about particular pre-designated questions*, action research aims at *timely, voluntary, mutual, validity-testing, transformative action at all moments of living*. This is a move away from a primarily reflective science about "bedrock truth" and toward critical inquiry-in-action by individuals, groups, organizations and the wider community toward creating a better world

In their 1978 chapter, Susman and Evered pointed to a crisis in organizational science in that "the findings in our scholarly management journals are only remotely related to the real world of practicing managers" (p. 582). The positivist approaches to science which have dominated our perspective on research "are deficient in their capacity to generate knowledge for use by members of organizations" (p. 585). Action research is future oriented, collaborative, implies system development, generates theory grounded in action, and is agnostic and situational, and as such, is clearly *not* supported by a positivist view of science. Susman and Evered argued that the conditions in which all of us try to learn in everyday life are better explored through a range of alternative philosophical viewpoints; Aristotelian praxis, hermeneutics, existentialism, pragmatism, process philosophies and phenomenology all point toward methods for improving validity under action conditions. Susman and Evered concluded:

We hope that this article will enable others to assess the scientific merits of action research. We believe that action research is both ascientific in terms of the criteria of positivist science and relevant in terms of generating good organizational science. As a procedure for generating knowledge, we believe it has a far greater potential than positivist science for understanding and managing the affairs of organizations. (p. 601)

But still another transformation—this time toward the *action turn*—is necessary to reach a full understanding of the action research that Susman and Evered called for so long ago. In making the action turn we re-vision our view of the *nature* and *purpose* of social science. Drawing on a participatory paradigm for research (Heron & Reason, 1997; Reason & Bradbury, 2000), we argue that the primary purpose of inquiry is not simply—or even primarily—to contribute to the fund of knowledge in a field, not to deconstruct taken-for-granted realities, nor even to develop emancipatory theory—although the latter can be valuable. Rather, it is to forge a direct link between intellectual knowledge and moment-to-moment personal and social action, so that inquiry contributes *directly* to the flourishing of human persons, their communities, and the ecosystems of which they are part (including the question of when foregrounding or backgrounding inquiry contributes best).

Since its origins in Kurt Lewin's social psychological experiments in the 1940s, action research emerged as a critique and alternative to more traditional views of social science (Argyris, 1970; Reason & Bradbury, 2001). Most social science studies seek to make causal links between predictor and dependent variables based on data or events that occurred in the past. In contrast, action research aims not only to understand past events, but also present phenomena, particularly the ongoing dynamics of human interactions, as well as future intentions and the forward design of joint organizing. Inquiry conducted in the present and for the future by co-participants remain underexplored in empirical scholarship to date. How do we generate valid information about a present situation when we are an "interested party" and action is urgent? How do we generate data about our practice and monitor its congruity or incongruity with strategies? How do we determine what we truly wish for the future? By employing action research strategies associated with present or future-oriented practical and transformational learning in action, we can re-vision the research method options available to research/practitioners and their rigor.

Second, action research differs from most social science studies that are aimed at aggregating data about many individuals, organizations, or events and attributing generalizable causal links among the variables studied, irrespective of the particulars. Thus, a kind of anonymous, third-person knowledge is sought, and it is then communicated to other anonymous third-persons, usually in the impersonal, third-person voice of a journal article. In contrast, action research seeks information that is shared directly with a client system, collaboratively interpreted to inform action, actions are jointly considered and implementation studied to enhance improvement efforts.

Qualitative studies achieve rigor through rich, in-depth accounts of one or more cases, and quantitative studies by generating statistically significant results. Although explanatory and predictive studies are clearly valuable to understanding the social world, other areas of reality have been relatively neglected. Action research begins to fill this void by emphasizing methods to obtain first-person, subjective data about oneself while in action to act differently in the moment. Action research generates second-person data from team members in order that they can learn to appreciate their multiple perspectives and through doing this, change how they work together. The methods that constitute the core of action research are not impersonal disembodied methods, but rather personal embodied disciplines of simultaneous research and practice among others "on line" (Heron, 1996; Reason, 1994; Sherman & Torbert, 2000; Torbert, 1991).

Empirical positivism concerns two territories of experience – the outside world territory where we find data; and the cognitive territory where we generate theory and systematic methods for testing the validity of the "fit" between data and theory. Research methods are, in effect, guidelines for how to sequence one's attention between the outside world territory of experience (to collect data) and the cognitive territory (to theorize, design, analyze, and write-up the study). One collects, analyzes and reports on data long after the actions are taken. This kind of science does not train us to study social dynamics in the present when we are co-participants in the ongoing action, let alone how to allow the data generated through this kind of study to influence our actions in the present. By contrast, educators Duenkel and Pratt, (2013) describe ecological education and action research as a transformative blend for formal and nonformal educators. Duenkel and Pratt analyze data from four cohorts of students engaged in capstone action research projects and describe their transformative learning experiences that highlight constructive encounters with layers of hegemony. Awakening to relations of power, however, facilitated a profound shift within these teacher-researchers, engendering a newly felt sense of empowerment. Integral to the transformative learning process were opportunities for reflection which fostered greater balance between the personal and professional, learning to accept feelings of confusion and disorientation, and evolving a deeper understanding of ecological education. This is a profoundly different kind of rigor and relevance. We think this is particularly apt in an era of such social complexity. Perhaps the best action related to the question of rigor of action research is to pause before we attempt to predefine what is recognized as meaningful inquiry, and simply encourage generative and transformative learning for action that benefits the whole system and the people the system serves.

CONCLUSION

Learning and doing action research in the neo-liberal world of contemporary higher education is often not supported by the larger institutional context. As Greenwood (2012) suggests, understanding how action research can be practiced, taught, and learned in contemporary universities requires understanding the dominant structures that organize higher education in the 21st century. He draws attention to the neo-liberal and Taylorist structures affecting higher education which make it difficult to step outside the margins. We caution that action research be viewed in the larger context in order to demonstrate how it can contribute to an improved future for higher education and for the organizational contexts in which it is used. Tosey and Marshall (2017) explore the demise of some of these inquiry based graduate programs in the UK and note that "On inquiry-based programmes, activities such as team teaching, working with small learning groups, and spending time with participants helping build learning communities are seldom credited adequately in workload models focused on formal, standardised contact hours" (p. 6). Moreover, "It seems to us that HRD practitioners with the self-awareness, understanding of individual, group and organizational dynamics, and inquiry skills described above are needed more than ever in an increasingly complex, unpredictable, and fractious world characterized by 'wicked' problems. . . . . Inquiry-based pedagogies seem acutely needed for practice, yet paradoxically are disappearing from HE" (p. 7).

For Argyris (1980), what social scientists do to ensure internal and external validity almost guarantees their findings will not be implementable – they may be replicable by other scientists conducting research, but the real world problems that led to the research in the first place will not be addressed by the knowledge produced under the tightly controlled conditions of normal science. What is needed instead, he argues, is research that produces "actionable knowledge (Argyris, 1996). Actionable knowledge makes clear the causal process that leads to action: If I do X, then Y will happen. Action research is a form of intervention research – focused on improving how we intervene in human systems in order to change them. Watkins (1981) stressed that in action research, the interventionist must produce or elicit the phenomena they hope to study. Clearly another barrier to using action research is the degree of interpersonal skills required of the researcher.

Similarly, Hackman (1985) was concerned that traditional laboratory research methods ignore the effects of context and the environment and assume a static state in the organization being studied. He commented, "Organizations do not hold still while we negotiate entry, make our intervention, and wait for an appropriate time to collect follow-up data." (p. 10). So, like others who sought research that is useful to practice (Lawler, 1985; Argyris, 1996), Hackman suggested that research that advances practice may also contribute to theory: "It may be that the best way to generate advances in basic theory is to do research that seeks solutions to real problems, and to keep one's eyes open for fundamental conceptual issues as one proceeds" (Ibid, p. 26).

Van de Ven and Johnson (2006) emphasize engaged scholarship, (including action research) that

adopts a pluralistic methodology that advances knowledge by leveraging the relative contributions and conceptual frameworks that researchers and practitioners bring to bear on a given problem or question. . . . The models that better fit the problems they were intended to solve are selected by users, and the gradual winnowing down of plausible rival models or hypotheses by the scholarly community produces an evolutionary conception of the growth of scientific knowledge. (p. 817)

We argue tthat, freed of some of the constraints of traditional social science research, action research may help build theory precisely because it has the time to develop iterative tests of theory in different contexts, and the capacity through multiple data collection approaches to better capture the complexity of organizational contexts.

Small wonder that the academy is reluctant to embrace such a challenging form of research. Where the academy has ceded ground to action research is in professional doctoral programs such as the UGA doctoral program described above. These scholar-practitioner programs focus on improving practice through research. It is unlikely that these small initiatives will challenge traditional social science practices, but they do represent a powerful means for changing human systems and a rich alternative to traditional research approaches.

REFERENCES

Alderfer, C. P. (1967). An organizational syndrome. *Administrative Science Quarterly*, *12*(3), 440–460. doi:10.2307/2391314

Argyris, C. (1968). Conditions for competence acquisition and therapy. *The Journal of Applied Behavioral Science*, *4*(2), 147–177. doi:10.1177/002188636800400201

Argyris, C. (1970). *Intervention theory and method*. Addison-Wesley.

Argyris, C. (1980). *Inner contradictions of rigorous research*. Academic Press.

Argyris, C. (1982). *Reasoning, learning, and action: Individual and organizational*. Jossey-Bass.

Argyris, C. (1995). Action science and organizational learning. *Journal of Managerial Psychology*, *10*(6), 20–26. doi:10.1108/02683949510093849

Argyris, C. (1996). Actionable knowledge: Design causality in the service of consequential theory. *The Journal of Applied Behavioral Science*, *32*(4), 390–406. doi:10.1177/0021886396324004

Argyris, C., Putnam, R., & Smith, D. (1985). *Action science: Concepts, methods, and skills for research and intervention*. Jossey-Bass.

Avison, D. E., Lau, F., Myers, M. D., & Nielsen, P. A. (1999). Action research. *Communications of the ACM*, *42*(1), 94–97. doi:10.1145/291469.291479

Bolman, L. (2013). Chris Argyris: The iconoclast. In *The Palgrave handbook of organizational change thinkers* (pp. 17–36). doi:10.1007/978-3-319-52878-6_29

Chandler, D., & Torbert, B. (2003). Transforming inquiry and action: Interweaving 27 flavors of action research. *Journal of Action Research, 1*(2), 133–152. doi:10.1177/14767503030012002

Checkland, P., & Holwell, S. (1998). Action research: Its nature and validity. *Systemic Practice and Action Research, 11*(1), 9–21. doi:10.1023/A:1022908820784

Coghlan, D., & Brannick, T. (2014). Doing action research in your own organization. *Sage (Atlanta, Ga.).*

Dickens, L., & Watkins, K. (1999). Action research: Rethinking Lewin. *Management Learning, 30*(2), 127–140. doi:10.1177/1350507699302002

Ellinger, A. D., & Watkins, K. E. (1998). Updating the critical incident technique after forty-four years. In *Academy of Human Resource Development 1998 Conference Proceedings* (pp. 285-292). Academy Human Resource Development.

Fisher, D., Rooke, D., & Torbert, B. (2003). *Personal and organizational transformations through action inquiry*. Edge/Work Press.

Flanagan, J. C. (1954). The critical incident technique. *Psychological Bulletin, 51*(4), 327–358. doi:10.1037/h0061470 PMID:13177800

Foster, P. (2012). Collaborative developmental action inquiry. In D. Coughlan & M. Brydon-Miller (Eds.), *Sage encyclopedia of action research management* (pp. 64–73). Sage.

Glassman, M., Erdem, G., & Bartholomew, M. (2013). Action research and its history as an adult education movement for social change. *Adult Education Quarterly, 63*(3), 272–288. doi:10.1177/0741713612471418

Graham-Cagney, A., & Watkins, K. (2014) *Insider Action Research Doctorates: Creating Transformative Learning Spaces in Professional Development in Higher Education.* An unpublished paper prepared for International Conference on Doctoral Education, Orlando, FL.

Hackman, J. R. (1985). Doing research that makes a difference. *Doing research that is useful for theory and practice, 126*, 148.

Heron, J. (1996). Co-operative Inquiry: Research into the human condition. *Sage (Atlanta, Ga.).*

Kwon, C. K., & Nicolaides, A. (2017). Managing diversity through triple-loop learning: A call for paradigm shift. *Human Resource Development Review*, 1–15.

Lawler, E. E. (1985). *Doing research that is useful for theory and practice*. Lexington Books.

Lewin, K. (1936). *Principles of topological psychology*. McGraw-Hill., 10. doi:10.1037/10019-000

Lewin, K. (1944). The dynamics of group action. *Educational Leadership*, *1*(4), 195–200.

Lewin, K. (1946). Action research and minority problems. *The Journal of Social Issues*, *2*(4), 34–46. doi:10.1111/j.1540-4560.1946.tb02295.x

Lewin, K. (1947). Frontiers in group dynamics II. Channels of group life; social planning and action research. *Human Relations*, *1*(2), 143–153. doi:10.1177/001872674700100201

Masters, J. (1995). The history of action research. *Action Research Electronic Reader*, *22*, 2005.

Miles, M. B., Huberman, A. M., & Saldana, J. (2013). Qualitative data analysis. *Sage (Atlanta, Ga.)*.

Nicolaides, A. (2015). Generative learning: Adults learning within ambiguity. *Adult Education Quarterly*, 1–17.

Reason, P., & Bradbury, H. (Eds.). (2001). *Handbook of action research*. Sage.

Reason, P., & Torbert, W. (2001). The action turn: Towards a transformational social science. *Concepts and Transformation*, *6*(1), 1–37.

Rudolph, J. W., Morrison, J. B., & Carrol, J. S. (2009). The dynamics of action-oriented problem solving: Linking interpretation and choice. *Academy of Management Review*, *34*(4), 733–356.

Sherman, F., & Torbert, W. (Eds.). (2000). *Transforming social inquiry, transforming social action*. Kluwer Academic Publishers.

Strauss, A., & Corbin, J. (1990). Open coding. Basics of qualitative research. *Grounded Theory Procedures and Techniques*, *2*, 101–121.

Susman, G. I., & Evered, R. D. (1978). An assessment of the scientific merits of action research. *Administrative Science Quarterly*, 582–603.

Torbert, W., Livne-Tarandach, R., McCallum, D., Nicolaides, A., & Herman-Baker, E. (2010). Developmental action inquiry: A distinct integral theory that actually integrates developmental theory practice and research. In S. Esbjörn-Hargens (Ed.), *Integral research in action* (pp. 413–430). SUNY Academic Press.

Torbert, W. R., & ... (2004). *Action inquiry: The secret of timely and transforming leadership*. Berrett-Koehler Publishers.

Tosey, P., & Marshall, J. (2017). The demise of inquiry-based HRD programmes in the UK: Implications for the field. *Human Resource Development International*, 1–10.

Van de Ven, A. H., & Johnson, P. E. (2006). Knowledge for theory and practice. *Academy of Management Review*, *31*(4), 802–821.

Watkins, K., & Wilson, J. (2001). Chris Argyris: The reluctant adult educator. In P. Jarvis (Ed.), *Twentieth century thinkers in adult and continuing education* (pp. 139–205). Routledge.

Watkins, K. E. (1991). *Validity in action research. Annual Meeting of the American Educational Research Association.*

Watkins, K. E., & Brooks, A. (1994). A framework for using action technologies. *New Directions for Adult and Continuing Education*, (63): 99–111.

Chapter 8
Valuing Indigenous Research Paradigms in the Context of Language Acquisition

Jennifer Morrison
https://orcid.org/0000-0003-4910-6925
Texas Tech University, USA

ABSTRACT

The purpose of this chapter is to justify the value of Indigenous research paradigms, specifically in the context of research on language acquisition. This argument has implications not only for research on language acquisition and the practice of language instruction but also for qualitative research, more broadly. Specifically, depending on the context of a given research project, it may be critical for educational researchers to value and respect Indigenous epistemologies and worldviews; otherwise, educational research endeavors may be adding to knowledge at the expense of devaluing research participants and local communities.

AN INDIGENOUS PARADIGM

To begin a discussion of an Indigenous methodological paradigm, one must first appreciate the over-arching premise of *relationality*. As Weber-Pillwax (2003) explained, "The foundation of Indigenous research lies in the reality of the lived Indigenous experience. Indigenous researchers ground their research knowingly in the lives of real persons as individuals and social beings, not in the world of ideas" (as cited in Wilson, 2008, p. 60). In this paradigm, "reality *is* [emphasis added]

DOI: 10.4018/978-1-7998-7600-7.ch008

relationships or sets of relationships" (Wilson, p. 73). As Chilisa (2012) points out, this relationality applies not only to interpersonal relations but also to our interconnectedness with the land, spirituality, and the cosmos. Within an Indigenous paradigm, everything is connected; we are related to our research, the participants in the research, the ideas inherent in the research, the process of conducting the research, and the process of sharing the research. To illustrate, Wilson (2008) quoted his friend, Peter, who described Indigenous knowledge as a fish net: One could investigate the knots in the net to try to understand why it holds fish, but it is the strings between the knots that make it strong. The strings here represent the relationships with world around us. Relationality is the lens through which to view an Indigenous paradigm.

Like the net, the elements of a research methodology rooted in an Indigenous paradigm will not follow a linear path but are circular and interconnected (and sometimes overlapping). The visual representation of Wilson's circle is reproduced below in Figure 1. Wilson (2008) explains that the epistemology that undergirds Indigenous research methodologies is essentially equivalent to Indigenous ontology. In other words, the understanding of reality and the understanding of knowledge (and the creation of knowledge) are inseparable:

In an Indigenous ontology there may be multiple realities, as in the constructivist research paradigm. The difference is that, rather than the truth being something that is "out there" or external, reality is in the relationship that one has with the truth... Thus, there is no one definite reality but rather different sets of relationships that make up an Indigenous ontology. Therefore, reality is not an object but a process of relationships, and an Indigenous ontology is actually the equivalent of an Indigenous epistemology. (p. 73)

Wilson (2008) explains that Indigenous ontology and epistemology can be seen clearly in the languages of many Indigenous people. Familial words illustrate this concept. In Cree, which is Wilson's native language, the word used between great-grandparent and great-grandchild are the same, implying a non-hierarchical relationship. Another example is found in Cherokee, where the word for *cousin* can imply an extended relationship in the family or someone who is "kin" to me. In other words, the language describes the relationship between speakers; language does not simply label individuals. Wilson (2008) sees these linguistic illustrations as clear examples of an Indigenous ontology that emphasizes relationships and of Indigenous epistemology in which speakers learn this worldview through learning the language.

Figure 1.

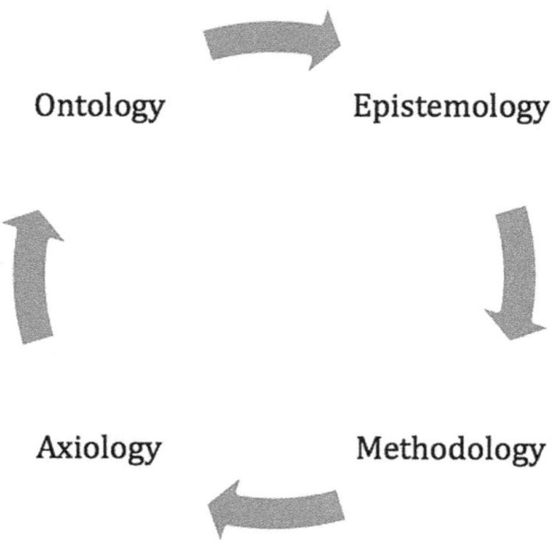

The previous concepts of Indigenous ontology and Indigenous epistemology lead to an understanding of Indigenous axiology, which focuses on accountability to one's relationships (Chilisa, 2012; Wilson, 2008). Being accountable means that one values the needs of those involved in the relationship and will not do anything to harm those relationships. This notion connects directly to methodology in research. Weber-Pillwax (2003, as cited in Wilson, 2008) named "three Rs" that are crucial to Indigenous methodology: respect, reciprocity, and responsibility. Chilisa (2012) named four Rs: "relational accountability, respectful representation, reciprocal appropriation, and rights and regulations during the research process" (p. 22). It is clear that, as Wilson (2008) claimed, the elements in an Indigenous paradigm are based on relationships between a person and the entire world (including the cosmos), and these elements cannot be separated into non-related entities. If separation happens, the paradigm collapses.

So far, we have been examining an Indigenous paradigm that could apply to all Indigenous people. However, Kovach (2009) reminds us that tribal ontologies and epistemologies are unique and must be treated as such. Still, published works explaining such worldviews are rare unless they are older texts often written by positivistic anthropologists who likely did not approach the understanding and telling of such information with relational methods (Smith, 2012; Kovach, 2009). For Indigenous peoples, unique worldviews are expressed through oral histories, containing some stories that are sacred and allowed to be shared only with tribal members (Teuton, 2012; Wilson, 2008). Other stories are acceptable to be shared with anyone who chooses to listen. While Indigenous storytellers acknowledge some problems with writing down stories that are meant to be received orally, they also admit that doing so can help to preserve cultures and languages (Iseke, 2013; Teuton, 2012). As Cherokee Elder and storyteller Hastings Shade said, "Sharing is probably the main thing as far as a *value* that we have. To me, sharing what we know…we're sharing with the other part of the world or the other people that says, 'What is a Cherokee? What's their *belief*? What do they *do*? What do they *know*?" (Teuton, 2012, p. 15).

For example, for traditional Cherokees, understanding their place in the world as beings who are tasked with balancing and connecting the Sky World, the Middle World, and the Under World is indelibly connected to the stories that are used to express that understanding (Teuton, 2012). Teuton expounded on this worldview: "Human beings in particular have the responsibility to create dynamic balances between the Sky World and the Under World through our application of knowledge" (p. 21). Thus, stories and storytellers are the means by which this worldview is shaped and also learned, emphasizing their roles in maintaining relationality and building community.

THE CONCEPT OF COMMUNICATIVE COMPETENCE IN LANGUAGE ACQUISITION

While the field of second language acquisition theory applied to language teaching and language learning is quite broad, there have been certain generally accepted overarching principles in the field that are based on decades of language acquisition research. From Noam Chomsky's linguistic theories on generative grammar to Krashen and Terrell's Natural Approach acquisition model, the field presents an array of ideas about how languages are learned and can be learned (Brown, 2014).

Although there seems to be as many models of second language acquisition as there are researchers theorizing about these concepts, most of the current models have been influenced by the concept of communicative competence (Hymes, 1972; Canale & Swain, 1980), which was a reaction to Chomsky's claim of linguistic (grammatical) competence being the foundation of language function. Hymes (1972) and Canale & Swain (1980) argued for a more comprehensive notion of competence, keeping the notion of Chomsky's grammatical competence but also adding the parameters of discourse, strategic, and pragmatic competence. While others have added to and refined these initial concepts about communicative competence (for example, Bachman, 1990; Bachman & Palmer, 1996), these ideas about competence in a second language impacted teaching methodology, leading to what is now known as Communicative Language Teaching, or CLT (Brown & Lee, 2015).

Brown and Lee (2015) summarized many of the commonly-accepted principles of CLT in their book *Teaching by Principles: An Interactive Approach to Language Pedagogy*. They categorize these principles of language learning and teaching into 8 main components, each of which affects the second language learning and teaching process: agency, identity and investment, self-regulation, automaticity, reward, languagculture, and transfer. Drawing on these principles of language teaching, the ultimate goal of CLT is to create an interactive learning environment for language learners, leading towards communicative competence. These ideas are illustrated in Figure 2.

Figure 2.

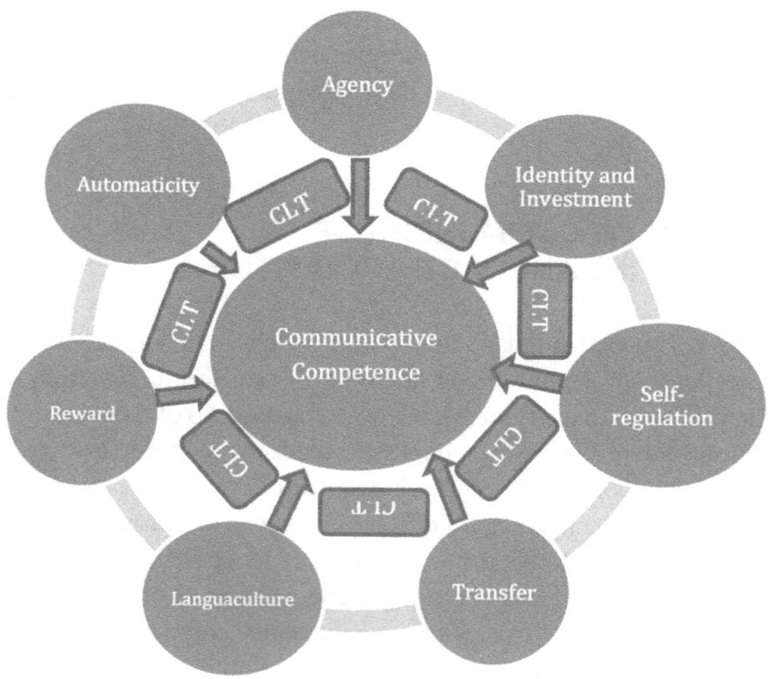

The legitimacy of such views and purposes of language learning and teaching, however, were brought into question in the late 1980s when issues of critical pedagogy came to the forefront in language teaching, specifically through theorists in applied linguistics and TESOL. Pennycook (1989) argued that the ideas about language learning and teaching above stem from a positivistic view of language, one that takes the stance that language teaching can be reduced to methods and methodologies that will result in the production of language for particular tasks and which views itself as apolitical. In his scathing review of a *methods-focused* approach to teaching English, Pennycook (1989), also said this: "First...all education is political, and second...all knowledge is 'interested'" (p. 590), a concept he supported by citing Giroux's and Foucault's notions of social power. He continued: "Education is fundamentally political since it is constantly involved in the (re)production of social and cultural inequalities...and of particular forms of culture and knowledge" (pp. 590-91). Stemming from his argument that all education is political, Pennycook (1989) suggested that the language teaching profession needed to carefully examine itself and the basic tenets upon which its practice had been built in the following areas: that education is political; that societal forces affect research paradigms, even within language acquisition that there are political and pedagogical implications of the concept of "communicative competence"; that the field often trivializes content; and that we must examine what method means to the field.

Ultimately, Pennycook (1989) concluded that language teaching "methods" are often influenced by economic and governmental powers, and that they should not be seen as prescribed modes of conveying knowledge about language and culture. He also warned against prescriptively using popular methods to achieve language production goals rather than descriptively looking at what is actually happening in classrooms with language learners. Rather, he suggested listening to teachers and students to find out what works best for them in their unique learning situations. In addressing international teaching contexts, Pennycook cites a study by Burnaby and Sun (as cited in Pennycook, 1989), which "suggests that we must examine the teaching context very carefully before promoting communicative approaches, since they may be inappropriate for a number of reasons" (p. 611).

Around twenty years after Pennycook's seminal criticism of language teaching method, the issue is still being discussed. In 2008, Canagarajah said, "No sensible professional can practice [English Language Teaching] today without being alert to the heterogeneity of English varieties, the conflicting claims of community and identity, the values behind methods and materials, and unequal classroom relationships and roles" (p. 213). Although Canagarajah (2008) was addressing the English teaching field, specifically, his mention of values inherent in chosen methods and materials is crucial to answering the question about how the theories and methods used in all language teaching are influenced by Western-based ideologies and powers, specifically by economically-driven factors, such as textbook publishing.

One example he used to illustrate his point is that of process-based writing, which has been the accepted *best practice* in the academy since the late 1980s. Echoing Pennycook's mention of the same issue, Canagarajah (2008) explained, "In recent pedagogical discourse, a process-oriented, student-centered, task-based approach has been treated as the most effective for teaching/learning. The ELT research and professional circles have treated this as 'the optimum interactional parameters within which classroom language learning can take place' (Holliday, 1994, p. 54)" (Canagarajah, 2008, p. 215). This is one specific example of how a particular language teaching method influenced by certain social powers has continued to hold on to dominance despite attempts to address concerns of cultural appropriateness and situational effectiveness.

In sum, the inherent positivistic Western-based influences on much of language pedagogy are not often enough problematized. Researchers are addressing the issues of appropriateness of methodology for certain cultures and settings while navigating outside of a positivistic paradigm. Current issues in language teaching, such as race, gender, privileged language varieties, and decolonization of language teaching, are all under scrutiny as the field moves forward (Canagarajah, 2008).

AN INDIGENOUS PARADIGM AND LANGUAGE ACQUISITION TEACHING AND RESEARCH

Connecting these concepts back to language teaching that values an Indigenous worldview, it is crucial to realize that methods and methodologies founded in a Western-based paradigm are not necessarily a template for success with all communities; specifically, the positivistic, Western-based ideologies that underlie commonly accepted methods may significantly clash with an Indigenous worldview. As May (2013) has shown, certain Indigenous tribes, such as the Navajo and Maori, have achieved revitalization successes through their own language education models, created by Indigenous people for their own purposes. The benefits of such programs are clear:

Within the space of 40 years, [these language programs] have been instrumental in addressing and, in some cases, actively ameliorating, a long, entrenched history of the active state-sponsored marginalization and diminution of indigenous languages, leading to their extensive language shift and loss. In so doing, research on these programs has focused primarily on their pivotal role in this process of indigenous language revitalization, as well as the wider politics of indigenous self-determination underpinning them. (May, 2013, p. 64)

Thus, a conversation between Indigenous methodologies and traditional Western-influenced methodologies must start with this realization: Indigenous understandings of learning and teaching must be foundational, and their purposes for language teaching can be quite different from traditional language teaching goals (Hinton, 2011). The conversation must be entered into with the acknowledgement that commonly accepted language teaching methods from the academy may be useful suggestions for practice, but they should not be automatically assumed to be the only or the best ways to teach. This article proposes that the most productive and ethical manner in moving the conversation forward is to acknowledge the significance of an Indigenous worldview and to apply this paradigm to methodologies for research and teaching.

This issue is not isolated merely to language teaching, learning, and research; rather, acknowledging an Indigenous worldview within the context of language is critical to identity formation and community formation for Indigenous peoples. In the words of Marie Battiste, a prominent Mi'kmaq scholar from the University of Saskatchewan:

Languages are the means of communication for the full range of human experiences and critical to the survival of the culture and political integrity of any people. Aboriginal languages provide a direct and powerful means of understanding the legacy of tribal knowledge. They provide the deep and lasting cognitive bonds that affect all aspects of Aboriginal life. Through sharing a language Aboriginal people create a shares belief of how the world works and what constitutes proper action. (Battiste, 1998, p. 18).

Battiste goes on to quote the voice of a tribal elder from Sioux Valley First Nations in Canada:

Our Native language embodies a value system about how we ought to live and relate to each other...It gives a name to relations among kind, to roles and responsibilities among family members, to ties with the broader clan group...There are no English words for these relationships...Now, if you destroy our languages you not only break down these relationships, but you also destroy other aspects of our Indian way of life and culture, especially those that describes man's connection with nature, the Great Spirit, and the order of things. Without our languages, we will cease to exist as a separate people. (as cited in Battiste, 1998, p. 18)

Identity, therefore, is inextricably bound to one's language, and one's language helps to develop one's worldview (Whorf, 1956; Wong Fillmore, 1991). When these languages are lost and fail to be cultivated with respect, Indigenous community members lose a part of their identity and possibly ways of understanding the world (McCarty, Romero, & Zepeda, 2006).

RESEARCH ON INDIGENOUS LANGUAGE ACQUISITION AND REVITALIZATION

McCarty, Romero, and Zepeda (2006) conducted an extensive study, gathering narratives of more than 150 adults and children involved in Navajo revitalization school programs, focusing on "language attitudes and ideologies" and "the role of Indigenous languages within their lives, their aspirations, and their senses of self" (p. 31). The narratives revealed three themes exhibiting the complexities inherent in Indigenous language identity: "the politics of shame and caring in school, the hegemony of English, and the iconic bonding of English with whiteness" (p. 37). For students, identities of shame vacillated between being ashamed of speaking Navajo and not being able to speak it when pressed to do so by teachers. Competing narratives on caring emerged both from teachers who expressed that students didn't care about the language and from students who said that teachers didn't care about them. Bishop (2011) found some of the same themes in narratives from Maori students in New Zealand who expressed their feelings that teachers viewed them from a deficit orientation rather than a positive orientation. Students in the Navajo study also expressed that they struggled with self-hate based on centuries of colonial oppression. Along with these findings, some parents expressed a belief that learning Navajo was not important because "The world speaks English." (p. 40). This frustrated the students who were, indeed, trying to learn the language.

Kouritzin (1999), in her case study work with people from language groups across the globe, recorded these same kinds of identity struggles with people who gave up a native language for English. She explained that most of the storytellers in her book experienced "tension and struggle between two or more competing cultural identities" (p. 99) and that they had trouble viewing themselves as a synthesis of those cultures and languages. Grande (2000) adds to this discussion by relating her own story of losing her ability to think metaphorically as she entered the academic world and academic English became a primary mode of communication. She expressed that doing so was necessary for academic survival, but she laments that this is the case.

Kouritzin (1999) suggested that people who find themselves between these two linguistic worlds must find a way to see themselves as a synthesis of the two worlds and languages, and that doing so is crucial for their languages and cultures to survive. She said,

In fact, Woolard (1989, cited in Ben Raphael, 1994) argues that the cultural hegemony dominating our thinking, which I take to mean the view of languages and cultures as bounded and fixed, may lead to the death of subordinate languages. Trying to maintain a "pure" form of the minority language, resisting the influence of the dominant language, may be less effective than a working corruption of the minority language, allowing for adaptation to the dominant language. (p. 98)

Although Indigenous peoples face a large hurdle in revitalizing their languages, finding ways to reshape views of those languages from within is crucial for what Watahomigie (as cited in McCarty, Romeo, & Zepeda, 2006) called "reverse brainwashing" (p. 43). Kouritzin's (1999) suggestion of developing a synthesized view of one's identity can work, but for that to happen, one must value all pieces of that identity. One critical factor for success in this process is the presence of caring adult mentors who view young people as valuable resources in the process and who understand that young people want to be taught the language and culture (McCarty, Romero, & Zepeda). Bishop (2011) expands on this idea by suggesting training teachers in non-deficit thinking, where teachers can "create learning contexts in classrooms where young Maori people were able to be themselves as Maori; where Maori students' humour was acceptable, where students could care for and learn with each other, where being different was acceptable and where the power of Maori students' own self-determination was fundamental to classroom relations and interactions" (p. 49). This is an excellent goal for teachers, and, in theory, also a worthwhile disposition for educational researchers to possess.

APPLYING AN INDIGENOUS RESEARCH PARADIGM

The central tenet of an Indigenous research paradigm is relationality, the worldview that all things are connected and related to one another, including to what we might call research (Chilisa, 2012; Kovach, 2009; Wilson, 2008). This notion of relationality informs an Indigenous research methodology, which first builds upon relations with the research partners and then uses methods that honor those relationships. Indigenous researchers have emphasized how narratives, specifically storytelling, can serve as a method that respects relational research approaches (Iseke, 2013; Chilisa, 2012; Wilson, 2008) and functions to create a "collaborative process" (Kovach, 2010, p. 43). Such stories thus serve as the foundation for building relationships with the participants who are co-researchers working together for the common good of the tribe (Chilisa, 2012; Kovach, 2009; Wilson, 2008). Indigenous story is the means by which tribal knowledge and worldview is shared (Iseke, 2013; Wilson, 2008), and learning from these stories requires a "respectful stance and reverence for the sharing of stories" (Iseke, 2013, p. 560) on the part of the research; furthermore, a researcher should understand that certain stories – such as sacred ones – cannot be written down (Iseke, 2013; Wilson, 2008).

Kovach (2009) supported the idea that Indigenous methodologies (informed by an Indigenous paradigm) "can be situated within the qualitative landscape because they encompass characteristics congruent with other relational qualitative approaches (e.g., feminist methodologies, participatory action research)" (p. 25). Kovach also emphasized, however, that there are also inherent differences which must be addressed so that Indigenous methodologies do not become subsumed under a larger umbrella of a Western-influenced qualitative paradigm. Acknowledging some the apparent similarities, Kovach (2009) and Wilson (2008) explained that the main difference between Indigenous methodologies and Western-based qualitative methods lies in specific tribal epistemologies.

Kovach (2009) further problematized the marrying of Western-based qualitative research and non-Western Indigenous research paradigms in two specific areas: language and knowledge itself. The two issues are intricately related, as Kovach asserted,

Indigenous knowledges have a fluidity and motion that is manifested in the distinctive structure of tribal languages. They resist the culturally imbued constructs of the English language, and from this perspective alone, Western research and Indigenous inquiry can walk together only so far. (p. 30).

Wilson (2008) also addressed this concern specifically regarding Indigenous discourse, of which he gives a specific tribal (Cree) example. In Cree discourse patterns, the storyteller cannot

draw conclusions for another or...make an argument. My role, based upon the guidelines of relationality and relational accountability, is to share information or to make connections with ideas. The ethic in place is that it is not right to make interfere with another's actions or thought process. (Wilson, 2008, p. 133)

These are precisely the methodological considerations to which educational researchers partnering in language acquisition research Indigenous populations must be sensitive.

CONCLUSION

The purpose of this article has been to argue that acknowledging and valuing Indigenous worldviews is essential for teachers and educational researchers who want to partner with Indigenous community members as students, teachers, and research partners. Indeed, respect for Indigenous epistemologies, ontologies, and axiologies should inform the research methodologies that educational researchers employ. This is especially critical in the context of teaching and research related to language acquisition, given that the development and use of language is essential to the development of one's identity and sense of community membership. Unless educational researchers understand these indigenous paradigms, educational research may come at the expense of the dignity and the identity of its participants, and, indeed, may produce knowledge that, in practice, serves to delegitimize the knowledge and values of local communities.

REFERENCES

Bachman, L. F. (1990). *Fundamental considerations in language testing*. OUP.

Bachman, L. F., & Palmer, A. S. (1996). *Language testing in practice: Designing and developing useful language tests*. OUP.

Battiste, M. (1998). Enabling the autumn seed: Toward a decolonized approach to aboriginal knowledge, language, and education. *Canadian Journal of Native Education, 22*(1), 16–27.

Bishop, R. (2011). *Freeing ourselves*. Sense Publishers. doi:10.1007/978-94-6091-415-7

Brown, H. D. (2014). *Principles of language learning and teaching: A course in second language acquisition*. Pearson.

Brown, H. D., & Lee, H. K. (2015). *Teaching by principles: An interactive approach to language pedagogy*. Pearson.

Canagarajah, S. (2008). The politics of English language teaching. In N. Van Deusen-Scholl & N. H. Hornberger (Eds.), Encyclopedia of language and education (2nd ed., pp. 213-237). Springer Science+Business Media. doi:10.1007/978-0-387-30424-3_16

Canale, M., & Swain, M. (1980). Theoretical bases of communicative approaches to second language teaching and testing. *Applied Linguistics, 1*(1), 1–47. doi:10.1093/applin/1.1.1

Fishman, J. A. (1996). What do you lose when you lose your language? In G. Cantoni (Ed.), *Stabilizing indigenous languages* (pp. 165–175). Northern Arizona University.

Hinton, L. (2011). Language revitalization and language pedagogy: New teaching and learning strategies. *Language and Education*, *25*(4), 307–318. doi:10.1080/09500782.2011.577220

Hymes, D. (1972). On communicative competence. In J. B. Pride & J. Holmes (Eds.), *Sociolinguistics* (pp. 1–21). Penguin.

Iseke, J. (2013). Indigenous storytelling as research. *International Review of Qualitative Research, 6*(4), 559-577.

Kouritzin, S. G. (1999). Face[t]s of first language loss. Routledge.

Kovach, M. (2009). *Indigenous methodologies: Characteristics, conversations, and contexts*. University of Toronto Press.

May, S. (2013). Indigenous immersion education: International developments. *Journal of Immersion and Content-Based Language Education, 1*(1), 34-69.

McCarty, T.L., Romero, M. E., & Zepeda, O. (2006). Reclaiming the gift: Indigenous youth counter-narratives on Native language loss and revitalization. *American Indian Quarterly, 30*(1-2), 28-48.

Pennycook, A. (1989). The concept of method, interested knowledge, and the politics of language teaching. *TESOL Quarterly*, *23*(4), 589–618. doi:10.2307/3587534

Teuton, C. B. (2012). *Cherokee stories of the Turtle Island Liars' Club*. University of North Carolina Press.

Whorf, B. (1956). Language, mind, and reality. In B. Carrol (Ed.), *Language, thought, and reality: Selected writings of Benjamin Lee Whorf* (pp. 246–270). M.I.T. Press.

Wilson, S. (2008). *Research is ceremony: Indigenous research methods*. Fernwood.

Wong Fillmore, L. (1991). When learning a second language means losing the first. *Early Childhood Research Quarterly*, *6*(3), 323–346. doi:10.1016/S0885-2006(05)80059-6

Chapter 9
An Emerging Methodological Approach to Anti-Racist Design and Implementation

Meredith Kier
College of William and Mary, USA

Deena Khalil
Howard University, USA

ABSTRACT

This chapter is about introducing critical race design (CRD), a research methodology that centers race and equity at the center of educational opportunities by design. First, the authors define design-based implementation research (DBIR) as an equity-oriented education research methodology where teaching and learning is informed by robust, iterative, evidence-based research conducted by multiple stakeholders. Next, they provide a brief overview of critical race theory in education (CRT) as a theoretical and methodological approach that aims to unpack and disrupt the structural inequities experienced by disenfranchised racial groups. They then describe how both education methodologies inform CRD, the authors' emerging anti-racist critical design methodology. Finally, they provide an example where they used CRD to design an online service-learning course that aims to situate the narratives of underrepresented STEM professionals as a curricular resource for nondominant adolescents.

DOI: 10.4018/978-1-7998-7600-7.ch009

INTRODUCTION

In the United States, there is an underrepresentation of women and professionals of color in science, technology, engineering, and mathematics (STEM) careers (Tsui, 2007). With a similar underrepresentation reflected in the K-12 STEM classrooms, traditional mathematics and science instruction needs to be more accessible to nondominant students (Tate, 1997; 2001). At the postsecondary level, there have been a number of commonly used strategies to recruit and retain underrepresented young adults in STEM courses and careers including summer bridge programs, mentoring, development of research internships, college-level tutoring, career counseling, academic centers, advising, and learning centers (Tsui, 2007). Evidence of this is seen in higher and adult education when pre- and post-graduates participate in service-learning experiences as both a means for continuing their education while simultaneously influencing and engaging with the community (Gasman & Commodore, 2014; Stanton, Giles, & Cruz, 1999). The combination of open access to higher education and federally funded programs in career and technical education (CTE) have, by design, provided similar opportunities to youth and young adults seeking to pursue alternate pathways to STEM careers (Elementary and Secondary Education Act [ESEA], 1965). Similarly, the goals of CTE have been introduced to the mainstream K-12 setting through College and Career Ready Standards, more specifically in the *Common Core State Standards of Mathematics* and the *Next Generation Science Standards* (Common Core State Standards Initiative, 2010; Darling-Hammond, Wilhoit, & Pittenger, 2014).

The inclusion of College and Career Ready Standards into K-12 education settings requires educators to forefront career pathways as opposed to assuming careers are a byproduct of learning (Herrington & Daubenmire, 2016; Stage, Asturias, Cheuk, Daro, & Hampton, 2013). However, including CTE in lesson designs can be a challenge as such programs may reinforce a system of education based on advantage (Tatum, 2003) where nondominant groups are overrepresented in the lower half of the hourglass global economy (Sassen, 2001). Thus, it is imperative for educators who are committed to equity to critically develop programs for nondominant students that enable them to access better paying careers in the top half of the hourglass economy (Ladson-Billings & Tate, 1995). In the United States, where white supremacy has historically marginalized Black students and continues to disenfranchise other students of color, we argue that the design and implementation of career and technical education need to be situated in critical and transformative research paradigms.

In this chapter, we outline an emerging methodology for programmatic interventions based on the tenets of Critical Race Theory (CRT; Ladson-Billings & Tate, 1995) and the principles of Design-Based Implementation Research (DBIR; Penuel, Fishman, Cheng & Sabelli, 2011). Both educational research methodologies stem from equity-oriented paradigms that aim to criticize the marginality of nondominant students while simultaneously empowering their agential knowledge production. Similarly, our emerging research methodology unifies CRT and DBIR in an approach that we have coined *Critical Race Design* (CRD). CRD aims to position nondominant learners as key foci for developing equitable learning ecologies. CRD centers on equity with tenets of CRT explaining *why* we take this approach to inquiry and action, and DBIR informing *how* we are designing P-20 education programs.

The following sections provide an overview of DBIR and CRT and a brief discussion on how they have informed K-12 and high education research. We then describe how both education methodologies informed CRD, and how principles of CRD may be utilized to interpret, disrupt, and reconstruct education programs to serve nondominant learners. Finally, we provide an example from our own project, *E-Communities*, to illustrate how our conceptualization of CRD originated and emerged during our design, implementation, and research of a service learning course for nondominant engineers who partnered with us to inspire teachers and students in one of the largest Black majority school districts in the country.

Defining Design-Based Implementation Research

Design-Based Implementation Research (DBIR; Penuel et al., 2011) is a design-based approach to inquiry and research that addresses policy level issues in schools through the implementation of equitable programs and research. DBIR draws from other pragmatic forms of inquiry including evaluation research, community-based participatory research, design-based research, and participatory-action research (Anderson & Shattuck, 2012; Fishman, Penuel, Allen, Cheng, & Sabelli, 2013) to bring implementations to scale. While these interdisciplinary methodologies are commonly integrated into school-based research, the combination of four key characteristics distinguishes DBIR from these other methodologies. According to Penuel et al., (2011), these characteristics include:

> *[i] a focus on persistent problems of practice from multiple stakeholders' perspectives; [ii] a commitment to iterative, collaborative design; [iii] a concern with developing theory related to both classroom learning and implementation through systematic inquiry; and [iv] a concern with developing capacity for sustaining change in system. (p. 332)*

In identifying the education problem to be addressed through design, multiple stakeholders come together to discuss critical issues within and around the education organization (i.e. school or district) and community. Stovall (2006) presents another goal of reform: "As Yamamoto suggests, the occupation of the socially conscious scholar is to participate in the activities that challenge hegemony at both the grassroots and intellectual levels," (p. 243). Participants in the DBIR process may include a wide array of stakeholders such as researchers, organizational leaders, practitioners, parents, community members, entrepreneurs, and students (Fishman et al., 2013; Penuel, Coburn & Gallagher, 2013). DBIR has been predominately used to identify key issues and questions regarding equity and access when strengthening capacity and connections between schools and communities (Lee & Krajcik, 2012; Penuel & Fishman, 2013). As trusting partnerships are formed between stakeholders and researchers, they, ideally, hone in on specific education questions that can feasibly be addressed given the organizational structure, time, commitment, and resources appropriated for research and programmatic design (Sabelli & Dede, 2013). Through planning and research, the community of stakeholders considers and decides upon research-based interventions and metrics to implement and iteratively assess selected strategies (Penuel & Fishman, 2013; Supovitz, 2008).

In this iterative process of designing and refining an intervention, researchers carefully document the successes and failures of each strategy, typically using mixed-methods for data collection and analysis in an effort to build a theory for design and a theory for action (Small, 2011). Rather than trying to emulate the outcomes of effectiveness studies, stakeholders use the findings to support the development of new questions or adapt interventions situated within the research contexts, ensuring the histories and voices of stakeholders, particularly nondominant school and community members, are heard (Lin, Schwartz, & Hatano, 2005). Typically, empirical research is not performed on the implementation of the intervention until there have been multiple iterations and the intervention design is being scaled-up for community sustainability (Penuel & Fishman, 2013). DBIR challenges current research methodologies as it builds knowledge through the process of iterative design; as knowledge is gained about the community and school context, it is used to adapt and revise strategies to achieve usable data for all stakeholders in a highly contextualized setting (Means & Harris, 2013).

Reviewing Tenets of Critical Race Theory in Education

CRT came to prominence in response to a need in the U.S. legal academy to connect conversations about race and racism as an experiential phenomenon to a theoretical and institutional one as well (Ladson-Billings & Tate, 1995). Instead of focusing on overt and willful acts of racism at the individual level, CRT unpacks the dominant structures in society that institutionally oppress communities of color, with the purpose of transforming societies and individuals (Tierney, 1993). CRT can be actively used as a method to change racial, political, social, and practical inequities (Milner, 2008). Sólorzano and Yosso (2002) outline five goals of CRT as an education research methodology: "i) the intercentricity of race and racism with other forms of subordination, ii) the challenge to dominant ideology, iii) the commitment to social justice, iv) the centrality of experiential knowledge, and v) the transdisciplinary perspective" (pp. 25-26). In an effort to blend the goal of unpacking institutional racism with actions that can disrupt it, critical race scholars have proposed several tenets to aid in utilizing CRT analytically. Here, we highlight three of these tenets-interest convergence (Bell, 1980; 2004), critique of liberalism (Crenshaw, 1988; 1995), and counter-storytelling (Delgado & Stefancic, 2012) to show how they can be aligned with DBIR by identifying racially inequitable spaces to be disrupted by designing programmatic interventions.

First, a necessary element in disrupting racism for nondominant community stakeholders (of color) is to outline how such a disruption can aid dominant stakeholders (primarily whites in the U.S.) as well; only through such an "interest convergence" (Bell, 1980; 2004) can systemic progress be achieved (Bell, 1980; 2004). Similarly, to critical legal theorists arguing that the civil rights of people of color will not be addressed unless it coincides with the interest of Whites, critical race educators argue that education reform in minoritized schools will not occur until such reform serves *all* students, particularly white students (Ladson-Billings & Tate, 1995). Previously, the goal of educating the masses was to have a literate citizenry that can function in an industrial system (Tyack, 1974). However, as the industrial economy gave way to a global economy, this "hidden curriculum," (Anyon, 1980, p. 67) of socially stratifying nondominant students in the U.S. risked the hegemony of the U.S. in the global economy. As a result, education policymakers aimed to raise high education standards for all (U.S. Department of Education, 1983) rather than limit access to schooling through exclusiveness and creditionalism (Loxley, & Seery, 2008; Ravitch, 2010). Debates have ensued over the convergence of equity and excellence--thus, mathematics and science literacy have been labeled the new civil rights frontiers (Tate, 2001), and a focus is placed on obtaining equal academic access and opportunities for underrepresented students in mathematics and science classrooms to ensure that all students are career ready.

Another tenet of critical race theory is the "critique of liberalism" (Crenshaw, 1988; 1995) which recognizes the fallacy of liberal ideologies such as institutions being meritocratic, objective, or neutral towards subordinated groups. This tenet critically exposes the permanence of racism within U.S. schools that marginalizes students of color (Howard, 2003; 2015; Ladson-Billings & Tate, 1995). For example, the disciplines of mathematics and science are often taught as subjects that are ahistorical, objective and race-neutral which leads to disengagement and disconnectedness with nondominant groups who witness few reflective narrative representations within the disciplines (Battey, 2013; Brown et al., 2016). CRT's critique of liberalism calls for one to question contemporary understandings of knowledge, intelligence, capability, and achievement as meritocratic in nature (Crenshaw, 1988). Despite the appearance of meritocracy or equal access, critical race theorists assert that access to mathematics and science are not based on neutral measures, and have led to a leaky school-to-career STEM pipeline among nondominant populations. Research indicates that STEM persistence is still informed by historic racial biases, and that bias and conflict continue to serve as "gatekeepers in providing access to science careers" (Brown et al., 2016, p. 146). Similarly, Battey (2013) found that while access to mathematics was often symbolically framed in a more neutral and meritocratic way, institutional and social constructions of race were major factors in limiting minoritized students' access to opportunities in mathematics. In one study, Denson, Avery, and Schell (2010) found that African American high school students had negative attitudes toward engineering because of implicit and explicit messages received of job discrimination, social stereotypes, and micro-aggressions. Further, Gándara writes (2006), "minoritized students are more likely to be assigned to low curriculum tracks independent of their test scores than white students (Lucas, 1999)" (p. 231).

For example, algebra is a course that is a major gatekeeper to entry into the college preparatory track; thus, when teachers track students of color into lower level mathematics courses, they are ultimately limiting college and career access (Gamoran & Hannigan, 2000; Paul, 2005).

Finally, as inequitable structures perpetuate institutional racism by norming and othering, stories from minoritized communities can disrupt pervasive narratives that aim to validate the norms and premises of whiteness (Delgado & Stefancic, 2012). The personal narratives of those from nondominant communities can be liberating and empowering as it situates their everyday experiences as epistemological resources (Lynn, 2004; Yosso, Parker, Solorzano, & Lynn, 2004), that can "challenge mainstream narratives that may be readily accepted as objective truths" (Horsford, 2010, p. 295). Stories from communities of color can serve as the basis for "1) curricular structures, processes, and discourses; 2) challenge dominant social and cultural assumptions regarding culture and intelligence, [and] 3) direct the formal curriculum towards goals of social justice and the hidden curriculum toward Freirean goals of critical consciousness" (Yosso, 2002, p. 98). Critical scholars have shown that short story descriptions of race and racialized experiences are successfully recounted with the intent to explore racial categorizations of individuals and how racism plays out in youth experiences (Denner, Cooper, Lopez & Dunbar, 1999; Diversi, 1998; Stinson, 2008; Kier, 2013). For example, scholars challenge the dominant narrative of STEM fields as exclusionary to people of color by providing learners with counter-stories of successful STEM professionals of color who have persevered in their field (Baber, Fletcher, & Graham, 2015; Harper, 2010; Stinson, 2008).

CRITICAL RACE DESIGN: AN EMERGENT ANTI-RACIST APPROACH

CRD is an anti-racist research and design methodology that connects critical race scholarship with the conceptualization, research, and iterative implementation of design-based perspectives and best practices organized for communities of color. This framework is an emerging methodology because it utilizes two, well-established methodological frameworks that position community stakeholders and nondominant learners as key to generating new knowledge of research, design, and praxis. The overarching goal of CRD is to attend to minoritized learners' broader social contexts and address their systemic marginalization and socialization within their learning ecologies. CRD emerged through the premise of conceptualizing and constructing the necessary structural supports for minoritized learners' ecologies based on and built through their unique funds of knowledge (González, Moll, & Amanti, 2005; Yosso, 2005). The strength of CRD as a methodology is that it i) links critical theory and organizing action, utilizing CRT by privileging race as a core driver that confronts inequitably racialized learning experiences (Ladson-Billings & Tate, 1995), and ii) DBIR privileges design as a propeller for developing learning experiences to be more equitable. Penuel and Fishman (2013) claim:

Addressing the needs of students from disadvantaged communities requires research on strategies for supporting effective adaptation of interventions to relate every day and scientific forms of meaning making, as well as research on strategies that attempt to create more equitable conditions for all students to learn. (p. 283)

Proponents of DBIR urge for large-scale, school-based interventions to be guided by research- and equity-based theories that guide inquiry and ways of knowing. We argue that CRT can and should be utilized to identify systemic and endemic racial inequities when researching and designing minoritized learning ecologies. Below we illustrate our conceptualization of CRD and show where CRT tenets drive the ontology, epistemology, axiology, and methodology of our anti-racist design-based implementation research. More specifically, we argue that these three tenets of CRT----"interest convergence" (Bell, 1980, p. 94), "critique of liberalism" (Crenshaw, 1988), and counter storytelling (Delgado, 1989) --are necessary tools to disrupt inequities and design racially equitable learning environments for learners of color (Dixson & Rousseau, 2005; Ladson-Billings, 2005; Lynn & Parker, 2006).

Empowering Nondominant Learners by Converging Multiple Stakeholder Interests

The first principle of DBIR is to bring multiple stakeholders together to jointly define the problems occurring in the teaching and learning contexts, and devise potential equitable solutions (Penuel et al., 2011). However, in STEM educational contexts, there are arguably competing interests among stakeholders informed by varying levels of power and privilege--those who argue for "equity" versus those who argue for "excellence" (Ravitch, 2004, p. 63). Because critical race scholars posit that all structures in social contexts are situated to uphold dominant groups' hegemonies, structures that benefit the disempowered or underprivileged only emerge when they align or converge with the interests of the dominant group (Bell, 1980). Thus, we propose that the first principle of CRD is for researchers to characterize and confront the power differentials among stakeholders, including researchers, policymakers, educators, parents, learners, and their corresponding interests. We encourage critical discussions between all stakeholders to collaboratively identify and understand contextual issues and competing interests behind the issues. Such discussions may take the form of interviews or focus groups within and between stakeholder groups in an effort to closely document convergent and competing goals for the development and implementation of school-based programs. When tensions inevitably arise between competing interests, CRD is used as a methodology to align the interests, voices, and needs of nondominant learners with dominant standards of rigor, so that students' funds of knowledge become central to the design of their learning ecologies. This occurs through stakeholders collaborating *with* learners, and learners taking an active role in the way that they learn. We recommend that researchers include prompts and questions that facilitate difficult conversations between stakeholders about race, the racial history of the surrounding community, and contemporary understandings of racialized access to educational and career experiences. This CRD tenet urges researchers to encourage, document and analyze frequent and transparent conversations to balance interpretations and competing interests with what it means for the most vulnerable and for the professed main beneficiaries--the learners.

Promoting Sustainability by Countering Race 'Neutral' Education Systems

A key component of DBIR is to have a goal of developing capacity, scaling up, and creating sustainable change (Penuel et al., 2011). Through CRD, we posit some headway may be possible if stakeholders consciously critique the 'neutrality' of large-scale education interventions that have historically failed learners of color, and iteratively redesign programs through a race-conscious approach aimed at disrupting education systems built around whiteness (Leonardo, 2002). CRD's foci of racializing all research and implementation efforts is a marked attempt at juxtaposing top-down education reforms that reinforce a system of education based on 'neutral' notions of meritocracy that privilege White norms. As the permanence of institutional racism within urban systems in U.S. is inevitable, it is important to continuously challenge it within each level of implementation through collecting and analyzing data frequently and revising programs to make them more contextually effective for stakeholders. Thus, we suggest that sustainability and scalability of programs is only possible through iterative questioning, race-conscious 'norming,' and generating solutions from the bottom up that are authentic to the context of nondominant learners. Researchers and program developers can apply CRD in design and implementation of programs by using a race-conscious approach to identify *who* key stakeholders are, and *what* is perceived and evaluated as knowledge. Thus, CRD would use a race-conscious approach when identifying stakeholders to develop a critical consciousness around how the sociopolitical infrastructure and socio-cultural histories of minoritized schools and communities are systematically disenfranchised. Recruiting stakeholders must be ongoing, and their strengths must be continuously assessed to redefine effective intervention strategies.

Countering Dominant Narratives by Iteratively Designing Programs

As a final principle of CRD, we suggest that when stakeholders come together to iteratively design and refine implementation, the personal narratives of nondominant stakeholders be analyzed as epistemological resources to counter racist narratives, dominant norms, and deficit practices. In this CRD principle, we call for nondominant stakeholders to share their stories and to build a community of trust that is authentic to the context of their lived experiences. By bringing together CRT's counter-storytelling tenet with iterative design cycles, stakeholders can share their positionality and narratives to co-resist, co-construct, co-practice, and co-refine. CRD's blended theory and methodology may be used to design programs and strategies for schools that emphasize the importance of iteratively, continuously, and systemically drawing upon learners' experiences, expertise, and interests as primary epistemological resources in science and mathematics classes. This allows for theory and practice to be multi-directional where a shared community-focused paradigm can counter top-down bureaucratic policies that are entrenched with Western Eurocentric notions and norms of excellence. Furthermore, we put forth that if learners have opportunities to co-develop their own learning ecologies using their experiences, then they can be truly situated as stakeholders.

CRITICAL RACE DESIGN IN ACTION: E-COMMUNITIES

We seek to use CRD, like other critical emergent education research methodologies, to include the truths of nondominant groups who have been oppressed and systematically marginalized in education research because of racism, classism, sexism, etc., (Kovach, 2005) and revise strategies of programmatic design and implementation. Critical emergent methods are inclusionary and attend to the historical context of participants' identities via research actions that validate their experiences and voices and counters the positivist euro-centric 'objective' norms of white patriarchal capitalistic supremacy (Denzin & Lincoln, 2008; Donnor, 2013; Gillborn, 2013). In the following paragraphs, we describe how our conceptualization of CRD was implemented in a project funded by the National Science Foundation (NSF), *E-Communities: A Collaboration between STEM Educators, Engineers, and Underserved Youth.* We postulate that when CRD principles drive research and implementation, we can develop racially equitable ecologies that build bridges between communities and schools.

Study Overview

E-Communities is a design-based implementation and research grant (Penuel et al., 2011) that is currently funded by the National Science Foundation for from 2015-2018. We (the mathematics and science teacher educators and researchers) have designed (YR 1) and completed one year of implementation (YR2), and will be making revisions to the research and implementation plan to identify and contextualize best practices for the stakeholders we are working with (YR3). The objectives of this project are to build the capacities of teachers to implement engineering design tasks in a culturally specific manner (Gay, 2010; Leonard, 2008) and to increase the interest and identity of students to pursue science, technology, engineering, and mathematics careers. We designed professional development systematically and iteratively based on the context and needs of the individuals (teachers, students, and engineers) of a predominately Black, underserved school district. We designed the following activities to build the communication and collaboration between the members of E-Communities and implement the objectives of the grant. We conducted two, 16-hour synchronous online service-learning courses, one with teachers and one with engineers to identify the needs and early capacities of each stakeholder to contribute to a future space where teachers and engineers can come together to design opportunities for students. The data collected from these courses guided our understanding of how stakeholder groups could share their unique funds of knowledge (González, Moll, & Amati, 2005) and access students' interests and values to co-construct engineering design challenges (EDC; Wheeler, Whitworth, & Gonczi, 2014) that are culturally-specific (Leonard, 2008) and connect students to engineering careers.

Study Context: Blended Service-Learning Communities in E-Communities

Our design-based implementation research study (Penuel et al., 2011) takes place in one of the wealthiest majority Black counties in the United States. Abundant career resources surround this community, including satellite offices of most corporations and government agencies. Despite the prevalence of these entrepreneurial opportunities that promise significant property tax-bases, the per pupil expenditure of the county's schools dramatically vary, with most schools significantly under-funded relative to schools located in similarly affluent districts less associated with Blackness. The district serves approximately 125,000 students, and 93% are students of color. We, the mathematics and science teacher educator on the project, worked closely with an informal engineering educator to recruit twenty experienced middle school math and science teachers (19 females, 1 male) from five geographically proximal schools that serve predominantly Black and Brown learners (approximately 70% Black; administrators from the schools and district,twenty volunteer engineers from local government agencies and businesses, an advisory board, and a project evaluator. We sought to develop a deep understanding of the individuals from each stakeholder group and design our program accordingly. This chapter focuses on how the voices of our engineer participants informed modifications to our design.

Research Study Design: *E-Communities'* Critical Race Design

In the following paragraphs, we illustrate how our CRD principles guided our methodology of iteratively researching and designing an online service-learning course to engage our engineering participants. We used CRD to identify the role that engineers would play in the program, to understand how the narratives of engineers could be centered in developing EDCs, and to revise our design based on the research and evaluation of the engineering course. In preparation for the engineering course, we brought stakeholders together to examine issues around learner engagement in STEM and how these classes could be made more relevant to learners' lived experiences. When using CRD, the historicity, positionality, and interests of stakeholders within that context are critically important to the development of school-based programmatic interventions. Since researchers have found that service learning and mentoring encourage interest and involvement in STEM fields (see Kier, 2013), we strived to capitalize on the assets of the community's professionals to create a link between the county's abundant wealth and the school districts' diminished resources. Together, we addressed issues around connecting practicing engineers with classrooms. During conversations among stakeholders, we documented many competing interests that were at stake. These interests included those of nondominant learners who must be engaged in classroom activities, administrators who have clear goals for school-wide initiatives and professional development, teachers who have an obligation to teach the curriculum, engineers who want to contribute to the community of students, and teacher educators and researchers who desire to research and understand how engineering could be feasibly incorporated into formalized mathematics and science curriculum. Our research documentation during stakeholder discussions guided our support to design authentic ways for engineer stakeholders to engage nondominant students. For example, teachers discussed that they wanted to develop a personalized connection between the engineers and their students. Teachers wanted engineers to show students that they were real people whose careers had direct relevance to activities that occurred in class. Teachers wanted engineers to be open to Q&A sessions with students and be able to discuss personal inspirations for pursuing their educational pathways. Our goal for the online course with engineers was to have them describe how their careers could align to middle school students and to the concepts outlined in the national mathematics and science standards.

To understand how the personal narratives of our engineers could counter deficit narratives around *who* can do mathematics/science, we used our individual networks and social media presence to recruit engineers who reflected the racial diversity of the teachers and the learners in the district; this approach was guided by research that suggests that learners identify with professionals who look like them, thus contributing to interest development and goal setting (Eccles, 2009; Syed, Goza, Chemers, & Zurbriggen, 2012. Engineers were representative of a variety of careers including medical engineers, aerospace engineers, and telecommunications engineers. We shared with the engineers our goal for them to share stories with students that reflected how they persisted through various challenges to become STEM professionals, given their own personal experience with navigating institutionally racist school systems and Eurocentric curricula (Davidson & Foster-Johnson, 2001). We proposed that teachers use the narratives of engineers of color to create or adapt EDCs.

We conducted early focus groups with the engineers to find out their motivations for participating in the project. We found that all engineers were drawn to the notion of service-learning. Engineers were eager to serve teachers and learners in their community, support a larger goal of making learning more applicable and relevant for learners, and making careers more accessible for students. We designed the engineers' online course to allow them to articulate their own histories and backgrounds and develop a narrative about their unique way of thinking about their career and processes within it (Lynn, 2004; Yosso et al., 2004). We shared with them concepts taught in middle school curriculum to see how their work correlated. We asked them to create and share professional narratives from various facets of their professional lives, video their story, and develop an engineering design challenge that connoted, in a small way, a part of how they thought about the world on a daily basis, while making sure to ground it in popular middle school math and science concepts.

The final CRD tenet that is incorporated into E-Communities was promoting sustainability through iteratively redefining and repositioning stakeholder involvement to converge interests. When we held our first online course with engineers, we requested that they meet with us weekly, over six weeks, to collaborate in developing their personal video narratives and ideate engineering design tasks related to their jobs that could be shared with teachers. During this course, we meticulously documented the engagement and satisfaction of the engineers (using video analysis of synchronous online meetings) and analyzed the products created (using a rubric for depicting narratives and creating engineering designs). We found that there was great variation in the participation of the engineers and the quality of their products. By the end of the online course between researchers and engineers, nearly half of the professional engineers said that, due to competing interests on their time, they could not continue to engage. We found that some of the engineers felt that there was not a clear 'end goal,' and that they wanted more input from teachers and learners as they generated various portions of their story as they were unsure how to communicate real world problems for middle school learners. The lack of time, generative format, and the virtual setting of the online course made engineers feel disconnected from students. We also found that the engineers continued on the project had a connection to one of our *E-Communities* stakeholders or to the community and thus had a more contextualized interest to motivate their commitment.

Our feedback from engineers and teachers led us to pivot and redesign E-Communities' programmatic implementation approach with a focus on beginning with learners', rather than engineers', narratives. We realized that beginning with the engineers' stories may have seemed too didactic and we needed to allow the interests of engineers and teachers to converge properly around learners' worldviews and interests. We created open-ended worldview surveys (Cobern, 1996) for teacher and their students to fill out to provide anchors for the engineers to relate their stories to students. We also found that it was more beneficial for teachers to share information related to their school's context as it broadened engineers' understandings of *who* they were sharing career information with, *how* middle school resources have expanded in exponentially higher number of ways that allow further connections between career and curricula, and *why* their career-oriented narratives may directly influence the lives of learners. We asked our teachers to review engineer narratives and products and co-construct questions with their students to guide engineers in proficiently communicating aspects of their work. Teachers discussed with other teachers how they saw the engineer's story connect to their learners as well as the content, serving as an inspiration for an engineering design challenge. Then, as part of our professional development model, we physically brought the engineers and teachers face-to-face (F2F) one weekend to further co-construct EDCs and discuss other meaningful ways to empower minoritized students' STEM identities.

Our *E-Communities* program is purposeful in merging the interests of all stakeholders and prioritizing the inclusion of nondominant voices that are traditionally disempowered--the scale-up of this was done slowly and iteratively in order to continuously disrupt structural inequities that prevent nondominant learners' access to potential success in STEM fields. Through our iterative research and formative assessments, we found future iterations of our programmatic implementation may include: limiting the recruitment of engineers to immediate local community to provoke more spontaneous co-construction with teachers to supplement online service-learning opportunities; offering a more customized approach for developing engineers' stories based on the interests and values of nondominant teachers and learners in their local school district; and, recruiting professionals from organizations that incentivize their employees through compensated time or community involvement. Further, we gained knowledge about the value of blending online models with face-to-face opportunities to promote further collaboration between engineers and teachers. While engineers and teachers alike can contribute their own personalized professional narratives online, it is still important to physically bring stakeholders together in the space that utilizes designing instruction to further situate the learning ecology and provide spontaneous conversations that enable stakeholders to develop further trust and commitment.

A Note on Design-Based Research Data Analytics and Sustainability

Critical Race Design, as with all design-based methodologies, guides the parameters around research and programmatic practice, most notably for answering what work is being done where, and for whom (Means & Penuel, 2005). Design-based research methodologies focus on iteratively assessing and designing macro-dynamic systems (e.g., instructional organization, coordination) that relate to learners' micro-dynamic interactions (e.g., situated cognition, affective engagement). Thus, design research serves to establish the *goals* of organizing research and implementation, whereas existing quantitative and qualitative research methods are used for sampling, data collection, and data analysis. Iteratively assessing the data is necessary for re-theorizing and re-designing programmatic intervention. For example, while the CRD principles enabled us to outline our goals for the service learning course design, quantitative and qualitative methods were used to inform how we may improve our design and implementation. Thus, we facilitated focus group discussions with stakeholders to understand what fosters and hinders their creativity in planning EDCs, how they have traditionally engaged with mathematics and science concepts, and how career applications could support all stakeholders' understanding of the real-world application of STEM. We triangulated this data with artifacts such as storyboards of narratives and reflections about implementation to examine the processes of co-designing culturally specific engineering design challenges that incorporate nondominant learners' worldviews. We also observed and transcribed videos to reflect on teachers' and engineers' engagement, commitment, and questions--all valuable indicators of the sustainability of *E-Communities*. These methods and data points helped assess how we may iteratively redesign equity-oriented learning inputs rather than the traditional technicist outputs.

DISCUSSION

As an emerging methodology, CRD principles helped us to define our design and implementation goals around iteratively disrupting and developing equitable learning environments, like the service-learning program described above, to meet the needs of various nondominant stakeholders. In this chapter, we specifically focused on engineers as stakeholders in an education community. Based on their experiences with one iteration of a 16-hour service-learning online course, engineers provided necessary feedback that led us to revise how our programmatic intervention could streamline their collaboration with teachers while simultaneously aiding nondominant engineers' and students' identity building. Our CRD methodology positioned us to continuously modify our goals around programmatic design and instructional strategies to ensure we attended to the place, race, and space of stakeholders in an effort to empower the most marginalized learners.

While CRD draws on ideas that evolved from *E-Communities* and previous education research methodologies with minoritized participants, the emergence of CRD was not a choice but a necessity, given the context of our research in and *with* underserved communities of color. Just as past critical and transformative methodologies were formed out of necessity, we found ourselves needing to forge a pathway to designing and critically analyzing our programmatic interventions with few methodological models to follow (Christ, 2013). When problems in design arose, CRD allowed us to interpret learning contexts that are transgressive in nature, be open to doubts and uncertainties, and keep goals and methods transparent, authentic, and reflexive. We propose that a strength of CRD is that it draws both theoretically and methodologically on equity-centered approaches, with tenets of CRT and DBIR situating how we seek to critically unpack, iteratively disrupt, effectively analyze, and sustainably transform education research performed within nondominant communities. Further, the emerging methodology links theory to action and shows how we privilege race as a core driver for identifying inequitable racialized learning experiences (Ladson-Billings & Tate, 1995). It also privileges design as a tool that must be iterative, collaborative, and contextually-based to develop sustainable changes (Lee & Krajcik, 2012). Thus, CRD as a methodology attends to nondominant learners' broader social contexts, by recognizing their marginalization, socialization, and multiple identity formations in their learning ecologies (Solorzano & Yosso, 2001) while simultaneously affirming community wealth and funds of knowledge (González, Moll, & Amanti, 2005; Yosso, 2005). Thus, the iterative-ness of CRD can afford stakeholders the opportunities needed to co-identify problems, co-design interventions, co-analyze data solutions, and co-construct changes by carefully documenting decisions following each cycle to explore how critical research and design may inform the re-contextualization and re-presentation of minoritized learning ecologies.

IMPLICATIONS

Implications for Research

Similar to other equity-oriented participatory forms of social science research, CRD stakeholders are invested in both the research and enactment of both unpacking mechanisms of oppression and pursuing transformative equitable outcomes. This approach to research represents a response to calls in recent equity literature for more work that uses "critical theory located in a transformative paradigm that employs multiple methods in order to create a more equitable and socially just education enterprise" (Parsons, 2014, p. 182). For example, the critical positionality of counter storytelling is to reassert the value of minoritized communities' experiences that have long been suppressed in white ethnocentric structures. However, the inevitable tensions that may be posed by two seemingly opposite methodologies, one critical and the other transformative, can perhaps be reconciled by contextualizing the multiple nested systems of learning ecologies CRD-researchers need to be iteratively disrupting and transforming.

Thus, perhaps one implication for research is to consider how transformative research methodologies can be translated into "iterative"-formative methods across the layers of learning ecologies, as they are not "fixed" spaces with "fixed" stakeholders to be transformed, but are actually continuously changing spaces with ever changing stakeholders. Similar to the development of CRT in education over the last 22 years that has identified inequitable structures and practices, we believe CRD is a working methodology that will ignite further critical race design researchers to engage with it; particularly as there is a need to design more spaces that critically disrupt structures with more praxes that can attend to the lived experiences of various minoritized groups. In other words, because learning ecologies occur across various settings (Penuel et al., 2013), further teaching and learning research paradigms need to be moved from the 'situational' context of classrooms and schools to include stakeholders (including researchers) from community and public pedagogy spaces including virtual communities.

Further, as with all critical emergent methods, we believe that critically designing and implementing strategies that attend to nondominant communities across various settings requires that we, in teacher education research, must move beyond researching in theoretical silos to be more transdisciplinary, interdisciplinary, and participatory in the spaces we design so we can attend to various political, economic, and organizational dimensions that influence research. Our examples in this chapter illustrate how both CRT and DBIR directed our research and design in context specific interventions. Thus the delineations between traditional "divisions" in education research paradigms need to be more systemically connected, such that "across-boundary collaboration" can improve and focus "networked improvement communities" (Dolle, Gomez, Russell, & Bryk, 2013 as cited in Penuel et al., 2013) that can empower stakeholders to not only create spaces in schools, but to reconsider what education is.

Implications for Practice

The process of designing *E-Communities* through a CRD methodology has implications to founding other critical and transformative methodologies that can inform nondominant students' learning environments. While systems of power and privilege are inevitable, it is important for those interested in developing equitable, learner-centered programs to critically examine which stakeholders must be present to create programs that are change agents and that are sustainable (Briars, Larson, Strutchens, & Barnes, 2015). For example, when designing and re-focusing career and technical education programs, it is necessary to convene a diverse sample of educators, administrators, parents, learners, and employers to the proverbial table to identify real issues of access and equity in careers and career-related pathways (Archer, Dewitt, Osborne, 2015; Zaniewski & Reinholz, 2016). We recommend that researchers and program designers be a part of the community, or perform thorough research of the community, to identify community assets and resources. Community wealth and knowledge is essential to facilitating conversations between stakeholders at times when interests must converge for the sake of fair and equitable education (Davis & Veenstra, 2014). We also believe that time and care need to be placed into designing spaces for information, stories, and community history to be shared among nondominant stakeholders. Without the personal narratives of nondominant stakeholders, it is nearly impossible to articulate true challenges or begin to consider how to dismantle areas of inequity (Swap, Leonard, & Shields, 2001). In vocational education, it is important for learners to be empowered by receiving genuine support from stakeholders when identifying career-related resources. This cannot be achieved by educators alone. CRD demands that stakeholders, like educators or professionals from the community, step outside of their silos and reconceptualize challenges with true input from diverse perspectives. Likely through this model, stakeholders could develop better ways for minoritized youth to contextualize critical thinking skills that could directly benefit the community in which they live. We argue that CRD is a working methodology and must be used to attend to the processes and inputs of design, rather the immediate solution or outputs. Like the engineering design process, these inputs need to be prototyped and tested. More importantly, stakeholders must be willing to redesign programmatic elements when they are not working. This can be most easily achieved, again, by attending to smaller inputs of the design and refining them before scaling up. CRD embraces a continuous cycle of formative assessment to ascertain if elements of programmatic design are working in the context that it is being implemented and if the program is serving its intended stakeholder. While this study focused on designing an equitable learning space for urban middle school learners in STEM, the principals of CRD can be applied to diverse communities, nondominant learners, and areas of study.

LIMITATIONS AND FUTURE RESEARCH

CRD also has inherent challenges, as it demands that design, implementation, and research be performed through critical and transformative worldviews. As we explained earlier, theoretically, critical research paradigms do not necessarily include the possibilities of transformative research paradigms. Methodologically, similar to other participatory- and action-oriented research methodologies, the challenge is to maintain an authentic balance between scientific thinking, critical positioning, and continuous self-reflexivity. Further, quality design-based research is characterized as having “a commitment to theory construction and explanation while solving real-world problems” (Reeves, Herrington, & Oliver, 2005, p. 103). The theory of design may not easily translate to the theory of implementation, hence the need to continuously re-assess and re-strategize. Logistically, this poses a need for adequate time and resources to properly identify and re-identify the multiple layers of structural inequities, co-design programmatic details to continuously disrupt and address them, and implement learning strategies and data collection methods in real time. Sabelli and Dede (2013) explain these limitations (time and resources) can be compounded when the goal of the research is for equitable access to learning spaces, thus stakeholders must commit to setting a realistic timeline for goals and many opportunities for formative assessment. Further limitations include recruiting and retaining community stakeholders in service learning spaces. Thus, in proposing CRD as an anti-racist methodology to design-based implementation research, we recognize the limits of reconciling critical and transformative paradigms, of aligning theories of design with theories of action, and of sustaining capacity and scalability with continuous reiteration.

Future research is needed to understand best practices and approaches for leveraging the expertise of community professionals in industry and government and bridging them to spaces in education. We need to continue to develop and test approaches that most effectively capture and share nondominant stakeholders' community wealth. We hope to inspire other scholars working with minoritized populations to examine *why and how* they are disrupting education inequities and constructing targeted solutions in order to best inform the design of their research and implementation that may best serve nondominant communities. It ensures that educational researchers, program developers, and practitioners continuously identify, confront, and apply the views of stakeholders, the contextual history, and the necessary design for successful program implementation. For research and action in any learning context, it is important that we, as educational researchers, remain critically conscious of our learners' worldviews and needs. CRD as an emergent methodology calls for the inclusion of nondominant learners' experiences, as education stakeholders often overlook this when shaping learning ecologies and defining measurable outcomes. It is important that researchers recruit and engage local stakeholders who understand the marginalized contexts of nondominant learners commit to designing equitable educational contexts that build on nondominant students' strength-based assets.

ACKNOWLEDGMENT

This research was funded by an ITEST grant (2015-2018) from the National Science Foundation (Award # 1510347). The opinions expressed are those of the authors and do not represent the views of the National Science Foundation, Howard University, or College of William and Mary. The authorship of this manuscript is credited equally to both co-first authors. Each author equally contributed toward its conceptualization, data collection, analysis, and writing. For questions regarding the manuscript, please contact Dr. Deena Khalil, deena_khalil@ howard.edu or Dr. Meredith Kier, mwkier@wm.edu.

REFERENCES

Anderson, T., & Shattuck, J. (2012). Design-based research: A decade of progress in education research? *Educational Researcher, 41*(1), 16–25. doi:10.3102/0013189X11428813

Anyon, J. (1980). Social class and the hidden curriculum of work. *Journal of Education, 162*(1), 67–92. doi:10.1177/002205748016200106

Archer, L., Dewitt, J., & Osborne, J. (2015). Is science for us? Black students' and parents' views of science and science careers. *Science Education, 99*(2), 199–237. doi:10.1002ce.21146 PMID:28579645

Baber, L., Fletcher, R., & Graham, E. (2015). Black Men Attending Community Colleges: Examining an institutional approach toward equity. *New Directions for Community Colleges, 172*(172), 97–107. doi:10.1002/cc.20167

Battey, D. (2013). "Good" mathematics teaching for students of color and those in poverty: The importance of relational interactions within instruction. *Educational Studies in Mathematics, 82*(1), 125–144. doi:10.100710649-012-9412-z

Bell, D. (1980). *Shades of Brown: New perspectives on school desegregation.* Teachers College Press.

Bell, D. (2004). *Silent covenants: Brown v. Board of Education and the unfulfilled hopes for racial reform.* Oxford University Press.

Briars, D. J., Larson, M., Strutchens, M. E., & Barnes, D. (2015). A Call for Mathematics Education Colleagues and Stakeholders to Collaboratively Engage with NCTM: In Response to Martin's Commentary. *Journal of Urban Mathematics Education, 8*(2), 23–36.

Brown, B. A., Henderson, J. B., Gray, S., Donovan, B., Sullivan, S., Patterson, A., & Waggstaff, W. (2016). From description to explanation: An empirical exploration of the African-American pipeline problem in STEM. *Journal of Research in Science Teaching, 53*(1), 146–177. doi:10.1002/tea.21249

Christ, T. W. (2013). The worldview matrix as a strategy when designing mixed methods research. *International Journal of Multiple Research Approaches, 7*(1), 110–118. doi:10.5172/mra.2013.7.1.110

Cobern, W. W. (1996). Worldview theory and conceptual change in science education. *Scientific Literacy and Cultural Studies Project*, 15. http://scholarworks.wmich.edu/cgi/viewcontent.cgi?article=1014&context=science_slcsp

Common Core State Standards Initiative. (2010). *Common Core State Standards for Mathematics*. National Governors Association Center for Best Practices and the Council of Chief State School Officers.

Crenshaw, K. W. (1988). Race, reform, and retrenchment: Transformation and legitimation in anti-discrimination law. *Harvard Law Review, 101*(7), 1331–1387. doi:10.2307/1341398

Crenshaw, K. W. (1995). Mapping the margins: Intersectionality, identity politics, and violence against women of color. In K. Crenshaw, N. Gotanda, G. Peller, & K. Thomas (Eds.), *Critical race theory: The key writings that formed the movement* (pp. 357–383). The New Press.

Darling-Hammond, L., Wilhoit, G., & Pittenger, L. (2014). Accountability for College and Career Readiness: Developing a New Paradigm. *Education Policy Analysis Archives, 22*(86), 1–38.

Davidson, M. N., & Foster-Johnson, L. (2001). Mentoring in the preparation of graduate Researchers of color. *Review of Educational Research, 71*(4), 549–574. doi:10.3102/00346543071004549

Davis, D., & Veenstra, C. (2014). Community Involvement in STEM Learning. *Journal for Quality and Participation, 37*(1), 30–33.

Delgado, R. (1989). Storytelling for oppositionists and others: A plea for narrative. *Michigan Law Review, 87*(8), 2411–2441. doi:10.2307/1289308

Delgado, R., & Stefancic, J. (2012). *Critical race theory: An introduction*. NYU Press.

Denner, J., Cooper, C. R., Lopez, E. M., & Dunbar, N. (1999). Beyond "Giving Science Away": How University Community Partnerships Inform Youth Programs, Research, and Policy. *Society for Research in Child Development, 8*(1), 1–20. doi:10.1002/j.2379-3988.1999.tb00010.x

Denson, C. D., Avery, Z. K., & Schell, J. W. (2010). Critical inquiry into urban African-American students' perceptions of engineering. *Journal of African American Studies, 14*(1), 61–74. doi:10.100712111-009-9107-4

Denzin, N. K., & Lincoln, Y. S. (2008). *The landscape of qualitative research* (Vol. 1). Sage Publications.

Diversi, M. (1998). Glimpses of street life: Representing lived experience through short stories. *Qualitative Inquiry, 4*(2), 131–147. doi:10.1177/107780049800400201

Dixson, A. D., & Rousseau, C. K. (2005). And we are still not saved: Critical race theory in education ten years later. *Race, Ethnicity and Education*, *8*(1), 7–27. doi:10.1080/1361332052000340971

Dolle, J. R., Gomez, L. M., Russell, J. L., & Bryk, A. S. (2013). More than a network: Building professional communities for educational improvement. *National Society for the Study of Education Yearbook*, *112*, 443–463.

Donnor, J. (2013). *Education as the property of whites: African Americans continued quest for good schools. In Handbook of critical race theory in education*. Routledge.

Dym, C. L., Agogino, A. M., Eris, O., Frey, D. D., & Leifer, L. J. (2005). Engineering design thinking, teaching, and learning. *Journal of Engineering Education*, *94*(1), 103–120. doi:10.1002/j.2168-9830.2005.tb00832.x

Eccles, J. (2009). Who am I and what am I going to do with my life? Personal and collective identities as motivators of action. *Educational Psychologist*, *44*(2), 78–89. doi:10.1080/00461520902832368

Elementary and Secondary Education Act of 1965. (n.d.). *The Social Welfare History Project*. Retrieved from https://socialwelfare.library.vcu.edu/programs/education/elementary-and-secondary-education-act-of-1965/

Fishman, B. J., Penuel, W. R., Allen, A. R., Cheng, B. H., & Sabelli, N. O. R. A. (2013). Design-based implementation research: An emerging model for transforming the relationship of research and practice. *National Society for the Study of Education*, *112*, 136–156.

Gamoran, A., & Hannigan, E. C. (2000). Algebra for everyone? Benefits of college-preparatory mathematics for students with diverse abilities in early secondary school. *Educational Evaluation and Policy Analysis*, *22*(3), 241–254. doi:10.3102/01623737022003241

Gándara, P. (2006). Strengthening the academic pipeline leading to careers in math, science, and technology for Latino students. *Journal of Hispanic Higher Education*, *5*(3), 222–237. doi:10.1177/1538192706288820

Gardner, D. P., Larsen, Y. W., Baker, W., Campbell, A., & Crosby, E. A. (1983). *A nation at risk: The imperative for educational reform*. United States Department of Education.

Gasman, M., & Commodore, F. (2014). The state of research on historically Black colleges and universities. *Journal for Multicultural Education*, *8*(2), 89–111. doi:10.1108/JME-01-2014-0004

Gay, G. (2010). *Culturally responsive teaching: Theory, research, and practice*. Teachers College Press.

Gillborn, D. (2013). The policy of inequity. Using CRT to unmask White supremacy in education policy. Handbook of critical race theory in education, 129-139.

González, N., Moll, L. C., & Amanti, C. (Eds.). (2005). *Funds of knowledge: Theorizing practices in households, communities, and classrooms*. Routledge.

Harper, S. R. (2010). An anti-deficit achievement framework for research on students of color in STEM. *New Directions for Institutional Research, 148*(148), 63–74. doi:10.1002/ir.362

Herrington, D., & Daubenmire, P. L. (2016). No Teacher Is an Island: Bridging the Gap between Teachers' Professional Practice and Research Findings. *Journal of Chemical Education, 93*(8), 1371–1376. doi:10.1021/acs.jchemed.5b00700

Horsford, S. D. (2010). Mixed feelings about mixed schools: Superintendents on the complex legacy of school desegregation. *Educational Administration Quarterly, 46*(3), 287–321. doi:10.1177/0013161X10365825

Howard, T. C. (2003). Culturally relevant pedagogy: Ingredients for critical teacher reflection. *Theory into Practice, 42*(3), 195–202. doi:10.120715430421tip4203_5

Howard, T. C. (2015). *Why race and culture matter in schools: Closing the achievement gap in America's classrooms*. Teachers College Press.

Kier, M. W. (2013). *Examining the effects of a STEM career video intervention on the interests and STEM professional identities of rural, minority middle school students*. Dissertation Study. North Carolina State University.

Kovach, M. (2005). Emerging from the margins: Indigenous methodologies. In L. Brown & S. Strega (Eds.), *Research as resistance: Critical, Indigenous, & anti-oppressive approaches* (pp. 19–36). Canadian Scholars' Press.

Ladson-Billings, G. (2005). What's the matter with the team? Diversity in teacher education. *Journal of Teacher Education, 56*(3), 229–234. doi:10.1177/0022487105275917

Ladson-Billings, G., & Tate, I. V. W. (1995). Toward a critical race theory of education. *Teachers College Record, 97*(1), 47–68.

Lee, O., & Krajcik, J. (2012). Large-scale interventions in science education for diverse student groups in varied educational settings. *Journal of Research in Science Teaching, 49*(3), 271–280. doi:10.1002/tea.21009

Leonard, J. (2008). *Culturally specific pedagogy in the mathematics classroom*. Routledge.

Leonardo, Z. (2002). The souls of white folk: Critical pedagogy, whiteness studies, and globalization discourse. *Race, Ethnicity and Education*, *5*(1), 29–50. doi:10.1080/13613320120117180

Lewin, K. (1997). *Self-hatred among Jews (1941)*. American Psychological Association. doi:10.1037/10269-012

Lin, X., Schwartz, D. L., & Hatano, G. (2005). Toward teachers' adaptive metacognition. *Educational Psychologist*, *40*(4), 245–255. doi:10.120715326985ep4004_6

Loxley, A., & Seery, A. (2008). *Some philosophical and other related issues of insider research. In Researching education from the inside*. Routledge.

Lucas, S. (1999). *Tracking inequality: Stratification and mobility in American schools*. Teachers College Press.

Lynn, M. (2004). Inserting the 'Race' into Critical Pedagogy: An analysis of 'race-based epistemologies'. *Educational Philosophy and Theory*, *36*(2), 153–165. doi:10.1111/j.1469-5812.2004.00058.x

Lynn, M., & Parker, L. (2006). Critical race studies in education: Examining a decade of research on U.S. schools. *The Urban Review*, *38*(4), 257–290. doi:10.100711256-006-0035-5

Means, B., & Harris, C. J. (2013). Towards an Evidence Framework for Design-Based Implementation Research. *Yearbook of the National Society for the Study of Education*, *112*, 350–371.

Means, B., & Penuel, W. R. (2005). Research to support scaling up technology-based educational innovations. *Scaling up success: Lessons from technology-based educational improvement*, 176-197.

Milner, H. R. IV. (2008). Critical race theory and interest convergence as analytic tools in teacher education policies and practices. *Journal of Teacher Education*, *59*(4), 332–346. doi:10.1177/0022487108321884

Parsons, E. C. (2014). Unpacking and critically synthesizing the literature on race and ethnicity in science education. Handbook of research in science education, 167-183.

Paul, F. (2005). Grouping within algebra 1: A structural sieve with powerful effects for low-income, minority, and immigrant students. *Educational Policy*, *19*(2), 262–282. doi:10.1177/0895904804274056

Penuel, W. R., Coburn, C. E., & Gallagher, D. J. (2013). Negotiating problems of practice in research–practice design partnerships. *Yearbook of the National Society for the Study of Education, 112*, 237–255.

Penuel, W. R., & Fishman, B. J. (2013). Large-scale intervention research we can use. *Journal of Research in Science Teaching, 49*, 281–304. doi:10.1002/tea.21001

Penuel, W. R., Fishman, B. J., Cheng, B. H., & Sabelli, N. (2011). Organizing research and development at the intersection of learning, implementation, and design. *Educational Researcher, 40*(7), 331–337. doi:10.3102/0013189X11421826

Ravitch, D. (2004). *The language police: How pressure groups restrict what students learn*. Vintage.

Ravitch, D. (2010). *The death and life of the great American school system: How testing and choice are undermining education*. Basic Books.

Reeves, T. C., Herrington, J., & Oliver, R. (2005). Design research: A socially responsible approach to instructional technology research in higher education. *Journal of Computing in Higher Education, 16*(2), 96–115. doi:10.1007/BF02961476

Sabelli, N., & Dede, C. (2013). Empowering design-based implementation research: The need for infrastructure. *Yearbook of the National Society for the Study of Education, 112*, 464–480.

Sassen, K. (1991). The polarization lidar technique for cloud research: A review and current assessment. *Bulletin of the American Meteorological Society, 72*(12), 1848–1866. doi:10.1175/1520-0477(1991)072<1848:TPLTFC>2.0.CO;2

Small, M. L. (2011). How to conduct a mixed methods study: Recent trends in a rapidly growing literature. *Sociology, 37*(1), 57–86. doi:10.1146/annurev.soc.012809.102657

Solórzano, D. G., & Yosso, T. J. (2001). From racial stereotyping toward a critical race theory in teacher education. *Multicultural Education, 9*, 2–8.

Solórzano, D. G., & Yosso, T. J. (2002). Critical race methodology: Counter-storytelling as an analytical framework for education research. *Qualitative Inquiry, 8*(1), 23–44. doi:10.1177/107780040200800103

Stage, E. K., Asturias, H., Cheuk, T., Daro, P. A., & Hampton, S. B. (2013). Opportunities and challenges in next generation standards. *Science, 340*(6130), 276–277. doi:10.1126cience.1234011 PMID:23599466

Stanton, T. K., Giles, D. E. Jr, & Cruz, N. I. (1999). *Service-Learning: A Movement's Pioneers Reflect on Its Origins, Practice, and Future*. Jossey-Bass.

Stinson, D. W. (2008). Negotiating sociocultural discourses: The counter-storytelling of academically (and mathematically) successful African American male students. *American Educational Research Journal, 45*(4), 975–1010. doi:10.3102/0002831208319723

Stovall, D. (2006). Forging community in race and class: Critical race theory and the quest for social justice in education. *Race, Ethnicity and Education, 9*(3), 243–259. doi:10.1080/13613320600807550

Supovitz, J. A. (2008). Implementation as iterative refraction. In J. A. Supovitz & E. H. Weinbaum (Eds.), *The implementation gap: Understanding reform in high schools* (pp. 151–172). Teachers College Press.

Swap, W., Leonard, D., & Mimi Shields, L. A. (2001). Using mentoring and storytelling to transfer knowledge in the workplace. *Journal of Management Information Systems, 18*(1), 95–114. doi:10.1080/07421222.2001.11045668

Syed, M., Goza, B. K., Chemers, M. M., & Zurbriggen, E. L. (2012). Individual Differences in Preferences for Matched-Ethnic Mentors Among High Achieving Ethnically Diverse Adolescents in STEM. *Child Development, 83*(3), 896–910. doi:10.1111/j.1467-8624.2012.01744.x PMID:22506789

Tate, W. (2001). Science education as a civil right: Urban schools and opportunity-to-learn considerations. *Journal of Research in Science Teaching, 38*(9), 1015–1028. doi:10.1002/tea.1045

Tate, W. F. (1997). Race-ethnicity, SES, gender, and language proficiency trends in mathematics achievement: An update. *Journal for Research in Mathematics Education, 28*(6), 652–679. doi:10.2307/749636

Tatum, B. (2003). *Why Are All the Black Kids Sitting Together in the Cafeteria?* (Rev. ed.). Basic Books.

Tierney, W. (1993). *Building communities of difference: Higher education in the twenty-first century*. Bergin and Garvey.

Tsui, L. (2007). Effective strategies to increase diversity in STEM fields: A review of the research literature. *The Journal of Negro Education*, 555–581.

Tyack, D. B. (1974). *The one best system: A history of American urban education.* Harvard University Press. doi:10.2307/j.ctv136c61j

Wenger, E. (2006). *Communities of practice a brief introduction.* http://www.ewenger.com/theory/

Wheeler, L. B., Whitworth, B. A., & Gonczi, A. L. (2014). Engineering Design Challenge. *Science Teacher (Normal, Ill.)*, *81*(9), 30–36. doi:10.2505/4/tst14_081_09_30

Yosso, T. J. (2002). Toward a critical race curriculum. *Equity & Excellence in Education*, *35*(2), 93–107. doi:10.1080/713845283

Yosso, T. J. (2005). Whose culture has capital? A critical race theory discussion of community cultural wealth. *Race, Ethnicity and Education*, *8*(1), 69–91. doi:10.1080/1361332052000341006

Yosso, T. J., Parker, L., Solorzano, D. G., & Lynn, M. (2004). From Jim Crow to affirmative action and back again: A critical race discussion of racialized rationales and access to higher education. *Review of Research in Education*, *28*(1), 1–25. doi:10.3102/0091732X028001001

Zaniewski, A. M., & Reinholz, D. (2016). Increasing STEM success: A near-peer mentoring program in the physical sciences. *International Journal of STEM Education*, *3*(1), 1–12. doi:10.118640594-016-0043-2

Chapter 10
Participatory Action Research in Adult Education:
Methods in Vivencia, Praxis, and Conscientization

Meagan Call-Cummings
George Mason University, USA

Melissa Hauber-Özer
https://orcid.org/0000-0001-6621-1204
George Mason University, USA

ABSTRACT

Participatory action research (PAR) is an embodied form of inquiry that engages those most affected by an issue or problem in creating knowledge and developing solutions. PAR epistemology intersects with a critical approach to adult education in its belief that programs, methods, and content must be relevant to learner needs and challenges and ought to lead to greater social justice. The purpose of this chapter is to offer a review of three critical, participatory inquiry methods that are connected to the ontological and epistemological anchors of PAR. The authors present readers with a useful description of how to enact these onto-epistemological anchors through these methods in diverse contexts. They conclude that these methods have great potential for critical educators to live out their own onto-epistemological commitments, better understand and meet learner needs, and facilitate positive social change.

DOI: 10.4018/978-1-7998-7600-7.ch010

INTRODUCTION

Participatory action research (PAR) is an embodied form of inquiry that engages those most affected by an issue or problem in creating knowledge and developing solutions (Gaventa, 1991; Wood & McAteer, 2017). As such, it explicitly seeks to upend traditional/positivist research paradigms and hierarchies, privileging the knowledge of the community – particularly those marginalized and oppressed by dominant social structures (Freire, 1990) in "a shared process of discovery" (Glassman & Erdem, 2014, p. 207). This process calls for a "critical and humanizing pedagogy" (Wood & McAteer, 2017, p. 253) entailing "inclusion, democracy, and respect for local knowledge emanating from lived experience of the specific context and culture" (p. 252). The goal of this pedagogy is not to reproduce the status quo but to democratize knowledge and society (Wood & McAteer, 2017).

PAR proponents attest that research must be conducted and disseminated with an in-depth understanding of the social and political context, which requires consulting local community members (Glassman & Erdem, 2014). This approach seeks not just to document human experience but "to improve the human condition" (Glassman & Erdem, 2014, p. 214) through problem solving and, ultimately, meaningful social change (Wood & McAteer, 2017). The resulting transformation must be horizontal, from and for the people, not from above, and power relations must be fundamentally reshaped rather than simply shifting the oppressed to the position of oppressor (Glassman & Erdem, 2014).

These ideas and ideals held by PAR practitioners are at once ontological, in that they connect to one's understanding of what can be known or learned, and epistemological, as they are rooted in one's conceptualization of how we know or come to know. PAR, then, is more than simply a method that can be transferred from one research design to another with little to no thought as to why it is engaged. Rather, PAR is an ontological and epistemological stance because it represents a major ontological and epistemological challenge to traditional, positivist approaches to knowledge creation (Cahill et al., 2019; Call-Cummings et al., 2019; Kindon et al., 2007; Torre & Ayala, 2009). Those who practice PAR take a firm stand against neoliberal knowledge production processes that, intentional or not, serve to re-produce and re-present systems and structures of dispossession, injustice, inequity, and violence.

Core Concepts

PAR is grounded in the core ontological and epistemological concepts of *vivencia*, *praxis*, and *conscientization* (Fals-Borda & Rahman, 1991). *Vivencia* goes beyond observation or documentation to sharing in the lived reality of community members, "a full experience of an event with its all possibilities, lived through direct participation" (Glassman & Erdem, 2014, p. 212). *Praxis* stems from this embodied experience, practical action rooted in reflection on "the conditions one faces in order to change them" (Freire, 1970, p. 33). It is flexible and responsive to local conditions with the goal of "liberation and [the] path to freedom" (Glassman & Erdem, 2014, p. 212). Freire (1970) defines research as *conscientization*, the "process in which men, not as recipients, but as knowing subjects achieve a deepening awareness both of the socio-cultural reality which shapes their lives and of their capacity to transform reality" (p. 27). Vio Grossi (1981, 1982) calls this *disindoctrination* – rejecting the systems and status quo imposed on oppressed groups.

Origins of PAR

Kurt Lewin (1946) is credited with developing action research (AR) in the 1940s as an attempt to reduce social conflict toward minority groups in the United States, and half a century later, William Foote Whyte (1994) designed a form of participatory action research to address problems in organizational structures (Glassman & Erdem, 2014). However, a very different strain of PAR surfaced in the Global South during the turbulent 1960s and 1970s, aiming to "throw off the intellectual, social, and material shackles of colonialism" (Glassman & Erdem, 2014, p. 207). This revolutionary, grassroots resistance to centuries of oppression sought not to restore social order but to change it from below; it emerged out of local communities from Colombia to India rather than under the leadership of one central figure (Glassman & Erdem, 2014; Pyrch, 2007).

Two Latin American intellectuals did, however, play key roles in developing and popularizing this liberatory method (Pyrch, 2007). Brazilian Paulo Freire's radical work in adult literacy education (Glassman & Erdem, 2014) and development of critical pedagogy (Freire, 1970) is often credited as crucial to PAR's onto-epistemological basis. Likewise, Colombian Orlando Fals Borda (1988, 2006) was instrumental in developing this "methodology of liberation" (Pyrch, 2007, p. 200) through which we can "investigate reality in order to transform it" (p. 203, citing Fals Borda, 1979). This is just as relevant in the 21st century in the face of globalization, neoliberalism, and unending war (Pyrch, 2007).

As a form of action research, PAR has a cyclical nature, consisting of a flexible process of data gathering, reflection, and action (Glassman & Erdem, 2014). Silver-Pacuilla (2004) describes this process, as codes emerge both from fieldwork and dialogic engagement with participants, themes are "spiraled" back to participants for analysis during ongoing data collection, and findings are presented to stakeholders to prompt change. PAR also fundamentally redefines the relationship between researcher and researched: the researcher acts as one of many co-researchers, becoming a co-participant in the lives of community members (Glassman & Erdem, 2014). The community, rather than the researcher, maintains ownership of data and shares equally in decisions about how to gather, analyze, and disseminate this knowledge (Chambers, 2002). Within the collective, university researcher(s) – often White and middle-class – and marginalized community co-researchers must negotiate power relations to enable authentic collaboration; however, privileged researchers can use their standing to bring attention to the knowledge of community members (Wood & McAteer, 2017).

THE SHARED COMMITMENTS OF PAR AND ADULT EDUCATION

PAR onto-epistemology intersects with the critical approach to adult education, particularly the belief that programs, methods, and content must be relevant to learner needs and challenges (Glassman & Erdem, 2014; Wood & McAteer, 2017) and ought to lead to greater social justice (Freire, 1970; Prins, 2007). This contrasts starkly with the common model of lifelong learning focused on preparing students for the labor market; instead of upholding the status quo, critical adult education seeks to transform hegemonic ideologies, social structures, and systems (Brookfield, 2018; Lucio-Villegas, 2018). Central to this tradition are Marxist ideology and Freire's critical pedagogy, which both demand accountability from the system and work to democratize education for the good of the learner rather than for those in power (Brookfield, 2018).

Critical adult education aims to increase learner autonomy and change future trajectories by facilitating active participation and drawing on bottom-up knowledge rather than transmitting 'approved' content from established authorities (Glassman & Erdem, 2014; Wood & McAteer, 2017). The educator takes the role of a facilitator of learning, encouraging critical reflection and advocating for the learners, rather than an objective source of knowledge (Brookfield, 2018). In addition, learners must develop a sense of solidarity and connectedness to identify shared challenges and work toward common goals (Prins, 2007). The authors of this chapter join numerous other scholars (e.g. Glassman & Erdem, 2014; Pyrch, 2007; Wells, 2014; Wood & McAteer, 2017) in asserting that PAR is an ideal way to gather and disseminate valuable community knowledge both *for* and *about* adult education.

PAR as Pedagogy in Adult Education

Freire's involvement with adult literacy campaigns in Northeast Brazil in the 1960s inspired his radical critical pedagogy (Lucio-Villegas, 2018), first outlined in *Pedagogy of the Oppressed* (1970). This pedagogy centers on dialogue, conscientization or awareness raising, and ultimate liberation from the oppressed/oppressor relationship (Lucio-Villegas, 2018). Critical pedagogy is often applied in adult education today through PAR praxis. For example, Gwon and Thongpriwan (2019) incorporated photovoice (Wang, 1999) into a university pre-nursing course to encourage students to reflect on their beliefs about the US healthcare system and consider other perspectives. Participants appraised this as an engaging, in-depth learning experience. Also in the discipline of healthcare, Janssen et al. (2013) find that PAR was a positive way for clinical physical therapists to learn about doing research.

PAR as Method in Adult Education

PAR has also been used extensively to conduct collaborative research *about* adult education, from adult literacy to ecology and cancer prevention. It is sometimes used as a method for needs analysis, as in Khachan's (2009) examination of Egyptian women's diglossic literacy needs. From the discipline of ecology, Ballard and Belsky (2010) demonstrate the use of PAR with a group of harvesters and land managers to investigate and solve an environmental problem affecting their livelihood (2010). Reporting on a participatory needs analysis conducted as part of a long-term partnership between a community literacy program and university writing center, Wells (2014) argues that collaborative, participant-driven inquiry is better suited to researching the "strategies, contexts, and constraints" (p. 51) of adult literacy programs than traditional methods.

PAR has also proven valuable for program and curriculum development in adult education. In Tulloch et al. (2017), PAR was used to develop, deliver, and evaluate an Indigenous language and literacy revitalization program for marginalized learners in Canada. The program was designed around a transformative bilingual education framework (Cummins, 2000) and a relational, empowering, and culturally responsive model built on participants' knowledge. Wood and McAteer (2017) describe a PAR process through which community members and teachers in South Africa designed an educational program for parents wanting to better support their children in school, while Cueva et al. (2018) worked with a group of Alaska Natives to develop and evaluate culturally relevant online learning modules about reducing risk factors and behaviors for cancer.

Several scholars have also demonstrated the value of PAR for program evaluation that includes the voices of adult learners and educators. For instance, Silver-Pacuilla (2004) engaged in PAR to learn about the challenges that women literacy learners with disabilities face in reaching their educational goals. The process spanned a year and included monthly focus groups, interviews, trust-building, reflection, and development of solutions. Participants shared their findings as recommendations to service providers, advice brochures for other women literacy learners, and public presentations (Silver-Pacuilla, 2004). Hanson (2014) reports on the use of PAR to learn about student experiences in a Canadian labor school for women which engages in feminist, Freirean transformative education, and Boughton and Durnan (2014) used PAR to evaluate mass adult literacy campaigns in Timor-Leste and Aboriginal Australia.

THREE PROMISING PARTICIPATORY INQUIRY METHODS

We have offered a quick yet thorough review of what participatory action research is and how it is anchored in the same onto-epistemological commitments to *vivencia*, *praxis*, and *conscientization* as critical adult education. *Vivencia*, or sharing in the lived experiences of those closest to a phenomenon, issue, or concern, is central to both PAR and critical adult education as the concept shifts our understanding of expertise from one that is exclusive and often disempowering to one that is inclusive and democratic. *Praxis*, or the interconnected process of action and reflection, connects PAR and critical adult education as it defines both processes as rooted in the reality of taking participant-drive, reflective action. Finally, *conscientization*, or awareness-raising, is at the heart of both PAR and critical adult education as both aim to move individuals or groups toward liberation from the oppressive structures, systems, and relationships that define injustice and inequity.

In the PAR and adult education courses we teach, we often hear the refrain, "Yes, we understand *what*, but *how*?" There have been many articles, chapters, and books dedicated to the *process* of participatory action research – the *what* (e.g. Fals-Borda, 1987; Kemmis et al., 2014; McIntyre, 2008). Some scholars or disciplines may use the term action research, others participatory action research, critical participatory action research, or community-based participatory research… the list goes on.

While we are grateful for those contributions and use them often in our own work, we are eager to move away from the silos of names and fields and to move back to the onto-epistemological roots of what we call participatory inquiry, which is inclusive of and welcoming to all inquiry traditions that place at a premium the experiences and knowledges of those closest to and most affected by an issue, problem, or phenomenon: that is, knowledge creation that deeply values and democratically engages the experiences and knowledges of those who are often strategically and systematically pushed to the margins of society. Built on the concepts of *vivencia*, *praxis*, and *conscientization*, participatory inquiry honors lived knowledge, engaging and leveraging it for meaningful social change. The remainder of this chapter will offer a review of three participatory methods that can be used both pedagogically and methodologically in any field (e.g., education, social work, or nursing) and for any substantive purpose, whether in the context of an adult education course, a research project on the treatment of those living in nursing homes, or a community organizing initiative around fair housing. They can be employed for both data collection and data analysis, as most often these processes are blurred.

These methods share a potential for facilitating the process of grassroots social change. We have used these tools in classrooms with adolescents, in church houses with adults, and in community settings with organizers[1]. Here, we describe three methods with particular potential for use in adult education settings. First, we discuss *Ripples of Change* (Chambers, 2002), which is an approach, rooted in the onto-epistemological concept of *vivencia*, to understanding the desires, goals, and experiences of a community as well as how the community sees change occurring. Second, we present *Priority Negotiations* (adapted from Chambers, 2002), which is an approach, rooted in the concept of *praxis*, to reflecting on, grappling with, and ultimately taking action around the priorities of multiple stakeholders related to social change. Finally, we offer *Force Mapping* (adapted from Chambers, 2002), which, rooted in the Freirean notion of *conscientization*, is a simple yet powerful way to facilitate the community's identification of sociocultural forces that both push and pull, help and hinder a movement for social change. All of these methods can be used as opportunities for blended data collection and analysis, as is common in participatory approaches to research, and are firmly anchored in a critical, participatory onto-epistemology and inquiry tradition that is at once pedagogically and methodologically rigorous.

Ripples of Change and *Vivencia*

Vivencia is grounded in the onto-epistemological concept of "empathetic involvement in processes and with the considered judgement of local reference groups" (Fals Borda, 2006, p. 33). Indeed, all of the methods we offer in this chapter fall under this concept, but we link it most closely with the activity Ripples of Change because of its potential as a first step toward illuminating, challenging, and transforming hierarchies of power anchored in racist, sexist, and xenophobic valuing of knowledge.

In this activity, a diverse research collective made up of all those who have a stake – or interest – in a particular community, issue, or problem is called to the proverbial table. For example, this collective could include a combination of adult learners, teachers, and administrators with or without the facilitation of university-based researchers. The authors of this chapter have done this with high school students and teachers concerned about school gun violence and often competing safety regulations and policies; community members, leaders, and policymakers involved in a local justice system working to better serve women of color; and a group of doctoral students hoping to transform their academic program into one that is empowering and supportive. It has been those groups – in which interests and agendas were the most diverse and the most contested – that this activity has yielded the most fruitful dialogue.

Once a group is gathered for this Ripples of Change activity, a facilitator draws three or four large, concentric circles in a place where all those who have gathered can have easy access (e.g., on a whiteboard). These circles represent the levels of change that members of the collective want to see related to their concern at various levels. These are referred to as "ripples" because, just like when a stone is thrown into a body of water, change occurs in ripples, often starting small or localized and then radiating outward to reach more people or to engage in broader ways. A few members of the collective are then asked to "label" these ripples or levels of change. It can be meaningful to ask those members of the collective who are the most marginalized or silenced in other situations to do this labeling in order to flip the traditional hierarchy of power on its head.

When using this method with high school students in an ongoing, long-term youth participatory action research (YPAR) project called *Courageous Conversations*, we, as facilitators, asked the student members of the collective to start our process[2]. We asked them to label the "ripples of change" that needed to occur to move the school community toward a desired end related to school gun violence. For them, the most immediate change became the innermost circle, which they labeled as individual change; that is, changes in individuals' understanding of and attitudes toward guns, gun violence, and gun regulations. They labeled the next circle moving outward as school-level change. For this research collective, members believed that if individuals within the school community (students, parents, teachers, and administrators) were to better understand the phenomenon of school gun violence, they could work to change school-level policies. Collective members labeled the largest, outermost circle "Community-level" and then, after some discussion, added a fourth circle and labeled it "Social/Policy change". As university researchers, we clarified with the students that they saw change starting from individuals changing their minds and then working as a group for school, community, and ultimately social (i.e., large-scale policy) change.

After labeling the ripples of change, all members of the collective were given index cards or note paper to record their individual ideas in response. In an attempt to flatten hierarchies of power once again, for our *Courageous Conversations* collective, we gave different colored papers to different categories of collective members. For example, students received blue papers, teachers yellow papers, and university researchers orange papers. We then gave ourselves a set amount of time to brain write[3] one idea or desired change per note paper and to place each in the circle that represented the appropriate level of change. The different colors facilitated a quick but useful validity check; for example, if everything under school-level change was orange, it was easy to tell that the university researcher members of the collective dominated those ideas. It was apparent that we needed to spend a little more time discussing the suggestions, questioning the assumptions that went into those ideas, and making sure there were not recommendations from other stakeholders that could contribute to our thinking.

Figure 1.

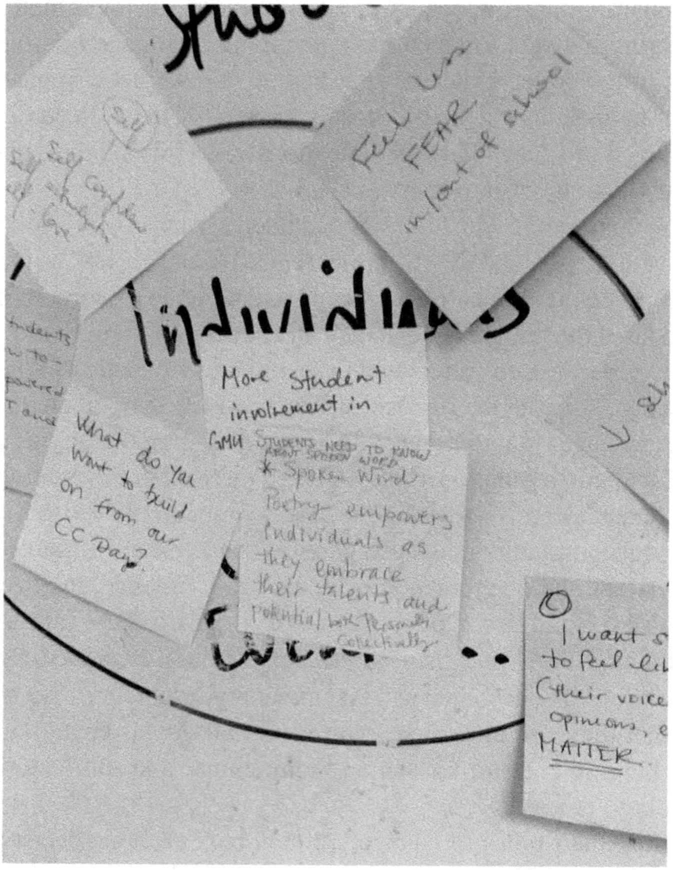

Once the allotted time was up, the same members of the collective who labelled the ripples or levels of change were asked to conduct a kind of thematic analysis, reading through all the note papers, placing similar papers or ideas next to each other, asking clarifying questions of the group, and, ultimately, identifying several main ideas to present to the rest of the group. The group engaged in dialogue to come to consensus about what changes were required and desired at each level of change. The changes our collective identified as necessary to achieving our ultimate goals became touchstones as the research collective moved on toward data gathering, analysis, and action. For example, after engaging in Ripples of Change, our data collection efforts focused more exclusively on how individual students saw and experienced changes in opinion about gun safety and gun regulations as they took part in dialogues ("Courageous Conversations") about recent and historical gun violence. As this project is ongoing, we have now come to a point where we have gathered and analyzed significant amounts of data focused on individual-level change that we are shifting our focus to school-level change – the second level or "ripple" of change identified by the collective as critical in the process of moving toward broad, social change related to gun violence.

Priority Negotiations and *Praxis*

The concept of *praxis* is used in many different ways and for varied means. Although often used as a kind of synonym for "practice," it is engaged in PAR as a critical, interconnected process of both self and collective action and reflection anchored in an intersubjective movement or reaching toward social change (McLaren, Ryoo, Crawford, & Moreno, 2010). Indeed, PAR's engagement with *praxis* is quite connected to Marx's use of the term, as he described the activities unique to humans through which we both create and change ourselves and our world (Brookfield, 2018).

The term *priority negotiations* can conjure up images of lawyers and teams of people with competing interests all vying for some limited amount of resources, but this is not at all the case here. A better term for this approach to *praxis* might be priority *dialogue*, for it centers the dialectical engagement Freire valued in his own work and writings (Brookfield, 2018; Lucio-Villegas, 2018). However, this engagement should be seen through the lens of critical pedagogy, which repositions traditionally dissimilar or even opposing groups – teachers and students – as equals and partners in collaborative, mutual learning (Brookfield, 2018; Lucio-Villegas, 2018).

Ideally, this activity should take place after some dialogic process like ripples of change has occurred and through which the multiple and potentially competing interests, ideas, and goals have been mapped. Once a research collective has arrived at a place where they agree on a vision for change, they should also tease out what their priorities are for moving toward that change. This activity has similar components as ripples of change in that it places a premium on dialogue. The goal is to come to an understanding of the various priorities of collective members, not to arrive at agreement or consensus. It is important to begin with the understanding that this consensus *may* happen in the process but is not the primary objective.

With the ripples of change or other idea mapping artifact placed where everyone can see it (often on a board in front of everyone or on the floor where people can walk around it), a few members of the collective are asked to rank desired changes as low, medium, or high priority. As in the ripples of change, it is usually beneficial to ask those who are typically in positions of less power or have fewer opportunities to voice their opinions to begin. This process might involve physically moving the original cards to another area or creating a list on sheets of paper. Facilitators may want to ask the other members of the research collective to list their own priorities as a way to compare or to start dialogue, but this is not necessary.

Once priority lists have been created, the group can engage in dialogue about what has been deemed low, medium, or high priority and why. It is important at this step for facilitators to help collective members engage in respectful and humble ways, so rules might need to be established. For example, each member might be given a set amount of time to voice their opinions, going around in a circle to make sure everyone gets an equal chance to speak, with an opportunity to "pass" if desired. Facilitators may also ask members to silently compare their priorities with those set by the smaller group and reflect on why there are differences before speaking. Alternatively, there may be a set of questions members of the collective are asked to reflect on or a fill-in-the-blank approach such as, "I appreciate the priority to ________________ because ______________;" followed by: "I think it's interesting that I did not prioritize ________________ because _________________:" or "I wonder if we should consider ______________."

Figure 2.

As we discussed earlier in this chapter, all of these activities are built on pedagogies and onto-epistemologies that value democratic knowledge creation, so it is useful to ensure facilitators are capable of engaging with respect for participants' knowledge and opinions and of encouraging respectful engagement among all collective members. Priority negotiation would be a valuable component of participatory needs assessment in determining with adult learners the order in which to address program design problems or incorporate curricular objectives. For example, the second author of this chapter, Melissa engaged in this process informally with a class of non-literate, beginner-level English learners who worked as custodians in university dormitories, using their first language, Spanish, so that they could express their ideas more easily. Although Melissa had assumed that establishing basics of print literacy should be the first priority, the learners expressed that learning how to respond to residents' greetings and questions would be more immediately useful. This was a simple but important form of praxis, listening to the learners' priorities and collaborating to adjust the curriculum according to what would begin to shift the power relations in their workplace.

Force Mapping and *Conscientization*

Grounded in post-Marxist critical theory and developed by Paulo Freire, the process of *conscientization* is one in which a person comes to an awareness of their own position in systems of power, oppression, and privilege; develops the ability to think critically about socio-cultural structures related to power; and builds the capacity to work toward the transformation of an oppressive reality. It can be difficult in the context of participatory inquiry to move through this process with a focus on identifying, teasing out, and confronting socio-cultural structures and systems of power because often research collectives are highly diverse. Members of the collective may identify as both oppressor and oppressed in various ways and this may cause anxiety about how or even if to engage in dialogue around difference, power, and oppression. For instance, in a class of doctoral students learning about PAR (see Call-Cummings et al., 2019), in which the first author of this chapter, Meagan, was the instructor and Melissa was a student, participants wrestled with the incongruence of their positions of privilege as highly educated individuals and the challenges they experienced related to gender, race, ethnicity, and ability. They questioned whether the problems they recognized in their PhD program warranted participatory inquiry since, all things considered, they could not really be considered oppressed.

Indeed, it can be tempting to gather a collective comprised of people who share similar identities and experiences as people may feel more at ease sharing concerns, voicing opinions, or working for change. While we do not argue that this approach is wrong (indeed, we have worked with and even formed participatory research collectives in these very ways), we encourage those interested in participatory inquiry to consider the potential for dialogue that lies in what Torre (2009) calls the contact zone: "a radically inclusive research site that would function as…a messy social space where very differently situated people could work together across their own varying relationships to power and privilege" (p. 110, citing Pratt 1992, p. 4). Ritterbusch (2019) discusses her experiences developing *sentipensar* (a kind of simultaneous thinking and feeling) as she engaged in PAR contact zones that brought together human rights activists, homeless drug users, and sex workers in Colombia. Cahill et al. (2019) present their experiences "resisting the 'school-to-sweatshop pipeline'" (p. 576) in Salt Lake City, Utah, suggesting that it was through the contact zone they intentionally created across "diverse standpoints within [their] multiracial intergenerational collective" that they "create[d] spaces of radical acceptance…holding the complexity of our interdependence" (p. 577) as necessary in their solidarity. Even those collectives that may seem relatively homogeneous, like the doctoral student class mentioned above, have divergent and intersectional experiences, identities, priorities, and needs that can lead to ambivalence or tension within the collective and can present opportunities for meaningful dialogue that moves the group toward a praxis of solidarity that, in turn, moves the group closer to goals of confronting and dismantling structural injustice.

Force mapping is an inquiry activity that can be useful to start or foster dialogue across difference – to enact one's onto-epistemological commitments to build an awareness and reach toward an understanding of people's varying experiences with systems of power and privilege, even in shared contexts. For this activity, members of the group are asked to identify forces that may play a role in the collective's work toward change. These forces may be abstract, such as systemic oppression, or highly concrete, such as limited funds for meetings. Through dialogue, the group works together to name these forces and to visually represent which way they push or pull the movement for change, as well as how strong the force is. Members draw on a whiteboard or other tool (like a large piece of paper) arrows that depict how the force pushes (against the movement or with/for it) and the strength of the force (larger arrows for stronger force, smaller arrows for weaker forces).

This process can help the group in several ways. First, as mentioned, it can build individual members' understanding of themselves and each other in relation to power and privilege, as well as other socio-cultural forces. Second, it has the potential to increase members' comfort and ability to articulate their thoughts on difficult, often taboo, topics. Relatedly, it can foster a dialogic space, which is a central component of successful participatory inquiry projects. Participatory research collectives often use varied and multimodal mapping strategies such as community mapping (Amsden & VanWynsberghe, 2005; Fang et al., 2006), body mapping (Sweet & Ortiz Escalante, 2017), and systems mapping (Saunders et al., 2016) as tools to better understand

and grapple with nuanced, multifaceted phenomena and to open up dialogue with stakeholders who often have disparate interests and agendas. Force mapping fits within this subset of dialogic participatory methods that are anchored by the vision of and commitment to *conscientization*.

This could have been a valuable exercise for the doctoral student collective. Over the semester, the participants engaged in ongoing, increasingly reflective dialogue regarding their personal and academic challenges, ultimately deciding that their concerns should be brought to the attention of administrators for the sake of other doctoral students. However, along the way a number of concerns emerged about managing expectations and possible repercussions that could come their way for critiquing the program. Visually mapping out the forces of power, the potential for change, and their roles in both could have aided in this process of *conscientization*, weighing the costs and benefits of conducting inquiry on their program, and deciding what types of data to collect and how to present it to stakeholders.

DISCUSSION AND CONCLUSION

Participatory inquiry is not easy. One does not just get out of bed one day and decide to try it out. It has been our experience that this is a long-term commitment and stance (Fals Borda & Rahman, 1991), one that comes with difficult decisions, ethical tensions, and very real negotiations of power and priorities. Both of us have, at more than one point in our academic lives, been told something to the effect of, "That's nice, but you have to get a job first." Or "Sounds great, but you need to make sure to set yourself up for tenure." We stand on the shoulders of Paulo Freire, Orlando Fals Borda, Gloria Anzaldúa, bell hooks, and many, many others to witness that staying true to our onto-epistemological commitments to *vivencia*, *praxis*, and *conscientization* as we take a stand against neoliberal knowledge production processes and for inclusive, dialogic knowledge creation is both trying and triumphant.

Along with many others (e.g. Cueva et al., 2018; Gwon & Thongpriwan, 2019; Khachan, 2009; Lucio-Villegas, 2018; Tulloch et al., 2017), we contend that participatory inquiry can and does produce useful knowledge about and for adult education. Moreover, it goes beyond research to create "serious and reliable knowledge upon which to construct power, or countervailing power, for the poor, oppressed, and exploited groups and social classes – the grassroots – and for their authentic organizations and movements" (Fals-Borda & Rahman, 1991, p. 3). As adult educators and early-career scholars, our work incorporates "this combination of liberating knowledge and political power" (Fals-Borda & Rahman, 1991, p. 3-4), guided by our onto-epistemological commitments to *vivencia*, *praxis*, and *conscientization*. We seek to share in the lived experiences of our learners and our participants, to take action with them toward improving their conditions, and to foster deep awareness about reality as well as the capacity for change (Freire, 1970; Glassman & Erdem, 2014).

Adult education ranges from literacy classes in church basements to terminal degree programs in Ivy League universities, and each context holds unique challenges as well as potential for positive change. In advocating for a critical approach to adult education pedagogy and research, we reject neoliberal models of lifelong learning that reproduce oppressive socioeconomic structures and embrace the possibility of democratic, transformative education that is accountable to the learner (Brookfield, 2018; Freire, 1970; Lucio-Villegas, 2018). In this chapter, we have offered three inquiry activities as possibilities, anchored in Freirean dialogical principles, for engaging in participatory inquiry for and with adult learners in any setting, formal or non-formal. These methods have great potential for critical educators to live out their own onto-epistemological commitments, better understand and meet learner needs, and facilitate liberatory education (Brookfield, 2018).

These methods, like the critical pedagogy in which they are grounded, are inherently political, transgressing the dominant social order and the hierarchy upholding it (Freire, 1996; Lima, 2018). Lima (2018) discards the term lifelong education in favor of 'permanent education' in order to emphasize "personal and democratic development" (p. 221) over economic and instrumental purposes. Adult education, in this perspective, is about the development of the person and the society rather than the transmission of marketable skills in order to gain economic power (Lima, 2018). More than just a shift in teaching style or change of curriculum, this demands a revolution in the purposes and practices – the praxis – of adult education, from the church basement to the ivory tower. Who better to lead this revolution than those already in the trenches?

REFERENCES

Amsden, J., & VanWynsberghe, R. (2005). Community mapping as a research tool with youth. *Action Research, 3*(4), 357–381. doi:10.1177/1476750305058487

Ballard, H. L., & Belsky, J. M. (2010). Participatory action research and environmental learning: Implications for resilient forests and communities. *Environmental Education Research, 16*(5–6), 611–627. doi:10.1080/13504622.2010.505440

Boughton, B., & Durnan, D. (2014). Cuba's "Yes, I Can" mass adult literacy campaign model in Timor-Leste and Aboriginal Australia: A comparative study. *International Review of Education / Internationale. Zeitschrift für Erziehungswissenschaft, 60*(4), 559–580. doi:10.100711159-014-9421-5

Brookfield, S. (2018). Critical adult education theory: Traditions and influence. In M. Milana, S. Webb, J. Holford, R. Waller, & P. Jarvis (Eds.), *The Palgrave international handbook on adult and lifelong education and learning* (pp. 53–74). Palgrave Macmillan. doi:10.1057/978-1-137-55783-4_4

Cahill, C., Alberto Quijada Cerecer, D., Reyna Rivarola, A. R., Hernández Zamudio, J., & Alvarez Gutiérrez, L. (2019). 'Caution, we have power': Resisting the 'school-to-sweatshop pipeline' through participatory artistic praxes and critical care. *Gender and Education, 31*(5), 576–589. doi:10.1080/09540253.2019.1582207

Call-Cummings, M., Hauber-Özer, M., & Dazzo, G. (2019). Examining researcher identity development within the context of a course on PAR: A layered narrative approach. *Qualitative Report, 24*(9), 2178–2196. https://nsuworks.nova.edu/tqr/vol24/iss9/6

Chambers, R. (2002). *Participatory workshops: A sourcebook of 21 sets of ideas and activities*. Earthscan.

Cueva, K., Revels, L., Cueva, M., Lanier, A. P., Dignan, M., Viswanath, K., Fung, T. T., & Geller, A. C. (2018). Culturally-relevant online cancer education modules empower Alaska's community health aides/practitioners to disseminate cancer information and reduce cancer risk. *Journal of Cancer Education, 33*(5), 1102–1109. doi:10.100713187-017-1217-4 PMID:28405897

Cummins, J. (2000). *Language, power and pedagogy: Bilingual children in the crossfire*. Multilingual Matters. doi:10.21832/9781853596773

Fals Borda, O. (1979). Investigating reality in order to transform it: The Colombian experience. *Dialectical Anthropology, 4*(1), 33–55. doi:10.1007/BF00417683

Fals-Borda, O. (1987). The application of participatory action-research in Latin America. *International Sociology, 2*(4), 329–347. doi:10.1177/026858098700200401

Fals Borda, O. (1988). *Knowledge and people's power*. Indian Social Institute.

Fals Borda, O. (2006). Participatory (action) research in social theory: Origins and challenges. In P. Reason & H. Bradbury (Eds.), *Handbook of action research* (pp. 27–37). Sage.

Fals-Borda, O., & Rahman, M. A. (1991). *Action and knowledge: Breaking the monopoly with participatory action-research*. Apex Press. doi:10.3362/9781780444239

Fang, M. L., Woolrych, R., Sixsmith, J., Canham, S., Battersby, L., & Sixsmith, A. (2016). Place-making with older persons: Establishing sense-of-place through participatory community mapping workshops. *Social Science & Medicine*, *168*, 223–229. doi:10.1016/j.socscimed.2016.07.007 PMID:27492575

Freire, P. (1970). *Pedagogy of the oppressed*. Herder & Herder.

Freire, P. (1996). *Pedagogia da autonomia. Saberes necessários à práctica educativa*. Paz e Terra.

Gaventa, J. (1991). Towards a knowledge democracy: Viewpoints on participatory research in North America. In O. Fals-Borda & M. A. Rahman (Eds.), *Action and knowledge: Breaking the monopoly with participatory action-research* (pp. 121–131). Apex Press. doi:10.3362/9781780444239.009

Glassman, M., & Erdem, G. (2014). Participatory action research and its meanings: Vivencia, praxis, conscientization. *Adult Education Quarterly*, *64*(3), 206–221. doi:10.1177/0741713614523667

Gwon, S. H., & Thongpriwan, V. (2019). PhotoVoice in a prenursing health care systems course. *The Journal of Nursing Education*, *58*(1), 64–64. doi:10.3928/01484834-20190103-14 PMID:30673098

Hanson, C. L. (2014). "I learned I am a feminist": Lessons for adult learning from participatory action research with union women. *Canadian Journal for the Study of Adult Education*, *27*(1), 1–16.

Janssen, J., Hale, L., Mirfin-Veitch, B., & Harland, T. (2013). Building the research capacity of clinical physical therapists using a participatory action research approach. *Physical Therapy*, *93*(7), 923–934. doi:10.2522/ptj.20120030 PMID:23559527

Kemmis, S., McTaggart, R., & Nixon, R. (2014). *The action research planner: Doing critical participatory action research*. Springer Science+Business Media.

Khachan, V. A. (2009). Diglossic needs of illiterate adult women in Egypt: A needs assessment. *International Journal of Lifelong Education*, *28*(5), 649–660. doi:10.1080/02601370903190086

Kindon, S., Pain, R., & Kesby, M. (2007). Participatory action research: Origins, approaches and methods. In *Participatory action research approaches and methods* (pp. 35–44). Routledge. doi:10.4324/9780203933671-13

Lewin, K. (1946). Action research and minority problems. *The Journal of Social Issues*, *2*(4), 34–46. doi:10.1111/j.1540-4560.1946.tb02295.x

Lima, L. C. (2018). Adult and permanent education in times of crisis: A critical perspective based on Freire and Gelpi. *Studies in the Education of Adults, 50*(2), 219–238. doi:10.1080/02660830.2018.1523087

Lucio-Villegas, E. (2018). Revisiting Paulo Freire: Adult education for emancipation. In The Palgrave international handbook on adult and lifelong education and learning (pp. 151-168). Palgrave Macmillan.

McIntyre, A. (2008). Participatory action research. *Sage (Atlanta, Ga.).*

McLaren, P. L., Ryoo, J. J., Crawford, J., & Moreno, D. (2010). Critical praxis. In C. Kridel (Ed.), *Encyclopedia of curriculum studies*. SAGE Publications, Inc., doi:10.4135/9781412958806.n90

Pratt, M. L. (1992). *Imperial eyes: Travel writing and transculturation*. Routledge. doi:10.4324/9780203163672

Prins, E. (2007). 'Aquí no somos unidos'/'We're not united here': Adult literacy and the obstacles to solidarity in postwar El Salvador. *International Journal of Qualitative Studies in Education: QSE, 20*(4), 401–431. doi:10.1080/09518390601176598

Pyrch, T. (2007). Participatory action research and the culture of fear: Resistance, community, hope and courage. *Action Research, 5*(2), 199–216. doi:10.1177/1476750307077324

Ritterbusch, A. E. (2019). Empathy at knifepoint: The dangers of research and lite pedagogies for social justice movements. *Antipode, 51*(4), 1296–1317. doi:10.1111/anti.12530

Saunders, S. G., Barrington, D. J., Sridharan, S., Meo, S., Hadwen, W., Shields, K. F., Souter, R., & Bartram, J. K. (2016). Addressing WaSH challenges in Pacific Island Countries: A participatory marketing systems mapping approach to empower informal settlement community action. *Habitat International, 55*, 159–166. doi:10.1016/j.habitatint.2016.03.010

Silver-Pacuilla, H. V. (2004). The meanings of literacy: A participatory action research project involving women with disabilities. *Women's Studies Quarterly, 32*(1/2), 43–58.

Sweet, E. L., & Ortiz Escalante, S. (2017). Engaging territorio cuerpo-tierra through body and community mapping: A methodology for making communities safer. *Gender, Place and Culture, 24*(4), 594–606. doi:10.1080/0966369X.2016.1219325

Torre, M. E. (2009). Participatory action research and critical race theory: Fueling spaces for nos-otras to research. *The Urban Review, 41*(1), 106–120. doi:10.100711256-008-0097-7

Torre, M. E., & Ayala, J. (2009). Envisioning participatory action research entremundos. *Feminism & Psychology, 19*(3), 387–393. doi:10.1177/0959353509105630

Tulloch, S., Kusugak, A., Chenier, C., Pilakapsi, Q., Uluqsi, G., & Walton, F. (2017). Transformational bilingual learning: Re-engaging marginalized learners through language, culture, community, and identity. *Canadian Modern Language Review, 73*(4), 438–462. doi:10.3138/cmlr.4052

Vio Grossi, F. (1981). Socio-political implications of participatory research. *Convergence (Toronto), 14*, 43–51.

Vio Grossi, F. (1982). Peasant participation, adult education, and agrarian reform in Chile. In B. Hall, A. Gillette, & R. Tandon (Eds.), Creating knowledge: A monopoly (pp. 153–174). Society for Participatory Research in Asia.

Wang, C. (1999). Photovoice: A participatory action research strategy applied to women's health. *Journal of Women's Health, 8*(2), 185–192. doi:10.1089/jwh.1999.8.185

Wells, J. M. (2014). Investigating adult literacy programs through community engagement research: A case study. *Community Literacy Journal, 8*(2), 49–67. doi:10.1353/clj.2014.0002

Whyte, W. (1994). *Participant observer: An autobiography*. ILR Press.

Wood, L., & McAteer, M. (2017). Levelling the playing fields in PAR: The intricacies of power, privilege, and participation in a university–community–school partnership. *Adult Education Quarterly, 67*(4), 251–265. doi:10.1177/0741713617706541

ENDNOTES

1 See www.researchforempowerment.com for several examples of projects that have engaged these methods.

2 To see a more detailed description of how this process works and to view pictures from such an activity, please visit www.researchforempowerment.com

\3 Brain writing differs from a typical session of brainstorming where often a group of people are asked to call out ideas or thoughts in a short amount of time to yield lots of ideas quickly. This can be fraught with problems, especially related to participants' fear of judgment. Brain writing is an alternative with similar goals of coming up with lots of ideas in a short amount of time but increases the likelihood of participation, and, therefore, more diverse ideas, because participants are asked to write their ideas down, pass them to another person to add, and then pass them to a third person to read aloud. Of course, this can be done in many different ways, but the core idea is to increase participation and idea generation by writing rather than yelling out ideas.

Chapter 11
Using Photo-Methods to Empower Participants in Education Research:
Creating Knowledge and Change

Michael L. Boucher, Jr.
Texas A&M University, San Antonio, USA

ABSTRACT

The use of photographs in ethnographic education research promises new insights and challenges to researchers who wish to do good science and good work in the communities under examination. The use of photo-elicitation is discussed as a method that can help alleviate what Foucault described as the analytical "gaze," allowing for discussions of difficult or taboo subjects like race, sex, gender, and dis/ability. The history, uses, and techniques are examined for different photo methods including photo-elicitation, photo-elicitation interviews, and photovoice. This chapter also contributes practical suggestions for using photos in ethnographic research and illuminates new research in the field. Using photos in the reviewed studies achieved positive results for participants and revealed new understandings of communities, culture, and individuals.

DOI: 10.4018/978-1-7998-7600-7.ch011

"There are always two people in every picture: the photographer and the viewer." — Ansel Adams – Landscape photographer and environmentalist (1902-1984)

"All photographs are accurate. None of them is the truth." — Richard Avedon – Fashion and portrait photographer (1923-2004)

"The camera is an excuse to be someplace you otherwise don't belong. It gives me both a point of connection and a point of separation." — Susan Meiselas – Documentary photographer (1948-)

INTRODUCTION: CULTURE AND ETHNOGRAPHY

While it has not always been the case, today there is an expectation that the people involved in an ethnographic study are participants in a co-equal relationship exploring culture, not objects under examination as in more experimental methods (Lather, 1986). Whether by a researcher who is an insider (emic) or an outsider (etic), historically and fundamentally, ethnographies are an exploration of culture (Geertz, 1973). Culture is centered when researchers write a "thick description" that includes the setting where the observations are performed and situates the individual in a cultural framework (Geertz, 1973 p. 6). Scholars explain that ethnography is a tool for "understanding the Other" (Patton, 2002 p. 84) or as "a dialogue with the Other" (Madison, 2012 p. 8). While place and culture are central to ethnography, ultimately, the project of the ethnographer is to explain the most complicated subject on earth, people. People are complex, illogical, inconsistent, unreliable, and erratic. According to Frankham and MacRae (2011), ethnography is, "an approach characterized by uncertainty and contradictions. It is a field defined by ambitious claims (holism, immersion, depth, rapport) accompanied by discussion of the impossibility of ever reaching those goals" (p. 34). Science is purported to be rational, systematic, logical, and detached. Our quest in social science is to use science to explain this infinitely complicated species of human beings, and as educators, our quest is to empower the people we study to know themselves, and their communities, better.

To assume the awesome responsibility of conducting scientific ethnographic work that does justice to the people involved, a committed researcher engages the whole person and does not assume that cultural characteristics are adequate to define the participants. The people in the cultures we study are sometimes indicative of the culture where they are immersed, and sometimes not (Quantz & O'Connor, 1988). As social scientists, we exist in that gap between the larger culture and the inner life of individuals. How well we define that gap is the difference between ethnographies that develop revealing and instructive understandings that challenge *a priori* notions about the participants, and a merely descriptive enterprise that reifies current worldviews. This chapter will outline some of the benefits, issues, and possibilities of using photos in ethnographic education research. These concepts come from previous studies and theoretical work (Boucher, 2014, 2017, 2018a, 2018b; Boucher & Helfenbein, 2015) To accomplish this, I will outline some of the history of the method as well as include three examples to illustrate the potential of the emergent method when used to empower participants.

THE IDEOLOGY BEHIND THE CAMERA

In *On Photography* (1977), Susan Sontag explained that photography, unlike painting was part of a larger project. She posed that the industrialization of film, cameras, and the images they created "… carried out a promise inherent in photography from its very beginning: to democratize all experiences by translating them into images" (p. 7). The publication of photos from faraway places indulged this notion of exploration from the comfort of home as photos from anthropological fieldwork were made available in magazines. Alexander Graham Bell became the president of the National Geographic Society in 1897 and moved the magazine from a mostly descriptive journal to the picture-rich format recognized today. "He told his editors to 'let the world hear from you as our representative'" (National Geographic Virtual Library, 2015). The work of the society became synonymous with Anthropology in the minds of the public throughout the 20th century and the photos of African, Middle Eastern, Central and South American people were fascinating to Americans and Europeans. While outwardly in the pursuit of science, these early explorers were less concerned with uncovering the ways of knowing that indigenous people hold, than revealing exotic, strange, and titillating characters that propped up their own ideology of White/European superiority. The historical entanglement of colonialism and this type of exploratory anthropology has been a topic in the anthropological literature and was summed up by Gogh (1968):

Until the Second World War most of our fieldwork was carried out in societies that had been conquered by our own government. We tended to accept the imperialist framework as given, perhaps partly because we were influenced by the dominant ideas of our time, and partly because at that time there was little anyone could do to dismantle the empires…Applied anthropology came into being as a kind of social work and community development effort for non-white peoples, whose future was seen in terms of gradual education, and of amelioration of conditions many of which had actually been imposed by their Western conquerors in the first place. (p. 13)

Gogh, who divided the world into White and "Non-White" peoples, lamented the loss of empires and the rise in "revolutionary movements" in the 1960s that hampered relationships used to gain access to subjects "as white liberals between the conquerors and the colonized" (p. 17). This attitude and practice of White researchers led to the creation of ethnographies whose sole purpose was to build a narrative of the backward Others and reassure Europeans and Americans of their cultural, racial, racial and scientific supremacy.

Accuracy Without Truth

Mead (1995) famously called Anthropology a "discipline of words" and yet today the discipline is being conducted in a world where photography has taken over as a means of communication in ways that prose and poetry did in the past (Pyyry, 2015; Rose, 2015). Self-portraits, or "selfies" dominate social media, but more important are the photos taken by everyday people at political revolutions, social justice protests and tragically, at police stops (Ly, 2014). Once thought to be incontrovertible, today photographs are questioned with the same scrutiny as a single testimony was in older times. Americans used to rely on professional photographers to bring the most dramatic or devastating moments of world history to their doorstep through daily newspapers, but now social media broadcasts life and death in real time, whether in Ferguson, Aleppo, or on a Minnesota street (Barnard & Cumming-Bruce, 2016; Eckert, 2021). Yet even with the unblinking eye of the camera, truth is often seen as elusive (Cobb, 2017).

Entering this new world, social scientists and critical ethnographers seek to better understand phenomena they observe and experience. Thomas (1993) referred to critical ethnography as a metaphor that "conveys a particular social imaginary" (p. 20). To Thomas, critical scholars do not debate over truth claims "but over metaphoric images that drive knowledge production" (p. 20). Thus, to think in a critical manner is to "unleash the researcher from the need to validate truth claims" and forgo the "domestication that comes from post-positivist objectivity" (p. 8). Carspecken (1996) explained, that critical researchers do not claim to have any final truths that can be tested and abstracted. Instead, they "agree with the pragmatists that any assertion will always be fallible" (p. 57). Carspecken posited that human fallibility makes critical research more reliable. Critical researchers understand that, when dealing with people, there is not solely one undeniable answer.

Sociologist Douglas Harper (2002) introduced the concept of photo-elicitation to a new generation of researchers and made a case for the use of photo-elicitation in ethnographic research. He explained:

> *Photo-elicitation is based on the simple idea of inserting a photograph into a research interview. The difference between interviews using images and text, and interviews using words alone lies in the ways we respond to these two forms of symbolic representation. This has a physical basis: the parts of the brain that process visual information are evolutionarily older than the parts that process verbal information. (p. 13)*

Harper theorized that images are more comforting and a more conversant experience than a traditional interview because of ways that the brain processes information and the types of images a researcher may use fall along a continuum from the most scientific to the most intimate. This is not a description of the artistic quality of the images, but a discussion of their emotional impact. Thus, the image itself can be more abstract or less artistic, but Harper's focus is on the content that will trigger more or less of a reaction. On the scientific end of the spectrum, Harper listed visual inventories or photos of artifacts; things that are detached from everyday life and experience. These photos engage the mind and the intellect but reveal little of the lived experiences of the participants. In the middle of the continuum are photos of work and home. These pictures of everyday life can chronicle the lived experiences of people as they portray themselves in public. These images are controlled, clean, and clothed in acceptability. At the other extreme are the emotionally raw, intimate portraits of friends, family and self. These reveal the internal struggles of the participants and uncover the contradictions between the lives that are portrayed in public and the familiarity of the daily lived experience. These are the naked, unshaven, unplanned, unkempt images that truly dominate our daily lives. These are images of the life that we do not show to anyone outside of our doors (Harper, 2002).

USING PHOTO-ELICITATION IN EDUCATION RESEARCH

Simply put, photo-elicitation is a technique to use photographs that elicit responses from participants in a study. While used most often in qualitative work, there are many ways to approach the use of photo in a study. Common methods seen in most of the literature are one of the following: 1) the participants take the photos and present the photos and their meaning to the researcher; 2) the participants are involved in another project that involves photography and use the photos to interpret their experiences with the project; 3) the pictures are taken by the participant or the researcher and are not seen by anyone but the researcher and participant. This flexibility in the method allows a researcher committed to just and ethical work with participants to create meaningful studies without demanding a specific ontology.

The best research starts with a great question. This question then dictates the choice of method. Other considerations in the choice of method include the purpose of the study and the needs of the participants. When the researcher seeks to do justice and empower the participants in a study, photo methods can decenter the researcher as the questioner and center the participant as inventor and creator of her/his stories. There are many ways that ethnographers can use photos in their studies but this is the general process as explained by Torre and Murphy (2015):

1. A topic for investigation is developed by researchers.
2. Researchers identify participants and invite them to the study.
3. Researchers or participants shoot photographs relevant to a question or topic.
4. After pictures are downloaded, the researcher uses the pictures to guide interviews and elicit responses.
5. Analyze the data and report findings.

There are many studies that do not use photographs to elicit dialogue, but instead rely on images as visual data that speak for themselves. (p. 8) The key component of photo elicitation interviews is that photographs are used during an interview. Photo method techniques can be used to empower participants, gaining new perspectives, new theorization by participants, more understanding, and richer, more consequential data and insightful new perceptions for the participants.

When it comes to data collection, photo-elicitation allows for a rich conversation that yields emotional, reflective responses from participants. This can be done in several ways, but most commonly, the pictures are shown to the participant and one simple question can start a long and lush conversation, "What is happening in this photo?" The power of this question cannot be overstated. This one question begins the discussion that can reveal theories, history, and reflection beyond what is often possible in a traditional interview. Especially when working with adolescents, the act of asking a teen a question is a revolutionary endeavor. Teens are seldom asked for their ideas, their opinions, and are even less often asked to teach adults about their world. That is one of the great reveals when using this method. Often, the participant will explain that no one has ever asked them to explain their lives and this research gives them the space to reflect and think about the phenomenon captured in that photo (see Gallo, 2002).

There are many ways to use photographs to better the ethnographic research project, aiding investigators to decenter the researcher and center the participant with a truth that only that person can tell. Reading the academic literature using typical ethnography as a method, the methodological pattern involves an ethnographer living among the people under study who gathers data through a triangulation of observation, self-reporting, and interviews (Carspecken, 1996; Creswell, 2012; Madison, 2012; Patton, 2002). Using this method, however, without any kind of intermediary between the researcher and the participant, the researcher runs the risk of seeing only the truth she or he wants to see and nothing more. These methods can all be enhanced, and the quality of data greatly extended, by using photos as a conduit to gain insight into the lived truths of the participants (Bridger, 2013; Crilly, Blackwell, & Clarkson, 2006; Croghan, Griffin, Hunter, & Phoenix, 2008; Gauntlett, & Holzwarth, 2006; Hergenrather, 2009; Jorgenson, & Sullivan, 2010; Leonard, & McKnight, 2014; Lomax, Fink, Singh, & High, 2011; Mandleco, 2013; Miller, 2015; Padgett, et. al., 2013; Prosser, & Loxley, 2007; Richard, & Lahman, 2014; Shaw, 2013; Wang, & Burris, 1994; Zartler, & Richter, 2012). In studies that use photos, they become the intermediary, releasing the researcher from her or his own tunnel vision and allowing the participant to avoid the 'inspecting gaze', as Foucault described it (1980, p. 155). For participants, talking about a picture adds to comfort and potentially eliminating the self-conscious interviews where the researcher writes notes and asks uncomfortable questions. Pictures also make the interview process more enjoyable by stirring participants to reveal their ideas, opinions, and life history. But most importantly, the participant is less like a subject under examination and more like a collaborator, sharing knowledge and teaching the researcher. Photo-elicitation empowers the participant to reflect on his or her own life, practice, space, or whatever the photo suggests.

Crilly, Blackwell, and Clarkson (2006) explained that using photo in interviews leads to data "difficult to achieve by verbal exchanges alone . . . This is because such stimuli bring factors external to the interview situation into view, prompting response to 'not now' moments, 'not here' events, and 'not present' actors" (pp. 341-342). In interviews that use images, participants are encouraged to theorize about whatever comes to mind, wherever the photo takes them. It is a tool that allows the conversation of the interview to flow naturally with less uncertainty from the participant.

To illustrate, there are some studies where the researcher provides photos and the participant is asked to react. This method can resemble a standard interview - however the participant in a photo interview can focus away from the researcher, leading to powerful insights from the participants (Boucher, 2014; 2018a; Boucher & Helfenbein, 2015). With appropriate time and investment in a constructive relationship with the researcher, the participant is freed from social convention to discuss uncomfortable topics by disassociating him/herself from the narrative. For example, a participant may discuss her or his perceptions of the physical or emotional characteristics of a stranger in a photo without looking for visual clues of approval and disapproval from the researcher (Walton & Niblett, 2013).

When the participants provide the photos, these photographs illustrate the lived experience in ways that a verbal interview cannot. Gallo (2002) used photos as part of an autodocumentary project to gain insights into the lives of immigrant and refugee workers. Her study, as part of a class to teach English and to advocate for worker's rights, gave workers disposable cameras and asked them to document how they found meaning in their work. She explained that this study was not only scientifically useful, but also empowering to the workers, "Too often they [immigrant workers] are ignored or belittled; portrayed as exotic, pitiable or comic. Therefore, it becomes an empowering exercise to show themselves through their own visual perspectives as multifaceted individuals expressing their joys and concerns" (p. 52). Allen (2011) used participant photos in a study that used photo-diary to help teens discuss their sexual lives and how that plays out in school. Allen argued that the use of photo allowed for a deeper discussion with participants:

By capturing and centering student embodiment, photos offer a detailed account of this lived materiality and opening for discussion. Photos also offer a sense of space and therefore insights about the spatial dimensions of sexuality which are abstract and less easily accessed through talk alone. (p. 488)

As these examples show, photographs can bring the larger outside world into the research study or, if the participant takes the photos, they can bring the participant to the world in a way that allows for them to remain in control. The use of photos, when done with fidelity to the needs of the participants, can empower them to theorize, discuss, and relay data that deepens our understanding of culture.

Using Participant Generated Images

In the field of education, Marisol Clark-Ibáñez's extensive work in photo-elicitation has made important contributions to the field of educational research and methodology (2004; 2007; 2008). She explained how the use of photo methodologies free the researcher to hear the authentic voices of children who are less likely to record their thinking through journals and other traditional ethnographic techniques. Clark-Ibáñez (2004) explained her use of photo-elicitation with children:

Using researcher-produced photographs is an excellent way to conduct theory-driven research. Researcher-photographers may capture taken-for granted aspects of the participants' community or life that prompt discussion. Yet they may limit themselves and miss an essential aspect of their research setting. In some cases, the interviewees alert researchers to omissions and questions that can be later included in the interview protocol. (p. 1509)

As Clark-Ibáñez (2004) indicated, using photo-elicitation gets at an important and elusive aspect of interviewing; investment in the process. Clipboard interviews can lead to overly simplistic, surface responses because the participant has no investment in the process. Often, traditional interviews from a prepared list of questions will yield one-word answers from participants or simply an "I don't know" response. In photo-elicitation, there is already a level of investment in the process. If the participants have taken the photos, they are eager to tell about them. If the researcher has done the photography, the participant is excited to see what the camera has captured. This curiosity is engrained in our nature. As visually centered beings, we are hard-wired to love images, especially ones that have a connection to ourselves. However, Clark-Ibáñez (2004) also cautions researchers from focusing too much on the images:

Finally, researchers must also be cautious of the tendency to capture the 'visually arresting' images (e.g., homeless person asleep near a school entrance) rather than what might be meaningful for the interview participants. (p. 1509)

In her foundational work on gender and "being bad," (2008) Clark-Ibáñez investigated the link between gendered expectations of *good* and *bad* in an urban elementary school. She found that the male participant, Dante, was likely to be punished for transgressing behavioral expectations, but the female participant, Pati, was punished for the both transgressing school rules as well as for "diverging from expectations of femininity" (p. 95). This study examined an urban elementary classroom where the teachers were mostly persons of color and the students were all Latinx or African American. In Ms. Belle's classroom, the *bad* students received much more frequent and harsher penalties. Dante (African American male) received 6 negative interactions per week and Pati (Bilingual Latina female) received 12 negative interactions per week compared to the average of 2 negative interactions per week for both males and females.

Through observations, Clark-Ibáñez learned that while both students had frequent negative interactions inside the classroom and in other school spaces, Dante was rewarded with special privileges by the principal while Pati was unsupported after a schoolyard fight where she was assailed by 8 boys. Clark-Ibáñez used photo elicitation to gain deeper insight into the apparent differences in how *bad* children were being perceived and how they perceived themselves. Clark-Ibáñez explained that in photo elicitation, "Children lead the pace, set the linguistic level of communication, and do not surpass their cognitive development" (p. 103). Clark-Ibáñez asked the *bad* students to take pictures of "what was important to them," a very open-ended assignment that yielded data that showed how the home and school are connected and the complicated "contours of their personal environments" (p. 104). Clark-Ibáñez's use of photo methods allowed her to see beyond the classroom observations and the typical clipboard interviews into the intimate lives of students' homes, families and friendships. She found that these *bad* students had rich lives outside of school that could have been tapped to help them inside the school. Knowing the lived experiences of these students could also have helped school personnel to reflect on their own conceptions of gender roles and attributes in order to improve the school experiences of all involved, students and the teacher. The use of photos with children allows for, "story-telling responses rather than a potentially intimidating 'question-and-answer' approach" that is effective with all ages, but comes even more alive when working with children (p. 103).

Using Participant and Researcher Generated Images

Kharon Grimmet (2016), sought to discover how adults with Intellectual disability (ID) perceived their Quality of Life (QoL). To accomplish this, she focused on the individual's personal development, career development, and self-determination using photography as the tool to aid these young adults tell their stories. Her participants took photos of their daily lives, both at home and at work over the course of 5 months. Her study highlights the power of using photos to study populations that may not disclose the inner workings of their experiences through traditional interviews (Grimmet, 2016).

Grimmet described that much of the literature on young adults with disability does not centre the experience of the person with ID and often, when it does, the person with ID is spoken *for* by parents or caregivers, not spoken *with*. Grimmet explained:

In an EBSCO search for research from the years 2000 – 2016 using the keywords: perspectives, individuals, disabilities, only 612 peer-review articles were identified. Of those 612 articles, only one peer-reviewed article described the perceptions of individuals with disability as it pertained to their health care experiences. The other 611 articles identified sought to understand the perspectives of parents, caregivers, family, support workers, teachers, nurses, employers, case managers, and healthcare workers. (in Boucher, ed. 2018 p.71)

Grimmet's participants described their work lives, their work relationships and their accomplishments through the photos that they took. Most were mundane snapshots of washrooms, factory floors, and self-portraits of the participant in front of the workplace. Despite their lack of meaning to outsiders, these images held deep meaning for the participants. In interviews, they discussed their work and home lives and a picture emerged of people who are in the process of asserting their independence against great forces within, and outside their bodies. The forces that seek to limit them are as diverse as families, employers, the state, and society, yet, within themselves, in fits and starts, the yearning is strong to be masters of their own actions and determine their own destinies.

Grimmet's study illuminates a vital part of the empowering potential of using photo. The participants were empowered to explain their lives, not just to a researcher, but also to their caregivers using the skills they obtained during the study. At some point, the participants became Gimmet's teachers, helping her to understand their lives, and she began to realize that in some cases, it was the researcher who had the disability:

I had never experienced a Catholic service and found myself alone as Faith's family was running late. Opening the [hymnal], I quickly discovered it was different from what I was accustomed to. Thankfully, Faith arrived with documents in hand that assisted me to, at least partially, participate in and understand the service.... During communion, she was adamant I cross my arms and receive a blessing from the priest. Looking around the very large sanctuary, I observed nobody else crossing their arms and questioned Faith many times before she grabbed my arm and swept me out from the pew and down the aisle toward the priest. I was scared, embarrassed, and frustrated as I walked down the aisle. I did not want to look like a fool, draw attention to myself, or get in trouble – not doing the right or expecting thing. I had to trust and rely on Faith, someone with an intellectual disability. Did she know what she was doing? (2016, p. 233)

Grimmet's reflection on this experience led her to the understanding, that in this case, she was the person with the *disability* and the participant with ID, Faith was the person who was the *caregiver*. She reflected, "I was the one who had no adaptive behavior skills, no prior knowledge or experience, no tools to use in this new environment. I had to trust a young lady who has a disability to get me through that experience…and I confess, in that moment, I did not expect her to successfully lead me through that situation" (2016, p. 233). As Grimmet explained, the photos, "helped prompt participants' in-depth responses by making people, places, events, and time a concrete visual instead of an abstract concept or memory" (2016, p. 73).

In the end, it is up to the researcher to decide the degree to which the images matter in the research project. At times, the images are throwaways, blurry, unflattering photos that are only good for eliciting a response. Other projects are more concerned with the art of the image. It can be a deeply personal and reflective portrait, or a photo of a beloved place or object. As with all studies, it is the responsibility of the researcher to create a safe, ethical, and robust proposal that will show care and justice to both the process and the participants.

Centering the Participant: Photovoice

An important contribution to using photos in Education research has been the introduction of Photovoice (Wang, 2006) Photovoice is used as part of participatory action research and encourages research where participants engage in environmental and community issues with the goal of engaging them in critically conscious action (Checkoway & Richards-Schuster, 2003; Rania, Migliorini, Cardinali, & Rebora, 2015). Photovoice, "encourages and leads participants to express and stand up for the issues that they consider important" (Rania, Migliorini, Rebora & Cardinalip, 2015 p. 383). According to Rania, Migliorini, Rebora and Cardinalip (2015) photovoice enables participants to speak about the strength of their communities by giving them tools to spread their voices to a larger audience, creates space for critical dialogues about community and personal issues, and facilitates opportunities for participants to interact with politicians and power structures. According to Wang and Burris (1994) the use of photos allows participants the ability to be experts on their own lives and experiences.

Hergenrather et al. (2009) summed up the purpose and strengths of the method, "Photovoice expands the representation and diversity of participant voices that assist to define and improve the realities experienced by community members, who, many times are not heard" (p. 694). Wang, (2006) found that the use of photo inspired participants to improve their communities based on their observations made through the study. As with other participatory research methods, photovoice centers the participants' experiences over that of experts who are not rooted in communities. Wang and Burris (1997) imbedded their participatory action research in feminist theory and in the work of Paulo Freire (1970/2000), "...problem-posing education starts with issues that people see as central to their lives and then enables them to identify common themes through dialogue" (p. 370). Wang (2006) stated that photovoice has three main goals: record and consider the community's strengths and needs, engage participants in critical dialogue; aid participants in reflection around the issues that matter to the community, and finally, to create and implement an action plan with local leaders (Rania, Migliorini, Cardinali, & Rebora, 2015).

The framework of photovoice pushes the researcher to move beyond merely the consumption of participant data for the purpose of publication. It demands a pedagogical ethos where participants *actually* participate in the research process. Choosing photovoice as a method is a choice to create opportunities for participants in order to change their own situations. It demands a teacher-researcher who will work to bring about that change.

USING PHOTO-ELICITATION TO "BREAKING THE FRAME"

Surveying the literature, the use of photos in ethnography can be described by two axes, 1) the number of people who will see the photos; and 2) the amount of control the participants have over the images. In studies that use photo-elicitation, there is a wide range of how public the study photos are. In some studies, participants take pictures that are used in a public performance to illustrate an aspect of the community and take the form of an art show, or a documentary (Brandt et al., 2016; Luttrell, 2010; Mahruf, Howes, & Shohel, 2007). Other times, the photos are taken by participants and only shown to the them as part of an interview process (Allen, 2011; Padgett, et. al., 2013). In some studies, the researcher takes photos and then the participants respond to the photos in an interview (Beilin, 2005; Boucher 2014; Boucher & Helfenbein, 2015; Mahruf, Howes, & Shohel, 2007). In others, the researcher uses found photos on the internet to use in interviews (Litner, 2005; Walton & Niblett, 2013). The decision of how much or how little control that a researcher turns over to her or his participants depends on the purpose of the study and the choice of method. For example, if the purpose is to get a perspective that only the participants give on a situation, then the researcher should hand over the camera to the people who are most likely to get that perspective.

However, there are times when the purpose of the study is to ask hard questions that rely on the researcher to facilitate deep reflection. These are often interview processes where the researcher asks the participant to see familiar things in a new way (Beilin, 2005). In this kind of study, the researcher is encouraged to "break the frame" of the participant by showing photos that allow for reflection on objects, places, or people that may not have seemed important before the study (Harper, 2010 p. 20). The phrase comes from framing theory, "the process by which people develop a particular conceptualization of an issue or reorient their thinking about an issue" (Chong & Druckman, 2007, p. 20). To "break the frame," a researcher shows familiar objects or scenes to the participants in an unfamiliar way. For example, a photo of a classroom from a higher advantage so the layout of the desks is clear, or photos that include the teacher, to show his or her interaction (Boucher, 2014; Samuels, 2004). Harper (2010) also explained that images are often able to transcend culture and can lead participants to new understandings of themselves and their cultures, "the idea behind breaking the frame is that photographs may lead an individual to a new view of their [sic] social existence. It is also possible to use images as bridges between worlds that are more culturally distinct" (p. 21).

I have found that this way of approaching photo-elicitation is especially useful for research with teachers in classrooms. In photos that are taken by the researcher in classrooms, the setting and the actions are familiar, but the photograph changes the viewpoint of the participant, thus breaking the frame. Used this way, photo-elicitation allows time and space for the teacher to see him or herself in a new way and will allow for a rich discussion of the event, their feelings about it, and the background of the interactions captured in the image (Boucher, 2014; Boucher & Helfenbein, 2015).

USING PHOTO-ELICITATION: THE THEORETICAL AND METHODOLOGICAL IMPLICATIONS

In framing the theoretical and methodological questions about how and when to apply photo-methods in research, I refer to the link between critical ethnography and photo-elicitation. While not always the case, the rise in the 1980s of critical ethnography sought to question the power structures inherent in the study of culture and then, not to exploit, but to empower participants (Carspecken, 1996; Madison, 2012: Simon & Doppo, 1986). To do justice, critical researchers who are interested in exploring culture through the eyes of the Other must be willing to subsume their own interests and work in solidarity with participants or risk the types of exploitation that characterizes so many early anthropological works (Anderson, 1989).

As critical ethnography examines power relations within and between cultures, so too should researchers who use photo methods be observant of the power relationships of both images and equipment used to create those images. iPads, iPods, cameras that are disposable or digital are all physical resources that may exclude participants. It is important to ask the implications of dispensing equipment to people who may not be able to afford necessities in life. But, what does it say that a researcher from an institution with multi-million-dollar endowments will avoid giving a camera worth a hundred dollars to a potential participant? This question reveals our values and biases. What if the equipment does not come back? Is it stolen? As with all qualitative work, the power relationships and positionality of researchers must be deeply interrogated. The equipment used to take photos, even if it never leaves the hands of the researcher is beyond the reach of our most vulnerable participants. That divide must be bridged for the participant to feel that the researcher can be trusted with the stories of that person's life.

Beyond the question of equipment, how researchers treat participants, their stories, and their images goes to the heart of who we are as scholars. Luttrell (2010) urged, when working with young people to make:

...a self-conscious effort to incorporate [participating children] as producers, interpreters, circulators, exhibitors and social analysts of their own and each other's images. (p. 234)

Empowering participants in a co-research relationship means treating our participant's stories, lives, and images with care. Insights can be obtained when participants generate their own meaning from the experiences and data. That does not mean that we will not be critical where criticism is needed, however, we must work to avoid supplanting our participant's stories with our own. Using photos allows for the creation of these co-research relationships in a non-threatening way. Luttrell continued:

We also need projects that open up opportunities to unsettle, fragment or dislodge other's gazes – if only for moments in time where young people can see themselves and be seen by others in alternative ways. (p. 234)

Photos enable people to cross boundaries of age, status, and location to bring individuals to a mutual understanding and shared experience and critical ethnography is well suited for photo methods because it values the stories of these individuals.

PRACTICAL IMPLICATIONS OF PHOTO-ELICITATION AND THE IRB

As with all research methods, the degree of intentionality the researcher brings to the committee or review board will be reflected in their reception of the research proposal. This is equally important when approaching the study as a critical ethnographer intent on examining power relationships. This critical attitude should also extend to our descriptions of method to the IRB. While using photos in research is an established method in some areas, it is relatively new in education. Thus, many boards are still new to the prospect and may be swayed by the inappropriate uses of photos seen in the news and may be concerned about using photos in school settings. The more that the researcher has reflected on the ethics of data handling, participant selection, and obtaining consent and assent from participants and stakeholders, the more likely it is that he or she will obtain permission and support from the committee. Understanding the positions of power as a researcher will help to adequately explain the purpose of the research and the participant protections put in place, making for a better proposal.

That said, there are a few practical tips that come from experience that can help new researchers through the process. The issue of taking pictures or using pictures in the classroom is a complicated one for many committees. There are two paths that can lead to success in doing educational research using photo. The first path is used when using photo-elicitation with the adults who work in a classroom or a school. In this case, it is wise to guarantee that your photos will not be seen by anyone accept the classroom teacher and yourself. In this pathway, the pictures themselves are not data. The data used in analysis is the audio-taped interview conducted while using the photos as a catalyst. In many districts, all students' guardians have signed waivers for photos to be taken "for educational purposes." These waivers are often included in an early packet of information that include *no shots; no school* provisions and were collected upon registration. The most important conversation is the one between the researcher and the school officials where the research is to be conducted. They must trust that their students will be treated with care, and their information securely and ethically. Usually, if the school officials are committed to the study, the IRB will not stand in the way.

The second pathway is one where students are directly involved as participants in the study. If the students are minors and the study involves talking with them and collecting data from them, consent forms will be required from all parents of minor children involved and it is important to be clear that all children are treated as participants who can give and withdraw assent at any time. The best bet with this situation is to keep the sample small and work with classrooms where the researcher has some personal investment. If this is your classroom, or a classroom that you are involved with, that will make this process much easier. If not, it will be very difficult to make sure that all students involved are covered. When it comes to adult classrooms and vocational classrooms, each student will have to give consent before data can be collected and in vocational classes where some are adults and others are minors, the parents of minors will have to give consent before any classroom photos can be taken.

USING PHOTO-METHODS IN EDUCATION RESEARCH

To better illustrate the power of using photo-methods with participants who would not normally be heard in academic literature, I will highlight two studies from authors that are part of an upcoming book, *Participant empowerment through photo-elicitation in ethnographic education research: New perspectives and approaches* (Boucher, ed., 2017). These studies are at both ends and the middle of the Harper's spectrum from the scientific to the intimate. In the first study, photos from the internet were used scientifically to uncover attitudes about bullying. In the second, deaf adolescents took intimate photos to explain transitional periods of their lives.

The Scientific End of Harper's Spectrum: Using Found Images

In the first study, Canadian researchers Gerald Walton and Blair Niblett (2013) add to the literature on bullying of gender variant adolescents in schools. They explain that the approaches to eliminate bullying focus on the paradigm of individual victims, perpetrators and bystanders and do not get to the core of the culture that allows, or even encourages bullying as a strategy for social acceptance and status. To summarize the literature, bullying continues despite the efforts of school officials and teachers because to kids, bullies are cool, and bullying is cool. To study this phenomenon, Walton and Niblett (2013) used photo-elicitation interviews with adolescents to understand the link between student attitudes toward prototypical youth and the participant's expectation that the prototypes would be bullied.

To accomplish this, the researchers gathered photographs of youth that represent different archetypes from internet searches and divided them into slides that reflected a diverse group of youths.

The slides were not arranged simplistically as 'one of these things is not like the other'. Rather, multiple forms of difference were evident in each slide. For instance, one set of images included one girl wearing a hijab, one who is Black, one White girl dressed in Emo style and one whose gender presentation is masculine. We designed the slides so that participants would debate their choice and rationale, which we asked them to do out loud. These slides were then shown to the participants who were asked to comment on the possibility that the pictured youth would be the victim of bullying and if so, what characteristics would invite the bullying. (p. 650)

In the study, the mostly White sample of adolescents discussed the act of bullying as a power relationship but were often hesitant to name the reasons why students would be bullied. Walton and Niblett referred to this as a spirit of "political incorrectness" where participants were concerned about their responses because they did not want to use terms and language to describe the adolescents in the pictures that would be viewed as racist by the adult researchers (p. 652).

Even with this hesitancy, the students reflected a framework of normative sexuality, gender, culture, and whiteness that reflected their understanding of who would be a victim of bullying. They focused on other aspects of the photographs, like being *chubby*, religious identifications, or boys who did not present stereotypically masculine traits. As one of their participants explained, "those whose appearance is not considered 'normal' or within the boundaries of social norms are usually treated as 'outsiders' who are typically not welcomed within youth cultures in schools" (p. 652). Using photos that were not tied to the participants in any way led to students' ability to explore their own thinking about bullying, but may have been an impediment to their discussion of race and gender.

The Intimate Side of Harper's spectrum: Using Personal Images

Dai O'Brien, a deaf researcher in the United Kingdom used auto-driven photo-elicitation to record the life and schooling experiences of deaf adolescent youth (O'Brien, 2013). The term auto-driven, in this case, means that the participants are in control of both the photography process and the subsequent interviews, "This is considered to give the participant the power to control the interview, but also to draw the researcher's attention to elements of their lives that the researcher may have considered to be inconsequential" (p. 157). He also pointed out an important caveat to turning over control of the project to participants:

A major disadvantage is that by handing over the control of each interview to the participant..., the researcher is not guaranteed data that can be compared between interviews. It is very possible that within a research project, no two interviews will cover the same topics, making it difficult to compare data between interviews, or build a unified picture of people's experiences. (p. 158)

As with all methods, this is a real consideration that must be addressed in the planning of the study. A strategy to hedge against a possible inability to obtain comparable data would be to make sure the sample size is large enough and chosen carefully for characteristics that reflect the phenomenon of interest (Creswell, 2007 p. 54). O'Brien had a sample size of 8, that eventually became 4, due to personal circumstances of his participants, but that sample was sufficient to gain rich, complex data.

According to the World Federation of the Deaf (2016), "Deaf people as a linguistic minority have a common experience of life, and this manifests in Deaf culture. This includes beliefs, attitudes, history, norms, values, literary traditions, and art shared by Deaf people." O'Brien's positionality as a deaf person and a member of Deaf culture makes him uniquely situated to build rapport and understand the experiences of his participants (Bourke & State, 2014). His discussion of the language used to build that rapport is especially instructive. His participants presented their interviews in British Sign Language (BSL), Sign Supported English (SSE) or English, depending on the preference of the participant, "careful language matching is essential in interviews with deaf people to ensure that an unproblematic communication environment is created" (p. 159). O'Brien's use of photo and his flexibility in language allowed his participants to discuss their feelings and theories about isolation and belonging to a community, their identity as members of a community, and independence in a community that relies on sound for communication.

O'Brien explained that mainstreaming in schools has had a negative effect the linguistic competence of deaf students and photo methods are a way that researchers can, "ensure that deaf people are not necessarily at a disadvantage in research interviews" (p. 170). His study highlights the ability of photo methods to amplify the voices of people who have been traditionally silenced in the literature.

Participants were simply asked to take photographs of places or things that were important to them as they grew up. If they wished to take photographs of people, they were encouraged to take photographs that represented the people they wanted to show. (p. 160)

The photos the students took were intimate glimpses of daily life. They were pictures of their rooms, their activities, and things that showed their passion for cars, love of film, youth clubs, and showed their senses of belonging, or a desire to belong. These intimacies allowed O'Brien to see deeply into these students lived experiences and ask questions about the meanings of these objects in the lives of his participants (O'Brien, 2013).

Each of these studies uses photos differently but were able to capture the life experiences of people who are often neglected in educational literature. These particular studies allow us to see that this technique can be used in many different spaces, not just traditional classrooms. It can be used to study the culture of a school as Clark-Ibáñez (2007; 2008) explained, or beyond high school as Grimmet's (2016) study shows or a workplace, as Gallo (2002) described.

To further illustrate, there are specific steps that I use in my work that I have found to be useful (Boucher, 2014; Boucher & Helfenbein, 2015). When it comes to hardware, in my studies I take the photos in classrooms using an iPad with a keyboard attachment. I have all of the sounds off on my iPad and sit out of the way of the teaching. I observe interactions between teachers and students, taking careful field notes using a word processing app for the iPad and when an interesting interaction happens, I snap a photo. These photos are not artistic, nor are they clear, and often they are dominated by the backs of people's heads, but in my work that is inconsequential. After the class, I will sit next to the teacher who is participating in the study, not across. That reduces my power in the dynamic and allows the participant to focus on the picture, rather than me scratching on a notepad, or looking and nodding. I ask teachers only a few questions and refer to the photos on the iPad. The questions that is most important is simply, "what is happening in this photo?" and that will start a discussion that allows teachers to talk about their teaching, theorize about uncomfortable subjects like race, gender, sexuality, and their own histories. When the conversation lulls, I will sometimes reflect the words of the participant, summarizing what I heard and ask for clarification. The photo is the gateway to a world inside the person observing the photo. The researcher, who is not gazing at the participant, shares the experience of the photo, and becomes the listener rather than the inquisitor.

CONCLUSION

As I tell my research students, all good research begins with at least one great question. When considering a method for investigating that question, a researcher must take into account the resources and setting of the study, who will participate, how to recruit participants, and many other factors. The research committee and the IRB will have many questions about the method that must be answered but the fundamental questions are simply; are you ethically, transparently, and systematically investigating a phenomenon? and; are you protecting the humanity, dignity, and privacy of your participants, keeping them from harm? The choices made for any study must answer these questions with fidelity. Within that framework, study after study has found that quality data can be gained from the application of methods that empower participants as co-researchers. Critical ethnography has shown a path and the use of photo enhances that work. As we advance through this era where images are at peoples' fingertips and in their pockets enabling everyone to feel like a reporter and investigator, social media has created an environment where many feel compelled to document and chronicle their daily experience. As researchers, we can harness this new power to tell the stories of marginalized, forgotten, and disenfranchised people, leading to a deeper understanding of the mosaic of humanity.

REFERENCES

Allen, L. (2011). "Picture this": Using photo-methods in research on sexualities and schooling. *Qualitative Research*, *11*(5), 487–504. doi:10.1177/1468794111413224

Anderson, G. L. (1989). Critical ethnography in education: Origins, current status, and new directions. *Review of Educational Research*, *59*(3), 249–270. doi:10.3102/00346543059003249

Barnard, A. (2016, December 13). Battle Over Aleppo Is Over, Russia Says, as Evacuation Deal Reached. The New York Times. https://www.nytimes.com/2016/12/13/world/middleeast/syria-aleppo-civilians.html?_r=1

Beilin, R. (2005). Photo-elicitation and the agricultural landscape: 'Seeing' and "telling" about farming, community and place. *Visual Studies*, *20*(1), 56–68. doi:10.1080/14725860500064904

Boucher, M. L. Jr. (2014). More than an ally: A successful white teacher who builds solidarity with his African American students. *Urban Education*, *51*(1), 82–107. doi:10.1177/0042085914542982

Boucher, M. L. Jr. (2017). The art of observation: Issues and potential of using photo-methods in critical ethnography with adolescents. *International Journal of Adult Vocational Education and Technology*, *8*(2), 1–14. doi:10.4018/IJAVET.2017040101

Boucher, M. L. (2018a). Interrogating whiteness: Using photo-elicitation to empower teachers to talk about race. In M. L. Boucher (Ed.), *Participant empowerment through photo elicitation in ethnographic education research: New research and approaches* (pp. 201–225). Springer.

Boucher, M. L. (Ed.). (2018b). *Participant empowerment through photo elicitation in ethnographic education research: New research and approaches*. Springer., doi:10.1007/978-3-319-64413-4

Boucher, M. L. Jr, & Helfenbein, R. J. (2015). The push and the pull: Deficit models, Ruby Payne, and becoming a "Warm Demander.". *The Urban Review*, *47*(4), 742–758. doi:10.100711256-015-0332-y

Bourke, B., & State, M. (2014). "Positionality: Reflecting on the research process" by Brian Bourke. *Qualitative Report*, *19*(18), 1–9.

Brandt, H. M., Freedman, D. A., Friedman, D. B., Choi, S. K., Seel, J. S., Guest, M. A., & Khang, L. (2016). Planting healthy roots: Using documentary film to evaluate and disseminate community-based participatory research. *Family & Community Health*, *39*(4), 242–250. doi:10.1097/FCH.0000000000000120 PMID:27536929

Bridger, L. (2013). Seeing and Telling Households: A Case for Photo Elicitation and Graphic Elicitation in Qualitative Research. *Graduate Journal of Social Science*, *10*(2), 106–131.

Carspecken, P. F. (1996). *Critical ethnography in educational research: A theoretical and practical guide*. Routledge.

Chong, D., & Druckman, J. N. (2007). Framing theory. *Annual Review of Political Science*, *10*(1), 103–126. doi:10.1146/annurev.polisci.10.072805.103054

Clark-Ibáñez, M. (2004). Framing the social world with photo-elicitation interviews. *The American Behavioral Scientist*, *47*(12), 1507–1527. doi:10.1177/0002764204266236

Clark-Ibáñez, M. K. (2007). Inner-city children in sharper focus: Sociology of childhood and photo-elicitation Interviews. In G. C. Stanczak (Ed.), *Visual research methods: Image, society, and representation* (pp. 167–196). Sage Publications.

Clark-Ibáñez, M. K. (2008). Gender and being "bad": Inner city students' photographs. In P. Thompson (Ed.), *Doing Visual Research with Children and Young People* (pp. 95–113). Routledge.

Cobb, J. (2017, June 22). Old Questions But No New Answers in the Philando Castile Verdict. The New Yorker. https://www.newyorker.com/news/daily-comment/old-questions-but-no-new-answers-in-the-philando-castile-verdict

Creswell, J. W. (2012). *Qualitative inquiry and research design: Choosing among five approaches* (3rd ed.). SAGE Publications.

Crilly, N., Blackwell, A. F., & Clarkson, P. J. (2006). Graphic elicitation: Using research diagrams as interview stimuli. *Qualitative Research*, *6*(3), 341–366. doi:10.1177/1468794106065007

Croghan, R., Griffin, C., Hunter, J., & Phoenix, A. (2008). Young people's constructions of self: Notes on the use and analysis of the Photo-Elicitation methods. *International Journal of Social Research Methodology*, *11*(4), 345–356. doi:10.1080/13645570701605707

Eckert, N. (2021, March 9). Video: The 9-minute encounter that rocked Minnesota and the world. Twin Cities. https://www.twincities.com/2021/03/08/george-floyd-video-derek-chauvin-trial-rocks-world3838220

Foucault, M. (1980). *Power-knowledge: Selected interviews and other writings, 1972-1977* (5th ed.). Pantheon Books.

Frankham, J., & MacRae, C. (2011). Ethnography. In Theory and Methods in Social Research (2nd ed.). SAGE Publications.

Gallo, M. L. (2002). Picture this: Immigrant workers use photography for communication and change. *Journal of Workplace Learning*, *14*(2), 49–57. doi:10.1108/13665620210419293

Gauntlett, D., & Holzwarth, P. (2006). Creative and visual methods for exploring identities. *Visual Studies*, *21*(1), 82–91. doi:10.1080/14725860600613261

Geertz, C. (1973). The interpretation of cultures: Selected essays (22nd ed.). Basic Books.

Grimmet, K. (2016). *Examining the quality of life of young adults with Intellectual Disability: Using photo elicitation to empower an unknown voice* (Doctoral thesis). ProQuest. (10195731)

Harper, D. (2002). Talking about pictures: A case for photo elicitation. *Visual Studies, 17*(1), 13–26. doi:10.1080/14725860220137345

Hergenrather, K. C., Rhodes, S. D., Cowen, C. A., Bardhoshi, G., & Pula, S. (2009). Photovoice as community-based participatory research: A qualitative review. *American Journal of Health Behavior, 33*(6), 686–698. doi:10.5993/AJHB.33.6.6 PMID:19320617

Jorgenson, J., & Sullivan, T. (2010). Accessing children's perspectives through participatory photo interviews. *Forum Qualitative Social Research, 11*(1), 8.

Lather, P. (1986). Issues of validity in openly ideological research: Between a rock and a soft place. *Interchange, 17*(4), 63–84. doi:10.1007/BF01807017

Leonard, M., & McKnight, M. (2014). Look and tell: Using photo-elicitation methods with teenagers. *Children's Geographies, 13*(6), 629–642. doi:10.1080/14733285.2014.887812

Lomax, H., Fink, J., Singh, N., & High, C. (2011). The politics of performance: Methodological challenges of researching children's experiences of childhood through the lens of participatory video. *International Journal of Social Research Methodology, 14*(3), 231–243. doi:10.1080/13645579.2011.563622

Luttrell, W. (2010). "A camera is a big responsibility": A lens for analysing children's visual voices. *Visual Studies, 25*(3), 224–237. doi:10.1080/1472586X.2010.523274

Ly, L. (2014). Can cell phones stop police brutality? *CNN*. https://www.cnn.com/2014/11/18/us/police-cell-phone-videos/

Madison, S. D. (2012). *Critical ethnography: Method, ethics, and performance* (2nd ed.). Sage Publications.

Mahruf, M., Howes, A. J., & Shohel, C. (2007). Transition from nonformal schools: Learning through photo elicitation in educational fieldwork in Bangladesh. *Visual Studies, 22*(1), 53–61. doi:10.1080/14725860601167200

Mandleco, B. (2013). Research with children as participants: Photo elicitation. *Journal for Specialists in Pediatric Nursing, 18*(1), 78–82. doi:10.1111/jspn.12012 PMID:23289457

Mead, M. (1995). Visual anthropology in a discipline of words. In P. Hockings (Ed.), *Principles of visual anthropology* (pp. 3–10). De Gruyter Mouton. doi:10.1515/9783110290691.3

Miller, K. (2015). Dear Critics: Addressing concerns and justifying the benefits of photography as a research method. *Forum Qualitative Social Research, 16*(27), 27.

National Geographic Virtual Library. (2015). *The history of the national geographic society*. http://gale.cengage.co.uk/national-geographic-virtual-library/history.aspx

O'Brien, D. (2013). Visual research with young d/Deaf people: An investigation ofthe transitional experiences of d/Deaf young people from mainstream schools using auto-driven photo elicitation interviews. *Graduate Journal of Social Science, 10*(2), 152–175.

Padgett, D. K., Smith, B. T., Derejko, K., Henwood, B. F., & Tiderington, E. (2013). A picture is worth . . .? Photo Elicitation interviewing with formerly homeless adults. *Qualitative Health Research, 23*(11), 1435–1444. doi:10.1177/1049732313507752 PMID:24122520

Patton, M. Q. (2002). *Qualitative research & evaluation methods: Integrating theory and practice* (3rd ed.). SAGE Publications.

Prosser, J., & Loxley, A. (2007). Enhancing the contribution of visual methods to inclusive education. *Journal of Research in Special Educational Needs, 7*(1), 55–68. doi:.2007.00081.x doi:10.1111/j.1471-3802

Pyyry, N. (2013). 'Sensing with "photography and 'thinking with' photographs in research into teenage girls" hanging out. *Children's Geographies, 13*(2), 149–163. doi:10.1080/14733285.2013.828453

Quantz, R. A., & O'Connor, T. W. (1988). Writing critical ethnography: Dialogue, multivoicedness, and carnival in cultural texts. *Educational Theory, 38*(1), 95–109. doi:10.1111/j.1741-5446.1988.00095.x

Richard, V. M., & Lahman, M. K. E. (2014). Photo-elicitation: Reflexivity on method, analysis, and graphic portraits. *International Journal of Research & Method in Education, 38*(1), 3–22. doi:10.1080/1743727X.2013.843073

Rose, G. (2013). On the relation between "visual research methods" and contemporary visual culture. *The Sociological Review, 62*(1), 24–46. doi:10.1111/1467-954X.12109

Samuels, J. (2004). Breaking the Ethnographer's frames: Reflections on the use of photo Elicitation in understanding Sri Lankan monastic culture. *The American Behavioral Scientist, 47*(12), 1528–1550. doi:10.1177/0002764204266238

Shaw, D. (2013). A new look at an old research method: Photo-elicitation. *TESOL Journal*, *4*(4), 785–799. doi:10.1002/tesj.108

Simon, R. I., & Dippo, D. (1986). On critical ethnographic work. *Anthropology & Education Quarterly*, *17*(4), 195–202. doi:10.1525/aeq.1986.17.4.04x0613o

Sontag, S. (1977). *On photography*. Penguin.

Thomas, J. (1993). *Doing critical ethnography*. Sage Publications. doi:10.4135/9781412983945

Walton, G., & Niblett, B. (2013). Investigating the problem of bullying through photo elicitation. *Journal of Youth Studies*, *16*(5), 646–662. doi:10.1080/13676261.2012.733810

Wang, C., & Burris, M. A. (1994). Empowerment through photo Novella: Portraits of participation. *Health Education & Behavior*, *21*(2), 171–186. doi:10.1177/109019819402100204 PMID:8021146

World Federation of the Deaf. (2016, October 24). *Deaf culture - WFD | world federation of the deaf*. https://wfdeaf.org/our-work/focus-areas/deaf-culture

Zartler, U., & Richter, R. (2012). My family through the lens. Photo interviews with children and sensitive aspects of family life. *Children & Society*, *28*(1), 42–54. doi:10.1111/j.1099-0860.2012.00447.x

ENDNOTE

This chapter is an updated reprint of Boucher, M. L. (2018). Using photo-methods to empower participants in education research. In Wang, V. (Ed.), Handbook of research on innovative techniques, trends, and analysis for optimized research methods (pp. 202-219). Hershey, PA: IGI Global. It also includes elements from Boucher, M. L. (Ed.) (2018). Participant empowerment through photo elicitation in ethnographic education research: New research and approaches. New York, NY: Springer. doi 10.1007/978-3-319-64413-4

Chapter 12
Case Study and Self Study as Means for Program Improvement in Teacher Education

Jahnette Wilson
University of Houston, USA

Samuel R. Brower
University of Houston, USA

Teresa Edgar
University of Houston, USA

Amber Thompson
University of Houston, USA

Shea Culpepper
University of Houston, USA

ABSTRACT

Proponents of the evidence-based movement in education maintain that decisions around policy and practice should be grounded in data outcomes. However, insufficient research exists on data use in teacher education programs as much of the research on data use is concentrated on K-12 programs. The purpose of this case study was to investigate the data use practices of an educator preparation program so as to facilitate program improvement efforts. The collective qualitative data described in this study was key to informing continuous improvement areas within this educator preparation program. Therefore, this case study offers insight as to how qualitative data can support and inform program improvement efforts.

DOI: 10.4018/978-1-7998-7600-7.ch012

INTRODUCTION

The evidence-based movement in education contends that decisions around policy and practice should be grounded in data outcomes (Moss, 2007; Trinder 2000). That said, little research exists on data use in teacher education programs as much of the research on data use is centered on K-12 programs. Further, there are no "universally accepted means" of assessing the performance of teacher preparation programs (CCSSO, 2016). Accountability and rigor in teacher education have been the focus of recent policy initiatives (Cochran-Smith & Villegas, 2015; CCSSO, 2016). With such an enormous focus on accountability, the importance of data-driven decision making for program improvement has gained significant attention over the last ten years (Aguerrebere, 2009; CCSSO, 2012; CCSSO, 2016, CCSSO, 2018, CAEP, 2013, Davis & Peck, 2018, Datnow & Hubbard, 2015, Mandinach & Gummer, 2015, Easton 2009, 2010). Data-driven decision making describes a "systematic collection, analysis, examination, and interpretation of data to inform practice and policy in educational settings" (Mandinach, 2012). With the rising external pressure of accountability, certain parameters associated with program data are required. For example, data collected for accountability purposes must be comparable from program to program. Therefore, the data must be standardized in an effort to capture a single measure to be used to assess the performance of the educators on program quality. However, standardized and quantitative data do not necessarily offer support and guidance to educators regarding what students do and do not understand (Wiess, 2012). .

However, there is no absolute consensus for assessing the performance of teacher preparation programs (TPP) (CCSSO, 2016). A 2013 report by the National Academy of Education (NAE) posits, "systems for evaluating TPPs use various types of evidence – each with its particular strengths and limitations – to make inferences about the quality of the preparation experience and its role in producing employable, high-quality teachers" (p. 1-2). Targeted attributes that are related to teacher preparation quality commonly measured include admissions and recruitment, quality and content of instruction, quality of student teaching field experiences, faculty qualifications, effectiveness and success regarding the preparation of quality teachers, and attrition rates.

A plethora of data is used in an attempt to evaluate the nuances of teacher preparation that are not necessarily observable or even quantifiable. Typical measures include grade point averages, percentage of minority students in a program, number of candidates in high-need areas, academic credentials, full-time vs part-time higher education faculty, textbooks, assignments, syllabi, required courses, fieldwork policies and hours, mentor qualifications, student teaching observation records, certification pass rates, surveys of graduates and P-12 principals, and even "value-added" measures of the impact on the learning of P-12 students. That said, Wiess (2012) suggests that data collected for accountability purposes must be comparable from program to program. Thus, input and output measures must be standardized so that data can be used to assess and compare performance across programs and institutions. However, standardized and quantitative data do not necessarily offer support and guidance to educators regarding what students do and do not understand (Feuer et al., 2013).

DATA USE: THE COMPLIANCY FACTOR

Accountability pressure has pushed higher education institutions of learning to move toward evidence-based program evaluation and improvement (Smith & College, 2009). Requirements of data collection and use have been routinely seen as responses to compliance rather than opportunities for inquiry related to program improvement (Peck et al., 2015). The 2013 NAE report discusses several different types of teacher preparation program (TPP) evaluation systems, all of which may be used to evaluate a single teacher preparation program. Each of these entities use differentiated but also overlapping data to assess teacher preparation programs (Feuer et al., 2013).

Through Title II of the Higher Education Act, the federal government requires institutions to report a vast amount of data for evidence that includes certification test results, as well as completion of student teaching requirements. National entities such as the Council for the Accreditation of Educational Preparation (CAEP) evaluate the quality of TPPs for accreditation purposes using a set of standards that require institutions to provide evidence, such as course syllabi, student teaching observation scores, and results of stakeholder surveys, including program graduates and their employers (Feuer et al., 2013).

State governments base accreditation on national entities such as CAEP, as well as the collection of quantitative data that includes the number of graduates approved for certification and fulfilling the state requirements of field work. Further, many states have started to develop annual program "report cards" to delineate how programs perform on various measures such as candidate completion rates, licensure examination pass rates, and placement and retention rates. These reports provide quantitative information that allow programs to be compared. Additionally, these more frequent reports reveal both program strengths and weaknesses thereby supporting a culture focused on using data for continuous program improvement (Feuer et al., 2013). More and more importance is being placed on these reports to inform program improvement, state accountability, and public information (CCSSO, 2016).

Additionally, certain media outlets and other independent organizations have developed rankings and ratings of higher education programs regarding the programs' quality. The National Council on Teacher Quality (NCTQ), in collaboration with U.S. News and World Report rate teacher preparation programs, is well known for ranking schools of education (Feuer et al., 2013).

Unfortunately, these accountability policies discussed, thus far, require data to be reported that is never actually used for improvement efforts by educators to change their practices (Dowd, 2005). Peck, Gallucci, and Sloan (2010) contend the compliance-oriented contexts of program evaluation can be problematic for teacher preparation programs (TPPs). Compliance mandates can require programs to direct resources in areas that are often based on implementation of policies rather than the meaningful work of program improvement measures. The threat of the loss of program accreditation for TPPs that do not meet these compliance-based measures reinforce the tendency to focus the majority of attention on compliance-based work rather than on gathering data that focuses on improved experiences for teacher candidates. The 2013 NAE report declares that teacher preparation programs may engage in self-evaluation to inform program improvement. It goes on to say that these self-studies should be based on evidence collected and use a model that connects research to practice as well as include collaboration amongst faculty and stakeholders beyond the walls of the institution (Feuer et al., 2013). Pinneagar and Erickson (2009) advocate for the use of self-study when examining accreditation and programmatic data. Educational researchers have established self-study as a research methodology to intentionally be used by teacher educators to improve their practice (Samaras & Freese, 2009; Hamilton, 1998). Self-study research follows in the lineage of other qualitative research methodologies; however, what distinguishes it from other methodologies is its orientation to collaborative improvement (LaBoskey, 2004; Loughran, 2005; Samaras & Freese, 2006; Davis & Peck, in press).

Self-studies focus on qualitative tools such as interviews to facilitate contextual data based on participants' lived experiences such that pertinent knowledge is uncovered to inform the study (Schultze and Avital, 2011). Interviews are recognized by the self-study community to be a strong data collection tool in a collective self-study as they allow researchers to engage with participants in a meaningful way through rich, spontaneous questions and answers, clarification, and verification opportunities (Kosnik et al., 2009). Further, qualitative data gathered through interviews offers the researcher in-depth insight to support interpretation of quantitative data (Rao & Woolcock, 2003). Data analysis of the qualitative data allows the researcher to attend to areas of interests grounded in theory as findings emerge from the collected data (Kosnik et al., 2009). LaBoskey (2004) sets forth four basic features for self-study research: (a) an aim towards improvement (b) interaction with colleagues, (c) the use of multiple methods; and (d) to make the work available to the professional community. Feldman (2003) argues that validity must be attended to so that the work can be made available and trusted by the academy. Validity is critical to defensible conclusions made based on the evidence provided for a TPP evaluation and program changes that are made following evaluation (Feuer et al., 2013); therefore, it is essential for the self-study researcher to consider the value of the work to others (Samaras & Roberts, 2011). Samaras and Roberts (2011) contend that sharing the self-study work "contributes to the accumulation of pedagogical, content, and issue-based knowledge and serves to build validation across related works" (p. 45).

SELF-STUDIES OF DATA USE PRACTICES

Self-studies that are conducted using data for program improvement and inquiry can be useful as the focus moves from compliance to one of inquiry. The inquiry process is highlighted by faculty making data-driven decisions regarding policy, coursework, and organizational changes (Peck et al., 2010). When conducted as self-study, data use for program improvement and inquiry may facilitate a culture of data-driven decision making, as opposed to compliance-oriented program evaluation (Feuer et al., 2013). Additional educational research confirms the value of faculty collaboration in teacher preparation programs (Spring, 2003). Although evidence-based decision-making is not necessarily a new idea, building a culture of evidence is becoming a valued practice in higher education (Dwyer et al., 2006). That said, building a culture of inquiry requires teacher preparation program leadership to provide opportunities for faculty to engage with data and use data to make programmatic and individual course decisions (Dowd, 2005). Thus, it is critical that teacher preparation programs offer structured time and space for data disaggregation and collaboration (Mace & Lublke, 2014).

Strategic structured social interactions are key to guiding and facilitating data use such that the collaborations among educators motivate and support capacity to use data to make instructional decisions. Additionally, these educators must have control over the adjustments to their own practice. That is, they must possess the authority to take action. Intentional and strategic support for evidence-based decision making for the purpose of program improvement moves the focus from data for accountability to data for improvement (Weiss, 2012). Engaging in the facilitation of data use practices to make individual and program level decisions will ultimately result in promising educational outcomes. Specifically, creating a culture that values evidence-based decision making has the potential to transform a teacher preparation program in all aspects, including its curriculum policies, and practices (Smith and College, 2009).

The purpose of this chapter is to investigate how one teacher preparation program used self-study to support their program in making sense of their data use practices given the amount of quantitative data collected throughout the program and used for accreditation reporting and compliance purposes. The researchers seek to answer the question: "*How can an institutional self-study in data use practices support a teacher preparation program in making sense of programmatic data*?"

CONTEXTUAL BACKGROUND OF THE STUDY

This case study was situated at a large urban university in the southern United States. Currently, there are approximately 900 students served in the teacher education program with over sixty teacher educators, including professors, instructors, and supervisors of student teachers. The program has a history of a commitment to excellence and program improvement as evidenced by the last three program iterations receiving national awards and accreditation. The current program iteration is CAEP-accredited and considered a distinguished program by the Association of Teacher Educators, which recognizes programs that exemplify collaboration, innovation, and high quality.

The teacher preparation program at this university is designed with attention to two main goals: 1) to create a culture of professional responsibility with both the *program* and *teacher candidates*; and 2) to create a culture of inquiry with both the *program* and *teacher candidates*. The teacher education program is committed to facilitating a collaborative culture to continue building the quality of the program and success of future teachers. That said, as is common in teacher preparation, the program is also steeped in compliance-oriented data reporting activities to maintain national and state level accreditation (Feuer et al., 2013).

A wealth of data has always been readily available and collected by program administration (i.e. certification test scores, stakeholder perception surveys, course grades, and lesson observation scores); however, the focus of collecting and analyzing these data was to monitor compliance and maintain accreditation status. Although useful for these purposes, these data collected did not readily lend itself to support faculty in understanding what their students actually knew and could execute in the field. Moreover, faculty operated, for the most part, as individuals and most improvement efforts were siloed and confined to impact their own course. Program faculty rarely spent time and resources to collaborate and collect data that crossed multiple course and program components. Nor did they attempt to focus on the impact their collective courses were having on the knowledge and skills of teacher candidates in their field work. This created a need and an opportunity for the researcher to use self-study as a means to examine data use practices within the teacher preparation program so as to facilitate improvement of the program's capacity to use data.

PHASE 1: AN EXAMINATION OF DATA USE PRACTICES

Participants

Participants in year one of the case study included 24 individuals in the teacher education program: teacher preparation leadership, faculty, instructors, staff, and educational preparation school partners, who are also part of the program. Teacher preparation leadership involved all individuals acting as leaders within the program: the Dean of the College of Education, the Associate Dean of Undergraduate Studies, the Curriculum and Instruction Department Chair, the Associate Chair for Teacher Education, and the Director of Student Teaching, as well as two individuals heavily involved in the data collected within the College of Education, including the Certification Officer of the College of Education. Teacher education faculty included a sample of those instructors of record who taught courses within the program. These instructors represented a range of longevity within the university. Some were new to the program and the college, while others had been with the university for a number of years. The majority of faculty who were interviewed were Clinical professors; however, there were three Tenured and Tenured-Track professors. The education preparation program partners identified as a principal and a cooperating teacher. The researcher, who was a faculty member within the teacher preparation program, also met with a focus group comprised of twelve supervisors who worked closely with teacher candidates.

Research Design and Data Collection

This case study, which was conducted in two phases, was designed, specifically, as a self-study of data use practices (Hamilton, 1998) where knowledge and learning were drawn from the teacher preparation program itself. The overall purpose of the study was to analyze data use practices used for program assessment and continuous improvement of the teacher education program. Research questions guiding the self-study were as follows:

- What ideas, beliefs, and feelings shape the way people engage data?
- What methods are used to collect, archive, and analyze the data?
- What program policies and practices affect data use?

These questions were specifically developed to gather qualitative data, rather than the more quantitative data that is typically collected within teacher preparation programs, as these data focus on inquiry into the perspectives of participants.

Year one of the self-study was an initial examination of data use practices within the teacher education program, and ultimately yielded invaluable results informing actionable steps that supported a concerted move toward an intentional data use plan for continuous improvement. This phase of the self-study was descriptive and based on the qualitative data drawn individual interviews, a focus group, surveys, and examination of pertinent artifacts such as CAEP and Inspectorate reports, agendas from data day and curriculum work, and identified data collected in the teacher preparation program. Through collaborative process, the researcher conducted interviews and focus-group discussions with teacher educators while triangulating findings with programmatic and accreditation data during the first phase of the study. Finally, the researcher took field notes during a follow-up meeting with the program's leadership team. The program's leadership team consisted of the Associate Dean of Undergraduate Studies, the Associate Chair of the Teacher Preparation Program, the Director of Student Teaching, and a faculty member within the program.

Data Analysis

The interviews and the focus group discussions were audiotaped and transcribed. The researcher then examined the transcripts of interviews and focus groups, as well as program documents (i.e. CAEP and Inspectorate reports; Data Day, Curriculum Work agendas, and an identified listing of data collected in the teacher preparation program). Low-inference coding was utilized to determine critical data within each transcription as low-inference tags are used to note observations "in terms that are concrete as possible, including verbatim accounts of what people say, for example, rather than researchers' reconstructions of the general sense of what a person said, which would allow researchers' personal perspectives to influence the reporting" (Seale, 1999, p. 148). The researcher coded the first three interview transcriptions, tagging evidence that appeared to relate to data use. Then, using the low-inference tags from the coded documents, the researcher grouped the tags into categories through inductive grouping procedures. The researcher continued to use the same grouping process to reduce the codes to a merged group of approximately twenty categories that were used to code and tag the remaining data. New codes were added as certain ideas were discovered within the data that seemed important, but were not included or addressed in the initial group of tags. For each coded document, a document summary was completed in which the researcher described the findings from that particular document that were related to the research questions. The document summaries were key in providing evidence for the themes and patterns that were found in the coded data. The researcher used the completed document summaries as a resource to summarize the emergent findings to present to the teacher preparation leadership team.

RESULTS

Phase 1: Initial Findings of the Self-Study

The first research question in the self-study focused on an examination of the ideas, beliefs, and feelings that shaped the way people engaged with data. The results of year one of the study revealed that most participants appeared to care about data and collaboration; moreover, they articulated the importance of collaboration and a shared vision. The majority of participants saw "data use" as a collective, collaborative practice, and expressed a desire to "see" data and become more "involved" with data use. One participant stated, "*I want to know more about what is collected and how to use it to improve my course*." There appeared to be pockets of ongoing collaborative data use practices throughout the program. Nonetheless, the differentiation of data use practices was clearly dependent upon roles: leadership, faculty, student teacher supervisors, and district partners. Moreover, there appeared to be no real consensus on which data mattered or were most useful as most individuals attended to their own practice, and did not necessarily "see" programmatically, a by-product of a focus on individual practice. Further, some participants appeared to distrust quantitative data and consistently called for a story to accompany the numbers presented in the data. As one participant remarked, "*I can make a number say anything I want it to say. I cannot make somebody else's story say something outside what they make it.*" Thus, there was a need to determine what overall data mattered as a program, as well as a continued, coordinated move toward collaborative data use and practices that would target continuous program improvement. This finding can clearly be seen in the words of a participant in the study: "*I would love to see our faculty collaboratively go after what we care about the most. We would need some agreement about data. I think it's a big deal for a program to self-study vs individuals.*

The second research question of the self-study of data use practices examined methods used to collect, analyze, and archive data collected throughout the program. These self-study data further indicated much program data were shown to be collected and available to faculty and staff. However, program collection methods had been somewhat cumbersome and not available to all stakeholders in a timely manner. Specifically, the constraints existing around the lack of a centralized data collection and management system continue to hold the program back with regard to the management of data. That said, the program had moved from "paper" to "online" data collection, analysis, and use, which appeared to allow for a more timely response to data use practices. However, as one participant pointed out, *"Out of the mountains and mountains and mountains of data that we have, how do we communicate it to everyone? How do we give it to everyone so that it is useful because they need to be able to access what they care about! It's hard to bring everyone together on it. We can't just throw all the data out there because everyone cares about different things.* Regarding values and beliefs about data use practices, the perceptions of data collection, archival, and analysis were much dependent upon the varied roles within the teacher preparation program. Another participant remarked, "*I don't know what is collected ... We are just out here in nebulousness.*" Some faculty did not know "how" to access these or from "whom." Thus, there appeared to be some disconnect between perception and reality regarding data use. Nonetheless, the interest in data cannot be denied or ignored. One participant commented, *"I would like to know more about what data is collected and how to use it to improve my course."* Another participant voiced this opinion, *"I think it would be helpful if we could really dig into the data ourselves versus someone presenting the data to us. It would mean more meetings, but I think we would get more out of the data. We would own the data more."*

Finally, the third research question of the self-study sought to examine the program policies and practices that impacted data use practices. Data analysis from year one of the study indicated policies regarding data use were not explicit and well-defined. However, certain roles in the program were more innately data-driven than others. This finding was exemplified within the following statement made during the focus group made up of supervisors of student teachers, "*I would say that this is definitely a data-driven position. We are constantly looking at the data we collect to make decisions for our instruction."* Moreover, non-tenured track promotion guidelines for clinical faculty had been developed and implemented within the department during the previous year. Within those guidelines, the following expectation was articulated: "competence in the field, evidence of ongoing scholarly and/or creative impact beyond the university, and evidence of interest in the welfare of the institutions of higher learning and leadership within the college," which could translate to collaborative, intentional data use practices. Further, coordinated, practice of collaborative work around data use appeared to be emerging in the program which would assist in negating the "silo-effect" of working individually. Supervisors felt strongly about collaborative data use, "*Faculty need to see what we collect. They really need to see the data."* Another commented, *"Somehow the information does need to be shared with those professors who are directly related to those in the field so they can support the needs of the students."* Finally, participants voiced a need for a more systematic data use plan for meeting accreditation requirements. As one participant commented, "*Never believe we need to be a different animal depending on who is coming to visit the zoo. No, I want to be just one kind of zebra. We need to be using data and evidence, and making changes occur based on evidence... based on outcomes.*"

Themes and Actionable Next Steps

The researcher worked with the leadership team to corroborate the emergent findings based on the analysis of these data. The group then collaboratively collapsed findings to four themes related to data use practices.

- Differentiated perceptions of data collection, analysis, and use are inherent within faculty and staff roles and responsibilities;
- A shared vision regarding data collection, analysis, and use should be the foundation for collaborative data use practices;
- Coordination of an intentional "data use" plan that includes time and space for the work will support building a culture that engages in "inquiry-based" data use practices, as opposed to one of compliance; and
- Context matters; Quantitative data is made more palatable and meaningful when accompanied by the qualitative story.

Finally, the researcher worked with the teacher preparation program leadership team to develop actionable next steps focused on program improvement efforts based on the findings and themes yielded by the self-study.

- Develop an improved communication plan focused on data collection, analysis, and use;
- Collaboratively construct programmatic outcomes with all stakeholders;
- Co-construct a data plan with stakeholders: "What data should guide programmatic decisions?"
- Use the collaboratively constructed programmatic outcomes and the data plan to inform continuous programmatic improvement efforts.

The group then delineated specific initiatives for the next academic year, which included: data-driven teacher education faculty meetings; online surveys with stakeholders regarding data use practices throughout the program; monthly data meetings with leadership; bi-annual data -driven stakeholder meetings; and bi-annual "data day" meetings with faculty and supervisors.

Phase Two: A Deeper Dive Into Data Use Practices

Year two of the self-study took place over the span of the year following year one of the study, and focused on a deeper dive into the teacher preparation program's data use practices. The researcher analyzed the data use practices resulting from initiatives set forth in the first phase of the self-study that were proposed for program assessment and continuous improvement. The focus was upon data collection around initiatives resulting from phase one of the data use self-study. Data collection included meeting agendas, meeting minutes, and online surveys that were completed by teacher education faculty, student teaching supervisors, P-12 administrators, mentor teachers, and teacher candidates. The second year of the self-study sought to further evaluate processes aimed at supporting and expanding data use practices throughout the program.

Year two of the self-study revealed a concerted, coordinated move toward collaborative data use that targeted continuous program improvement, which could clearly be seen in the data collection based on the initiatives that emerged out of year one of the self-study. These data indicated a program-wide shift to a more intentional and deliberate focus on data-driven decision-making, as well as collaboration among stakeholders. All stakeholders including leadership, faculty, supervisors, cooperating teachers, administrators, and even teacher candidates appeared to value these data collected during student teaching and seemed to be making a conscious effort to seek out relevant data. The online survey results revealed that participants appeared to be somewhat more aware of data collected within the program as the survey data showed stakeholders were using course evaluations, formative and summative student feedback, certification data, and student teacher evaluation data to make course level and program level changes and improvements.

Perhaps the most significant change noted in the data collected during year two of the study was that it appeared intentional data use for continuous improvement was becoming a more common practice in the teacher education program at this university. The institution of data day and the data-driven stakeholder meetings seemed to give rise to data-driven conversations and programmatic improvement efforts that would further support teacher candidate growth and practice. A culture of coordinated, collaborative work around data use seemed to have become the norm as evidenced by the plethora of data documented in the agendas and minutes from the various initiatives that came out of year one of the self-study. Meetings appeared to be grounded in data-driven tasks informed by programmatic improvement efforts. Program policies and practices over the past academic year were directly impacted by consistent data use practices.

DISCUSSION OF THE FINDINGS

The leadership within this teacher preparation program found that qualitative data collected during the self-study was critical to informing programmatic improvement. At the conclusion of the first year of the self-study, one member of the leadership team noted, "*This self-study was telling and useful. As a program, we need to take some steps back. But, that raises the question: Exactly what and how?*" Analysis of the data illuminated both areas of strength and areas of growth within the teacher education program of this university. Year two of the self-study clearly evidenced much to be celebrated around more consistent data use practices. Strides were made in beginning to build a culture of evidence within the teacher preparation program that was inquiry-based versus strictly compliance-oriented. During one interview from year one of the self-study, one participant in the study eloquently stated, "*Looking at data in and of itself is a waste of time unless you are responding to it.*" The leadership of this teacher preparation program took that to heart and determined to work on advancing a culture for continuous improvement grounded in data use practices.

We are continuing to learn more about the various policies and practices that support evidence-based decision making and program improvement in teacher education (Peck & McDonald, 2013, 2014). Self-study (Hamilton, 1998), where knowledge and learning is drawn from the study of a college's own teacher preparation practice has been shown to be useful in revealing robust themes in data use policies and practices (Davis & Peck, 2018). The qualitative data gathered in this case study and described within this article proved to be critical in the continuous improvement efforts of the teacher preparation program. The findings from the self-study of data use practices were invaluable to this institution, and provided the teacher preparation program with insight to its' own data use practices; specifically, how data use practices include the interaction of the individuals involved in the work of teacher preparation, the level to which tools for data use are available, and the organizational framework around data use practices. Through the data examined in this case study, the leadership of the teacher preparation program learned that collaborative, data-driven inquiry goes beyond checking off a box. Efforts grounded in collaboration provided for authentic use of data and informed program improvement program improvement efforts (Cuthrell & Lys, 2014). They used the qualitative data from the case study to inform a comprehensive approach to data use practices within the teacher preparation program such that all stakeholders have a part in the process of using data related to program outcomes in efforts to build a culture of evidence (Peck & Davis, in press).

IMPLICATIONS

As members of the leadership team of this teacher preparation program learned, there is merit and value in studying one's own institutional practices (Feuer et al., 2013). With an increasing call for accountability in teacher education (Cochran-Smith, et al, 2018), teacher preparation programs must engage in examining their own practices (CAEP, 2019) for continuous improvement. Thus, this case study may offer an illustration of how self-study data can provide a means to improve practice (Samaras & Freese, 2009; Hamilton, 1998). The qualitative data collected through interviews, focus groups, and artifacts informed a comprehensive approach to data use practices within the teacher preparation program, resulting in collective programmatic improvement (Kosnik et al., 2009). As proposed by Pinneagar and Erickson (2009), the leadership team of this teacher preparation program found the use of self-study to examine accreditation and programmatic data to be invaluable; thus, the usefulness and value of the findings within this case study of self-study may inform the academy how institutional self study in data use practices can support quantitative programmatic data in facilitating programmatic improvement initiatives.

REFERENCES

Aguerrebere, J. (2009). *Panel discussion at Teachers' use of data to impact teaching and learning*. Alliance for Excellent Education.

Cochran-Smith, M. (2009). "Re-culturing" teacher education: Inquiry, evidence, and action. *Journal of Teacher Education*, *60*(5), 458–468. doi:10.1177/0022487109347206

Cochran-Smith, M., Carney, M., Keefe, E., Burton, S., Chang, W.-C., Fernandez, M., ... Baker, M. (2018). *Reclaiming accountability in teacher education*. Teachers College Press. doi:10.1007/978-981-13-2026-2

Cochran-Smith, M., & Villegas, A. M. (2015). Framing teacher preparation research: An overview of the field, part I. *Journal of Teacher Education*, *66*(1), 7–20. doi:10.1177/0022487114549072

Council for the Accreditation of Educator Preparation. (2013). *2013 CAEP Standards*. http://caepnet.org/standards/introduction

Council of Chief State School Officers. (2016). *Accountability in teacher preparation: Policies and data in the 50 states and DC*. https://www.ccsso.org/sites/default/files/2017-10/50StateScan092216.pdf

Council of Chief State School Officers. (2018). *Measuring What Matters.* http://www.ccsso.org/sites/default/files/2018-02/Measuring%20What%20Matters.pdf

Council of Chief State School Officers Task Force on Teacher Preparation and Entry into the Profession. (2012). *Our responsibility, our promise: Transforming teacher preparation and entry into the profession.* Council of Chief State School Officers. https://ccsso.org/sites/default/files/2017-10/Our%20Responsibility%20Our%20Promise_2012.pdf

Cuthrell, K., & Lys, D. (2014). *Using data for program improvement.* American Association of Colleges of Teacher Education. https://secure.aacte.org/apps/rl/res_get.php?fid=1638&ref=rl

Datnow, A., & Hubbard, L. (2015). Teachers' use of assessment data to inform instruction: Lessons from the past and prospects for the future. *Teachers College Record, 117*(4), 1e26.

Davis, S., & Peck, C. (in press). Using data for program improvement in teacher education: A study of promising practices. *Teacher College Press.*

Davis, S., & Peck, C. A. (2018). Building capacity and commitment for data use in teacher education programs. In E. Mandinach & E. Gummer (Eds.), *Data for continuous programmatic improvement: Steps colleges of education must take to become a data culture* (pp. 68–92). Harvard Education Press.

Dowd, A. (2005). *Data don't drive: Building a practitioner-driven culture of inquiry to assess community college performance.* Lumina Foundation for Education. https://www.luminafoundation.org/files/publications/datadontdrive2005.pdf

Dwyer, C., Millett, C., & Paye, D. (2006). *A culture of evidence: A secondary assessment and learning outcomes.* Educational Testing Service.

Easton, J. Q. (2009). Using data systems to drive school improvement. Keynote address at the STATS-DC 2009 Conference, Bethesda, MD.

Easton, J. Q. (2010). Helping states and districts swim in an ocean of new data. Keynote address at the Educate with Data: Stats-DC 2010, Bethesda, MD.

Feldman, A. (2003). Validity and Quality in Self-Study. *Educational Researcher, 32*(3), 26–28. doi:10.3102/0013189X032003026

Feuer, M., Floden, R., Chudowsky, N., Ahn, J., & National Academy of Education. (2013). *Evaluation of teacher preparation programs: Purposes, methods, and policy options.* National Academy of Education.

Hamilton, M. L. (Ed.). (1998). Reconceptualizing teaching practice: Self-study in teacher education. Falmer.

Kosnik, C., Cleovoulou, Y., & Fletcher, T. (2009). The use of interviews in self-study research. In C. A. Lassonde, S. Galman, & C. Kosnik (Eds.), *Self-study research methodologies for teacher educators* (pp. 51–69). Sense Publishers. doi:10.1163/9789087906900_005

LaBoskey, V. K. (2004). The methodology of self study and it's theoretical underpinnings. In J. J. Loughran, M. L. Hamilton, V. K. LaBoskey, & T. Russell (Eds.), *International handbook of self-study of teaching and teacher education practices* (pp. 817–870). Kluwer. doi:10.1007/978-1-4020-6545-3_21

Loughran, J. (2005). Researching teaching about teaching: Self-study of teacher education practices. *Studying Teacher Education*, *1*(1), 5–16. doi:10.1080/17425960500039777

Mace, D. P., & Lueblke, P. (2014). *Creating organizational support for collaborative inquiry*. American Association of Colleges of Teacher Education. https://secure.aacte.org/apps/rl/res_get.php?fid=1638&ref=rl

Mandinach, E., & Gummer, E. (2015). Data-driven decision making: Components of the enculturation of data use in education. *Teachers College Record*, *117*(4), 1–8.

Mandinach, E. B. (2012). A perfect time for data use: Using data-driven decision making to inform practice. *Educational Psychologist*, *47*(2), 71–85. doi:10.1080/00461520.2012.667064

Moss, P. (2007). Evidence and decision making, the 106th yearbook of the National Society for the Study of Education. National Society for the Study of Education.

Peck, C. A., Gallucci, C., & Sloan, T. (2010). Negotiating implementation of high-stakes performance assessment policies in teacher education: From compliance to inquiry. *Journal of Teacher Education*, *61*(5), 451–463. doi:10.1177/0022487109354520

Peck, C. A., & McDonald, M. (2013). Creating "cultures of evidence" in teacher education: Context, policy, and practice in three high-data-use programs. *New Educator*, *9*(1), 1228. doi:10.1080/1547688X.2013.751312

Peck, C. A., McDonald, M., & Davis, S. (2015). *Using data for program improvement in teacher education: A study of promising practices*. American Association of Colleges of Teacher Education. https://secure.aacte.org/apps/rl/resource.php

Peck, C. A., & McDonald, M. A. (2014). What is a culture of evidence? How do you get one? And…should you want one? *Teachers College Record*, *116*, 1–27.

Pinnegar, S., & Erickson, L. B. (2009). Uncovering self-studies in accreditation review processes. In C. A. Lassonde, S. Galman, & C. Kosnik (Eds.), *Self-study research methodologies for teacher educators* (pp. 151–168). Sense Publishers.

Rao, V., & Woolcock, M. (2003). Integrating qualitative and quantitative approaches in program evaluation. In F. J. Bourguignon & L. P. da Silva (Eds.), *The impact of economic policies on poverty and income distribution: Evaluation techniques and tools* (pp. 165–190). Oxford University Press.

Samaras, A., & Freese, A. (2006). *Self-study of teaching practices.* Pete Lang Publishing, Inc.

Samaras, A., & Freese, A. (2009). Looking back and looking forward: An historical overview of the self-study school. In C. A. Lassonde, S. Galman, & C. Kosnik (Eds.), *Self-study research methodologies for teacher educators* (pp. 3–19). Sense Publishers.

Samaras, A., & Roberts, L. (2011). Flying solo. Teachers take charge of their learning through self-study research. *Journal of Staff Development, 32*(5), 42–45.

Schultze, U., & Avital, M. (2011). Designing interviews to generate rich data for information systems research. *Information and Organization, 21*(1), 1–16. doi:10.1016/j.infoandorg.2010.11.001

Seale, C. (1999). *The quality of qualitative research. Introducing qualitative methods series.* Sage.

Spillane, J. (2012). Data in practice: Conceptualizing the data-based decision-making phenomena. *American Journal of Education, 118*(2), 113–141. doi:10.1086/663283

Spring, J. (2003). *American education.* McGraw Hill.

Trinder, L. (2000). A critical appraisal of evidence-based practice. In L. Trinder & S. Reynolds (Eds.), Evidence-based practice, a critical appraisal (pp. 212-241). Blackwell Science. doi:10.1002/9780470699003.ch10

Weiss, J. (2012). Data for improvement, data for accountability. *Teachers College Record, 114*(11), 110307.

Chapter 13
The Democratic Potential of Narrative, Poetry, and Performance

Aaron S. Zimmerman
Texas Tech University, USA

ABSTRACT

This chapter will present an overview of three particular methodologies of arts-based research: narrative, poetry, and performance. This chapter will discuss the ways in which these methodological approaches to research may be effective means through which to capture and share the knowledge possessed by community stakeholders. This chapter has positioned community stakeholders as partners in arts-based research. When university faculty and community stakeholders form reciprocal, mutually beneficial partnerships, it becomes possible to create and disseminate the knowledge needed to support a democratic society.

There is reason to believe that arts-based research may be an effective way to facilitate the communication of knowledge within a deliberative democracy. In large part, the justification lies in the nature of arts-based research. Arts-based research is primarily concerned with advancing human understanding by persuading an audience to see the world through the eyes of another person via creative modes of exploration and representation (Diaz, 2002; Leavy, 2009). Art can spotlight the subjective nature of human experience (Belanger, 1998; Eisner, 2008), and by recognizing and emphasizing subjective, personal dimensions to experience, art has the power "to reveal what [previously] had not been noticed" (Barone & Eisner, 2012, p. 102).

DOI: 10.4018/978-1-7998-7600-7.ch013

This appreciation for the qualitative dimension of experience makes arts-based research a potent antidote to neoliberal epistemologies that portray standardized, quantitative measures as the only data worth knowing (Ball, 2016). As Eisner (1997) writes, "We have…concretized our view of what it means to know. We prefer our knowledge solid and like our data hard. It makes for a firm foundation, a secure place to stand" (p. 7). While neoliberalism would have us take for granted that "hard data" is the only knowledge worth knowing, Eisner (2008) reminds us that "the life of feeling is best revealed through those forms of feeling we call the arts" (p. 7). In other words, the arts invite qualitative and pluralistic definitions of what counts as knowledge (Gallas, 1991; Rasmussen, 2010; Sinclair, 2003). The arts insist that the world of innermost personal feeling is equally important (and equally real) as the world that is best represented through quantitative data and statistics.

"By connecting people on emotional and visceral levels, artistic forms of representation facilitate empathy, which is a necessary precondition for challenging harmful stereotypes…and building coalition/community across differences" (Leavy, 2009, p. 14). Since these (i.e., facilitating empathy, challenging stereotypes, understanding the lives of others) are precisely the experiences that individuals need in order to participate productively within a democratic society (Cipollone et al., 2018; Cooke, 2000; Gutmann & Thompson, 2004; Zeichner, 2020), arts-based research may, therefore, be a medium of research worth considering seriously. The remainder of this chapter will present three different arts-based research methodologies that may facilitate the sharing of knowledge between democratic citizens.

NARRATIVE

The narrative method involves "a collaborative method of telling stories, reflecting on stories, and (re)writing stories" (Leavy, 2009, p. 27; see also Clandinin & Rosiek, 2007; Pinnegar & Daynes, 2007). The process of constructing narrative involves an attempt to "access participants' life experiences and engage in a process of storying and restorying in order to reveal multidimensional meanings and present an authentic and compelling rendering of the data" (Leavy, 2009, p. 27). The more evocative and aesthetically compelling a constructed narrative is, the more its readers will be able to empathize with the narrative's protagonist(s).

This approach has clear relevance to the goal of producing and sharing the knowledge held by community stakeholders. Narratives may serve as a way to capture local knowledge in a complex, nuanced, evocative, and personally resonant manner, thus deepening the understanding between researchers, readers, and community stakeholders. Because a compelling narrative has the potential to transform one's outlook on the world and on the lives of others (Jones, 2003; Kim, 2006), it is logical to conjecture that life histories and lived experiences – when presented through a compelling and evocative narrative – will be more impactful than by merely providing researchers and practitioners with facts, demographics, and interview transcripts.

Indeed, narrative researchers are interested not only (or even primarily) in facts, but, rather, are concerned with the meanings that are captured and conveyed through narrative. Furthermore, these meanings are presented in rich, contextual detail (i.e., through moments of lived experience), and, thus, narrative inquiry is not focused on producing decontextualized knowledge that is generalizable. In this way, narrative inquiry allows researchers to capture the contextualized, authentic essence of dynamic human experience (Bailey & Tilley, 2002; Hardy, Gregory, & Ramjeet, 2009).

The epistemology of narrative inquiry is, in part, rooted in a particular philosophy of how human beings experience the world. As the philosopher John Dewey (1938) argued, human beings experience the world through a continuous flow of experience (rather than through discrete, disconnected moments). Relatedly, Dewey argued that human beings do not experience the world as a series of encounters with abstract forms or abstract ideas; rather, human beings encounter the world through singular moments that are contextualized in concrete detail in the real world. Abstract ideas may be helpful tools that human beings can use to understand the world upon reflection; however, our encounter with the world occurs within specific moments of lived experience (moments that cannot necessarily be reduced to abstract and generalizable principles). For this reason, a rich description of narrative experience (i.e., a description of the flow of experience as it occurs in real time and in context) may be one of the most valid and reliable ways to capture and represent knowledge about the human experience.

It is worth noting that Dewey's philosophy is recognizable in the methodological frameworks developed by narrative researchers. For example, Clandinin and Connelly (2000) have developed an approach to narrative inquiry that encourages researchers to attend to elements of personal and social interaction, of the continuity of time (past, present, and future), and of place and situation. That is to say, when researching lived experience, narrative researchers must pay particularly close attention to the manner in which human experience unfolds (i.e., through interactions with others, in time, and in particular places and situations). This approach is noticeably different than other approaches to social science (e.g., building statistical models with large scale data sets) which may seek to smooth away any particulars present in the data that is collected. Many social science methods (both quantitative and qualitative) rely on the process of categorization; narrative inquiry (and, arts-based methods, more generally) tend to eschew categorization and generalization in favor of a focus on particulars, uniqueness, and singularity.

As one example of how narrative inquiry might be applied to the context of democratic participation, community stakeholders might be asked to compose an autoethnographic narrative (Pelias, 2003, 2004). The purpose of an autoethnography is to situate an individual and his or her experiences within a particular society and culture and to facilitate an experience of resonance with the reader. The reader, by encountering the richly described lived experiences of the author, should come away from the narrative with a better, more empathetic understanding of what it is like to be that particular individual within that particular cultural context. This outcome is significant, given that empathy is a critical disposition for deliberative democracy (Miklikowska, 2012; Morrell, 2007; Rampal & Mander, 2013; Segal, 2011).

Authors of autoethnographic narratives are encouraged to describe their world from a subjective point of view:

As researchers working from critical perspectives such as feminism have challenged the subject-object, rational-emotional, and concrete-abstract dualism that have historically guided social scientific inquiry, a space has opened for researchers to create methods... [that] account for the researcher's subjectivity. (Leavy, 2009, p. 38)

As discussed earlier in this chapter, arts-based methodologies (including narrative) can challenge the neoliberal assumptions that "hard" data (i.e., numbers and metrics) are the only data worth knowing; these alternative forms of data (e.g., narrative data) are particularly valuable in cultivating empathy, understanding, and communication between individuals from different generations, cultures, or social strata. Because autoethnographic narratives intentionally place premier value on the subjectivity of the author and the particulars of the author's lived experience, these narratives may prevent researchers and consumers of research from reducing the lives of others into disembodied demographics. Indeed, narratives may help researchers, practitioners, and democratic citizens at large to see through the eyes of another.

POETRY

Poetry, like autoethnographic narratives, spotlight subjectivity; poetry, however, can be considered to be an even more intentionally evocative presentation of specific moments in the social world. As Leavy (2009) writes, "Sensory scenes created with skillfully placed words and purposeful pauses, poems push feelings to the forefront capturing heightened moments of social reality as if under a magnifying glass" (p. 63). As is the case with many arts-based methodologies, poetry may serve as a particularly valuable supplement within the political context of neoliberalism: specifically, in the midst of a political and educational milieu that prioritizes rubrics, standardized tests, and strictly measurable (i.e., quantifiable) outcomes (see Hurst, 2015; Regan & Jesse, 2019; Saltman, 2016), poetry can, in Leavy's words, "push feelings to the forefront."

By attending to linguistic features such as the length and rhythm of words, nuances in meaning, and the implications of silences, poems can represent individual moments of social reality in ways that heighten subjective interpretation and the conveying of feeling (Richardson, 1998). In this way, poetry can serve as a powerful alternative way of representing data, especially if the researcher wants his or her audience to understand the data in a particular way:

The representation of the data in poetic form is not simply an alternative way of presenting the same information; rather it can help the researcher evoke different meanings from the data, work through a different set of issues, and help the audience to receive the data differently. (Leavy, 2009, p. 64; see also Glesne, 1997; Rasberry, 2002)

Said in another way, poetry can be used to not only represent data creatively, but can also be used to reinterpret existing data, such as interview transcripts and field notes (Hanauer, 2010; Xerri, 2017). By examining words poetically, "we can surprise both ourselves and our audiences with new possibilities" (Cahnmann, 2003, p. 37).

As one example of how poetry might be applied to the context of community-engagement and democratic participation, researchers might ask community stakeholders to participate in a process of poetic transcription (Faulkner, 2009; Glesne, 1997; Leavy, 2009). Poetic transcription is a process through which an interview transcript is presented in poetic form. After conducting an interview and obtaining an interview transcript, a researcher will identify specific words and phrases from the transcript and use these words and phrases to form the raw material for a poem. Community stakeholders might be asked to participate in interviews and then be given the agency to create poetry based on these interview transcripts. This methodological approach, similar to others described in this chapter, "offers qualitative practitioners a way to address…commitments to participants' voice… Researchers committed to accessing subjugated voices might be especially inclined toward this interpretation style" (Leavy, 2009, pp. 75-76). Amplifying the voices of community members is, indeed, one of the central commitments of deliberative democracy, thus making poetry a particularly apt methodological choice in the context of learning from the knowledge possessed by democratic citizens.

PERFORMANCE

Performance is a methodology that can bring research to life, affording a real-time exchange of meaning between performer and audience (Leavy, 2009; Madison, 2003; Norris, 2000). The experience of crafting and presenting a performance can disrupt conventional ways of knowing and conventional ways of thinking about knowledge and the communication of knowledge (Gray, 2003). Specifically, performances communicate a knowledge that is fluid, integrated, negotiated, and experienced in-the-moment. The fact that performance affords an *experience* of knowledge (rather than a static, disembodied account of knowledge; e.g., knowledge communicated through the reading a textbook) makes performance a particularly powerful medium for strengthening interpersonal connection and inter-subjective understanding.

Just as a performance integrates multiple dimensions of experience and expression (i.e., word, number, image, gesture, and sound; see Norris, 2000), performance can build bridges between subjugated voices and mainstream perspectives (Langellier & Peterson, 2006). Because the act of performance is an act of direct and unfiltered communication, performance makes it impossible for subjugated voices to remain unheard. Thus, performance has been recognized as one of the most powerful means of exposing oppression and compelling critical awareness (Denzin, 2006; Erel, Reynolds, & Kaptani, 2017; Österlind, 2008; Valentine, 2006), and, in this way, performance can be understood as a political act. Performance may, therefore, be one of the most effective means by which to advance deliberative democracy as well as democratic participation.

Conceptualizing research and knowledge as *performance* positions research and knowledge as inextricably rooted in the social world. That is to say, performative research eschews research questions and research designs that are static or predetermined or that seek to produce generalizable knowledge. Rather, the questions asked and the knowledge produced through performative inquiry are situational, contextual, fluid, and, perhaps, even ambiguous: "Through performance…it is possible to say something in a meaningful way that still preserves the uncertainty, paradox, or ambiguity that characterises many aspects of the social world" (Douglas & Carless, 2013, p. 56). This highlights, once again, one of the unique and valuable features of arts-based research methodologies. Given art's ambiguous, flexible, and complex nature, arts-based research has the potential to reflect accurately the ambiguity, flexibility, and complexity characteristic of our social world.

It is also worth noting that while the nature of performative research is radically different from traditional methodologies of social science and knowledge creation, the epistemological and methodological stances towards knowledge production and knowledge dissemination characteristic of performative research may be critical within the context of a deliberative democracy. Indeed, such an epistemological, methodological, and political stance towards research is in harmony with Dewey's pragmatic stance towards knowledge: i.e., abstract knowledge exists only as a tool for solving the real-world problems of society, and such solutions must be perpetually developed, negotiated, and recreated through dialogue (Dewey, 1938; Honneth, 1998; Warren, 2017). Performing research (i.e., performing knowledge) provides the concrete means through which such dialogue and renegotiation of knowledge can take place.

As one example of how performance might be applied in the context of community-engaged research, community stakeholders might be asked to author an ethnodrama. An ethnodrama consists of writing up research findings in the form of a script. In the words of Saldaña (2005), an ethnodrama is a written script that consists of "dramatized, significant selections of narrative collected through interviews, participant observation field notes, journal entries, and/or print and media artifacts… [This is the process of] dramatizing the data" (p. 2). Ethnodrama is a particularly apt approach within the context of presenting research on the lives of others, given

> *the ability of dramatic performance to get at and present rich, textured, descriptive situated, contextual experiences and multiple meanings from the perspectives of those studied in the field…[This allows researchers] to explore the dimensionality, tonality, and multisensory experiences that occur within the field [and the community] in ways not enabled by traditional textual representation. (Leavy, 2009, 145)*

As with narrative and poetry, performance challenges the neoliberal assumption that the only knowledge worth knowing about communities and citizens can be expressed through quantitative measures and written reports; indeed, the art of performance insists that some knowledge must be experienced to be understood.

CHALLENGES RELATED TO DEMOCRATIC APPROACHES TO RESEARCH

One of the primary challenges related to the endeavor research within a deliberative democracy concerns deciphering what knowledge (i.e., whose knowledge) counts (Domínguez, 2019; Lewis, 2018; Matusov, 1996; Mignolo, 2009; Spivak, 1988; Zeichner, 2016). Specifically, it is not a trivial assumption that researchers (and university-based educators) when asked will casually relinquish their claim to epistemological authority regarding tenets of what should be taught in schools, what knowledge is worth knowing, how knowledge should be taught in schools, or how knowledge should inform public policy (Baquedano-López, Alexander, & Hernández, 2013; Guillen & Zeichner, 2018; Jurow, Horn, & Philip, 2019; Pinar,

2005; Schubert, 2004; Zeichner, 2009; Zygmunt & Cipollone, 2018; Zygmunt & Clark, 2016). These are not only epistemological but also deeply political and ethical questions. Therefore, one must acknowledge the deeply-rooted challenges related to the endeavor of democratizing education and democratizing research (Ansley & Gaventa, 1997; Ferman & Hill, 2004).

To further consider the conflicts related to the relinquishing of epistemological authority, in the context of teacher education, for example, teacher educators may object that community stakeholders "just tell stories" or that these community members are "not pedagogically skilled enough to teach" preservice teachers (Zeichner et al., 2016, p. 288). Conversely, community mentors may experience frustration with respect to their interactions with teacher educators and preservice teachers. For example, community mentors may discern a lack of willingness from teacher educators and/or preservice teachers to engage in difficult conversations regarding privilege; or, community mentors may observe that teacher educators and/or preservice teachers rely on deficit language when describing students and families. Guillen and Zeichner (2018) have documented one such interaction:

After one small group discussion…a number of mentors questioned one preservice teacher who shared his thoughts on college, which he believed was not for everyone. This disturbed a number of mentors…It did not appear to bother this preservice teacher when a mix of mentors from a variety of ethnic backgrounds challenged his statement. They were most concerned that for many in their communities, college was an educational goal that they had historically been discouraged or barred from accessing, and teachers were often the gatekeepers…[T]hey found it most frustrating that the preservice teacher did not understand their perspectives in attempting to highlight the dangers in his thinking. (p. 148)

Thus, one of the primary challenges related to using university-based knowledge to create and confirm the curriculum and pedagogy of the next generation of teachers is that such curricula and such methods of instruction may or may not incorporate the knowledge or commitments of local communities. If that is the case, then ostensibly democratic education may, in fact, only serve to reinforce epistemologies reified by hegemonic political agendas (Domínguez, 2019; Lewis, 2018; Zeichner, 2016). This observation further justifies the need to consider more participatory, community-engaged approaches to research and knowledge production. As this chapter has argued, arts-based approaches to research (including narrative, poetry, and performance) may be particularly appropriate in this respect.

Having acknowledged these potential challenges and resistances, it is worth concluding this chapter with a discussion of how to mitigate these potential epistemic tensions between community stakeholders and university-based researchers, university-based educators, and university-based teacher educators. Indeed, these tensions are considered to be a common feature of university-community collaborations (Bringle, Clayton, & Price, 2009; Huxham & Vangen, 2005; Jacoby, 2003). Unless these research partnerships are developed reciprocally, research outcomes (and, more importantly, trust between the research partners) can be jeopardized.

Given that community engagement is intended to reverse exploitative neoliberal trends in education (including in higher education), community-engaged researchers must remain mindful of community stakeholders' motivations and concerns throughout the research process (Howarth, Currie, Morrell, Sorensen, & Bengle, 2017). Towards this end, university-based researchers must plan community-engaged research projects in collaboration with community stakeholders; university-based researchers must debrief and reflect upon the research process with community stakeholders during each stage of the project; and, university-based researchers must be willing to adjust their plans based upon the input and the best interests of the community stakeholders as the research project develops. Thus, even though the authoring of narrative, poetry, and performance hold the promise of amplifying the voice community stakeholders, arts-based researchers must not impose these research endeavors upon community stakeholders.

DeLugan and colleagues (2014, p. 161) provide community-engaged researchers with a useful set of questions that serve as guideposts before initiating any given community-engaged research project. Arts-based researchers might ask the following questions before and during a research partnership with community stakeholders:

- How is the research project mutually beneficial to both the university-based researchers and the community partners?
- How do community partners understand the purpose and processes of these research projects?
- How can the community stakeholders and the teacher educators collaborate to identify shared goals?
- Does the plan for the research project fairly distribute work between community stakeholders and the university-based researchers? Are responsibilities, risks, and rewards evenly distributed?
- Have the researchers and the community stakeholders discussed potential difficulties or conflicts related to planning, implementation, or dissemination?

As discussed earlier, university-based researchers and educators may perceive themselves to be experts (Barnes, 2017; Zeichner et al., 2016) and may, in the process, discount the value of the knowledge held by community stakeholders. Some university-based researchers may see communities only as sights for data collection (Bringle & Hatcher, 2002), and, such a stance will inevitably foster feelings of patronization, mistrust, and alienation (Buys & Bursnall, 2007; Enos & Morton, 2003; Freeman, Brugge, Bennett-Bradley, Levy, & Carrasco, 2006). Thus, even when community-based arts-based research projects are initiated with the goal of learning more about the needs, goals, and concerns of a given community – e.g., "What does this community value?...What kinds of things do people want for their children?" (Cipollone et al., 2018, p. 719) – university-based researchers must approach these research projects with an attitude of reciprocity. Unless the community stakeholders have a say in the design of the research and unless they are clearly able to perceive how they themselves will benefit from this research partnership, the university-community partnership may suffer (Bracken, 2007; Howarth et al., 2017; Prins, 2005; Savan, Flicker, Kolenda, & Mildenberger, 2009; Strier, 2014).

To assist in the negotiation of university-community collaborations, Dumalo and Janke (2012, p. 163) suggest a series of discussion prompts that researchers can use as a framework for addressing the tensions that they may encounter during their research partnership with community stakeholders:

- How do we (i.e., researchers and community stakeholders) want to relate to each other?
- What roles and responsibilities do we each want to take?
- What work do we do together? What work do we do separately?
- How do we define what to expect from this university-community partnership?
- How do we determine what information, opinions, and ideas to share?
- What information and issues can we talk about? What information and issues should we not talk about?
- What information does each of us need in order for each of us to do the work?

When university-based researchers and community stakeholders openly engage in dialogue focused on these issues, university and community stakeholders can form and sustain trusted and productive research partnerships that have the potential to reify and disseminate local knowledge that can augment community voices in the democratic process.

CONCLUSION

This chapter has positioned community stakeholders as partners in arts-based research. When university faculty and community stakeholders form reciprocal, mutually beneficial partnerships, it becomes possible to create and disseminate the knowledge needed to support a democratic society. Narrative, poetry, and performance have the potential to shift "the center of gravity" (Guillen & Zeichner, 2018, p. 141) of knowledge production from the university to the community. By representing lived experiences and by giving voice to subjugated perspectives, art has the power to directly confront audiences with insider knowledge of unique social realities. Arts-based research, thus, helps to make room for "the intersection of social and political progress, the emergence of alternative theoretical and epistemological groundings, [and] overarching social justice-oriented research initiatives" (Leavy, 2009, p. 253).

One might say that the validity of arts-based research (and of the arts, more generally) rests in a given artwork's evocative power: "The evaluation of the knowledge generated in arts-based research includes...a creative examination of how art evokes responses and connections; [and] a careful inquiry into the methods that art uses to unsettle ossified thinking and provoke imagination" (Sinner, Leggo, Irwin, Gouzouasis, & Grauer, 2006, p. 1252). This is a particularly apt quote to consider, given that, to advance democracy, the disruption of ossified thinking is necessary. That is to say, within a democracy, individuals must be willing to listen and to adopt new perspectives. Furthermore, learning to listen and learning to think differently are dispositions cultivated not only by "experts" (e.g., university-based researchers) but also by (if not primarily by) the democratic citizens themselves (i.e., local community members). The arts may be a particularly powerful way to empower democratic citizens to share their expertise; and, furthermore, encountering the arts (for example, through narrative, poetry, and performance) may be a particularly powerful (and evocative) way for democratic citizens (including researchers) to recognize the significance of listening to new voices.

REFERENCES

Ansley, F., & Gaventa, J. (1997). Researching for democracy and democratizing research. *Change: The Magazine of Higher Learning*, *29*(1), 46–53. doi:10.1080/00091389709603114

Bailey, P. H., & Tilley, S. (2002). Storytelling and the interpretation of meaning in qualitative research. *Journal of Advanced Nursing*, *38*(6), 574–583. doi:10.1046/j.1365-2648.2000.02224.x PMID:12067396

Ball, S. J. (2016). Subjectivity as a site of struggle: Refusing neoliberalism? *British Journal of Sociology of Education*, *37*(8), 1129–1146. doi:10.1080/01425692.2015.1044072

Baquedano-López, P., Alexander, R. A., & Hernández, S. J. (2013). Equity issues in parental and community involvement in schools: What teacher educators need to know. *Review of Research in Education*, *37*(1), 149–182. doi:10.3102/0091732X12459718

Barnes, M. E. (2017). Encouraging interaction and striving for reciprocity: The challenges of community-engaged projects in teacher education. *Teaching and Teacher Education, 68*, 220–231. doi:10.1016/j.tate.2017.09.004

Barone, T., & Eisner, E. W. (2012). Arts-based research. *Sage (Atlanta, Ga.)*.

Belanger, S. L. (1998). Knowing through art. In W. Braud & R. Anderson (Eds.), Transpersonal research methods for the social sciences: Honoring human experience (pp. 166-168). SAGE.

Bracken, S. (2007). The importance of language, context, and communication as components of successful partnership. *New Directions for Community Colleges, 2007*(139), 41–47. doi:10.1002/cc.291

Bringle, R., Clayton, P., & Price, M. (2009). Partnerships in service-learning and civic engagement. *Partnerships: A Journal of Service-Learning & Civic Engagement, 1*(1), 1-20.

Bringle, R., & Hatcher, J. (2002). Campus-community partnerships: The terms of engagement. *The Journal of Social Issues, 58*(3), 503–516. doi:10.1111/1540-4560.00273

Buys, N., & Bursnall, S. (2007). Establishing university-community partnerships: Processes and benefits. *Journal of Higher Education Outreach & Engagement, 29*(1), 73–86.

Cahnmann, M. (2003). The craft, practice, and possibility of poetry in educational research. *Educational Researcher, 32*(3), 29–36. doi:10.3102/0013189X032003029

Cipollone, K., Zygmunt, E., & Tancock, S. (2018). "A paradigm of possibility": Community mentors and teacher preparation. *Policy Futures in Education, 16*(6), 709–728. doi:10.1177/1478210317751270

Clandinin, D. J., & Connelly, F. M. (2000). *Narrative inquiry: Experience and story in qualitative research.* Jossey-Bass.

Clandinin, D. J., & Rosiek, J. (2007). Mapping a landscape of narrative inquiry: Borderland spaces and tensions. In D. J. Clandinin (Ed.), *Handbook of narrative inquiry: Mapping a methodology* (pp. 35–75). Sage. doi:10.4135/9781452226552.n2

Cooke, M. (2000). Five arguments for deliberative democracy. *Political Studies, 48*(5), 947–969. doi:10.1111/1467-9248.00289

DeLugan, R. M., Roussos, S., & Skram, G. (2014). Linking academic and community guidelines for community-engaged scholarship. *Journal of Higher Education Outreach & Engagement*, *18*(1), 155–168.

Denzin, N. K. (2006). The politics and ethics of performance pedagogy: Toward a pedagogy of hope. In D. S. Madison & J. Hamera (Eds.), *The Sage handbook of performance studies* (pp. 325–338). Sage. doi:10.4135/9781412976145.n18

Dewey, J. (1938). *Education and experience*. Simon and Schuster.

Diaz, G. (2002). Artistic inquiry: On Lighthourse Hill. In C. Bagley & M. B. Cancienne (Eds.), *Dancing with data* (pp. 147–161). Peter Lang.

Domínguez, M. (2019). Decolonial innovation in teacher development: Praxis beyond the colonial zero-point. *Journal of Education for Teaching*, *45*(1), 47–62. doi:10.1080/02607476.2019.1550605

Douglas, K., & Carless, D. (2013). An invitation to performative research. *Methodological Innovations Online*, *8*(1), 53–64. doi:10.4256/mio.2013.0004

Eisner, E. W. (1997). The promise and perils of alternative forms of data representation. *Educational Researcher*, *26*(6), 4–10. doi:10.3102/0013189X026006004

Eisner, E. W. (2008). Art and knowledge. In G. J. Knowles & A. L. Cole (Eds.), *Handbook of the arts in qualitative research* (pp. 3–12). Sage.

Enos, S., & Morton, K. (2003). Developing a theory and practice of campus-community partnerships. In *Building partnerships for service-learning* (pp. 20–41). Jossey-Bass.

Erel, U., Reynolds, T., & Kaptani, E. (2017). Participatory theatre for transformative social research. *Qualitative Research*, *17*(3), 302–312. doi:10.1177/1468794117696029

Faulkner, S. (2009). *Poetry as method: Reporting research through verse*. Left Coast Press.

Ferman, B., & Hill, T. L. (2004). The challenges of agenda conflict in higher-education-community research partnerships: Views from the community side. *Journal of Urban Affairs*, *26*(2), 241–257. doi:10.1111/j.0735-2166.2004.00199.x

Freeman, E. R., Brugge, D., Bennett-Bradley, W. M., Levy, J., & Carrasco, E. R. (2006). Challenges of conducting community-based participatory research in Boston's neighborhoods to reduce disparities in asthma. *Journal of Urban Health*, *83*(6), 1013–1021. doi:10.100711524-006-9111-0 PMID:17103339

Gallas, K. (1991). Arts as epistemology: Enabling children to know what they know. *Harvard Educational Review*, *61*(1), 40–51. doi:10.17763/haer.61.1.l7518t5kh06710n4

Glesne, C. (1997). That rare feeling: Re-presenting research through poetic transcription. *Qualitative Inquiry*, *3*(2), 202–221. doi:10.1177/107780049700300204

Gray, R. E. (2003). Performing on and off the stage: The place(s) of performance in arts-based approaches to qualitative inquiry. *Qualitative Inquiry*, *9*(2), 254–267. doi:10.1177/1077800402250957

Guillen, L., & Zeichner, K. (2018). A university-community partnership in teacher education from the perspective of community-based teacher educators. *Journal of Teacher Education*, *69*(2), 140–153. doi:10.1177/0022487117751133

Gutmann, A., & Thompson, D. F. (2004). *Why deliberative democracy?* Princeton University Press. doi:10.1515/9781400826339

Hanauer, D. I. (2010). *Poetry as research: Exploring second language poetry writing*. John Benjamins. doi:10.1075/lal.9

Hardy, S., Gregory, S., & Ramjeet, J. (2009). An exploration of intent for narrative methods of inquiry. *Nurse Researcher*, *16*(4), 7–19. doi:10.7748/nr2009.07.16.4.7.c7157 PMID:19653542

Honneth, A. (1998). Democracy as reflexive cooperation: John Dewey and the theory of democracy today. *Political Theory*, *26*(6), 763–783. doi:10.1177/0090591798026006001

Howarth, J., Currie, M. A., Morrell, E., Sorensen, J., & Bengle, T. (2017). Challenges of building community-university partnerships in new poverty landscapes. *Community Development (Columbus, Ohio)*, *48*(1), 48–66. doi:10.1080/15575330.2016.1244696

Hurst, D. (2015). Neoliberalism, markets and accountability: Transforming education and undermining democracy in the United States and England. *Policy Futures in Education*, *3*(1), 3–15. doi:10.2304/pfie.2005.3.1.6

Huxham, C., & Vangen, S. (2005). *Managing to collaborate: The theory and practice of collaborative advantage*. Routledge.

Jacoby, B. (2003). *Building partnerships for service-learning*. Jossey-Bass.

Jones, K. (2003). The turn to a narrative knowing of persons: One method explored. *Narrative Studies*, *8*(1), 60–71. doi:10.1177/136140960300800109

Jurow, S., Horn, I. S., & Philip, T. M. (2019). Re-mediating knowledge infrastrctures: A site for innovation in teacher education. *Journal of Education for Teaching*, *45*(1), 82–96. doi:10.1080/02607476.2019.1550607

Kim, J. (2006). For whom the school bell tolls: Conflicting voices inside an alternative high school. *International Journal of Education & the Arts*, *7*(6), 1–19.

Langellier, K. M., & Peterson, E. E. (2006). Shifting contexts in personal narrative performance. In D. S. Madison & J. Hamera (Eds.), *The Sage handbook of performance studies* (pp. 151–168). Sage. doi:10.4135/9781412976145.n9

Leavy, P. (2009). *Method meets art: Arts-based research practice.* Guilford.

Lewis, J. S. (2018). Releasing a tradition: Diasporic epistemology and the decolonized curriculum. *The Cambridge Journal of Anthropology*, *36*(2), 21–33. doi:10.3167/cja.2018.360204

Madison, D. S. (2003). Performance, personal narrative, and the politics of possibility. In Y. S. Lincoln & N. K. Denzin (Eds.), *Turning points in qualitative research* (pp. 469–486). AltaMira Press.

Matusov, E. (1996). Intersubjectivty without agreement. *Mind, Culture, and Activity*, *3*(1), 29–45. doi:10.120715327884mca0301_4

Mignolo, W. D. (2009). Epistemic disobedience, independent thought and decolonial freedom. *Theory, Culture & Society*, *26*(7-8), 159–181. doi:10.1177/0263276409349275

Miklikowska, M. (2012). Psychological underpinnings of democracy: Empathy, authoritarianism, self-esteem, interpersonal trust, normative identity style, and openness to experience as predictors of support for democratic values. *Personality and Individual Differences*, *53*(5), 603–608. doi:10.1016/j.paid.2012.04.032

Morrell, M. E. (2007). Empathy and democratic education. *Public Affairs Quarterly*, *21*(4), 381–403.

Norris, J. (2000). Drama as research: Realizing the potential of drama in education as a research methodology. *Youth Theatre Journal*, *14*(1), 40–51. doi:10.1080/08929092.2000.10012516

Österlind, E. (2008). Acting out of habits–can theatre of the oppressed promote change? Boal's theatre methods in relation to Bourdieu's concept of habitus. *Research in Drama Education*, *13*(1), 71–82. doi:10.1080/13569780701825328

Pelias, R. J. (2003). The academic tourist: An autoethnography. *Qualitative Inquiry*, *9*(3), 369–373. doi:10.1177/1077800403009003003

Pelias, R. J. (2004). *A methodology of the heart: Evoking academic and daily life*. AltaMira Press.

Pinar, W. F. (2005). The problem with curriculum and pedagogy. *Journal of Curriculum and Pedagogy*, *2*(1), 67–82. doi:10.1080/15505170.2005.10411529

Pinnegar, S., & Daynes, J. G. (2007). Locating narrative inquiry historically: Thematics in the turn to narrative. In D. J. Clandinin (Ed.), *Handbook of narrative inquiry: Mapping a methodology* (pp. 3–34). Sage. doi:10.4135/9781452226552.n1

Prins, E. (2005). Framing a conflict in a campus-community partnership. *Journal of Planning Education and Research*, *25*(1), 57–74. doi:10.1177/0739456X04270370

Rampal, A., & Mander, H. (2013). Lessons on food and hunger: Pedagogy of empathy for democracy. *Economic and Political Weekly*, *48*(28), 50–57.

Rasberry, G. W. (2002). Imagine, inventing a data-dancer. In C. Bagley & M. B. Cancienne (Eds.), *Dancing the data* (pp. 106–120). Peter Lang.

Rasmussen, B. (2010). The "good enough" drama: Reinterpreting constructivist aesthetics and epistemology in drama education. *Research in Drama Education*, *15*(4), 529–546. doi:10.1080/13569783.2010.512187

Regan, P. M., & Jesse, J. (2019). Ethical challenges of edtech, big data and personalized learning: Twenty-first century student sorting and tracking. *Ethics and Information Technology*, *21*(3), 167–179. doi:10.100710676-018-9492-2

Richardson, M. (1998). Poetics in the field and on the page. *Qualitative Inquiry*, *4*(4), 451–462. doi:10.1177/107780049800400401

Saldaña, J. (Ed.). (2005). *Ethnodrama: An anthology of reality theater*. AltaMira Press.

Saltman, K. (2016). Corporate schooling meets corporate media: Standards, testing, and technophilia. *Review of Education, Pedagogy & Cultural Studies*, *38*(2), 105–123. doi:10.1080/10714413.2016.1155953

Savan, B., Flicker, S., Kolenda, B., & Mildenberger, M. (2009). How to facilitate (or discourage) community-based research: Recommendations based on a Canadian survey. *Local Environment*, *14*(8), 783–796. doi:10.1080/13549830903102177

Schubert, W. H. (2004). Curriculum and pedagogy for reconstruction and reconceptualization. *Journal of Curriculum and Pedagogy*, *1*(1), 19–21. doi:10.1080/15505170.2004.10411471

Segal, E. A. (2011). Social empathy: A model built on empathy, contextual understanding, and social responsibility that promotes social justice. *Journal of Social Service Research*, *37*(3), 266–277. doi:10.1080/01488376.2011.564040

Sinclair, N. (2003). Aesthetic Values in Mathematics: A Value-Oriented Epistemology. *International Group for the Psychology of Mathematics Education*, *4*, 199–205.

Sinner, A., Leggo, C., Irwin, R., Gouzouasis, P., & Grauer, K. (2006). Arts-based education research dissertations: Reviewing the practices of new scholars. *Canadian Journal of Education*, *29*(4), 1223–1270. doi:10.2307/20054216

Spivak, G. C. (1988). Can the subaltern speak? In C. Nelson & L. Grossberg (Eds.), *Marxism and the interpretation of culture* (pp. 271–315). University of Illinois Press. doi:10.1007/978-1-349-19059-1_20

Strier, R. (2014). Fields of paradox: University-community partnerships. *Higher Education*, *68*(2), 155–165. doi:10.100710734-013-9698-5

Valentine, K. B. (2006). Unlocking the doors for incarcerated women through performance and creative writing. In D. S. Madison & J. Hamera (Eds.), *The Sage handbook of performance studies* (pp. 309–324). Sage. doi:10.4135/9781412976145.n17

Warren, M. E. (2017). A problem-based approach to democratic theory. *The American Political Science Review*, *111*(1), 39–53. doi:10.1017/S0003055416000605

Xerri, D. (2017). Poetry education research as an anchorage of thought: Using poetry as interview stimulus material. *International Journal of Research & Method in Education*, *40*(4), 379–388. doi:10.1080/1743727X.2015.1119110

Zeichner, K. (2009). *Teacher education and the struggle for social justice*. Routledge. doi:10.4324/9780203878767

Zeichner, K. (2016). Advancing social justice and democracy in teacher education: Teacher preparation 1.0, 2.0, and 3.0. *Kappa Delta Pi Record*, *52*(4), 150–155. doi:10.1080/00228958.2016.1223986

Zeichner, K. (2020). Preparing teachers as democratic professionals. *Action in Teacher Education*, *42*(1), 38–48. doi:10.1080/01626620.2019.1700847

Zeichner, K., Bowman, M., Guillen, L., & Napolitan, K. (2016). Engaging and working in solidarity with local communities in preparing the teachers of their children. *Journal of Teacher Education*, *67*(4), 277–290. doi:10.1177/0022487116660623

Zygmunt, E., & Cipollone, K. (2018). A pedagogy of promise: Critical service-learning as praxis in community-engaged, culturally responsive teacher preparation. In T. D. Meidl & M. M. S. Dowell (Eds.), *Handbook of research on service-learning initiatives in teacher education programs*. IGI Global. doi:10.4018/978-1-5225-4041-0.ch018

Zygmunt, E., & Clark, P. (2016). *Transforming teacher education for social justice*. Teachers College Press.

Chapter 14
Grounded Theory Approach and the Process of Men Taking Responsibility in Domestic Violence Interventions

Zeynep Turhan
https://orcid.org/0000-0002-5343-9442
Bartin University, Turkey

ABSTRACT

Research on the effectiveness of domestic violence perpetrator interventions, where perpetrators are challenged to apply non-violent behavior and develop a healthy intimate relationship, is uneven and often inconsistent. Effective and accurate methodological tools and approaches can provide a better understanding of how the program providers offer successful practices to perpetrators of domestic violence. The goal of this chapter is to discuss the potential benefits of the grounded theory approach in a qualitative inquiry that might reduce methodological challenges in the examination of the process of men taking responsibility in interventions. This chapter contributes to the debate about the best way to examine perpetrators' behavioral change processes and identifies the benefits of implementing the grounded theory approach and reflexivity.

DOI: 10.4018/978-1-7998-7600-7.ch014

INTRODUCTION

Grounded theory approach is a research methodology that is often used in social sciences. The reason for applying grounded theory approach might be to develop a new theory. Theory is vital for reducing complex social problems and recognizing relevant aspects in broader contexts (Gibson & Hartman, 2014). Moreover, grounded theory provides adequate comparisons by focusing on specific properties across social elements (Gibson & Hartman, 2014). Behavioral change process among perpetrators of domestic violence can be better understood by applying theoretical sampling, a constant comparative method and coding exercises found in the grounded theory approach.

The main body of domestic violence perpetrator literature has concentrated on the rate of or factors around completion or drop-outs among perpetrators (Cuevas & Bui, 2016; Kalogo, 2015; Rondeau, Brodeur, Brochu & Lemire, 2001; Rosenbaum, Gearan, & Ondovic, 2001; Rothman & Coutinho, 2007; Waller, 2016). However, some researchers have explored the steps and process of behavioral change rather than solely focusing on the outcome. For instance, different stages of change and various conditions around the readiness to take new action have been investigated (Carbajosa, Catalá-Miñana, Lila, Gracia, & Boira, 2017; Chovanec, 2009; Daniels & Murphy, 1997). A few studies about changing behavior during psychotherapy also examined the stages of change and the effectiveness of some techniques (McConnaughy, Prochaska & Velicer, 1983; Prochaska & DiClemente, 1982). Therefore, a consideration of the different conditions and their stages is critical to the identification of motivational dynamics based on their relationship with program providers, family members and other social networks. This in-depth examination can improve our understanding of how they attempt to take responsibility and develop a healthier and more respectful intimate relationship. Importantly, participants' interactions need to be taken into account by the researchers who attempt to identify the conditions around taking responsibility for violence during interventions.

Researching male perpetrators' behavioral change process is very difficult because several complicated and sensitive conditions are often involved in this process. The difficulty of this examination is likely to be associated with the men's minimizations and denials of their violent behavior, blaming women, insufficient understanding of the consequences of violence on their family members (Dobash et al., 2000; Dutton & Hemphill, 1992; Ehrensaft & Vivian, 1999; Harne & Radford, 2008), restrictions on seeing their children (Alderson et al., 2013; Stanley et al., 2012), potential mental health or physical complications, masculine identity, tolerance of violence, patriarchal beliefs and gendered power relations (McCarry, 2007; Sayem and Nury, 2013).

Some perpetrators experience immigration-related stressors such as lack of housing and employment, poor language skills, dealing with two cultures, adaptation, lack of social support, racism and discrimination (Guru, 2006; Hancock & Siu, 2009; Williams, 1994), as well as stigmatizations about domestic violence in communities. From the available evidence around perpetrators of domestic violence, it is possible to state that a great number of problems can be interconnected with perpetrators' insufficient efforts to achieve non-violent behavior. A discussion of the methodological challenges and the benefits of applying the grounded theory approach in examining the degree that perpetrators take responsibility for their violence can contribute to understanding behavioral change process in interventions. Therefore, this chapter aims to explore how grounded theory approach might help examine the process of behavioral change among perpetrators of domestic violence.

BACKGROUND

Even though the literature on domestic violence perpetrator interventions often discusses the effectiveness of strategies for achieving behavioral change that challenge men to take responsibility, there is inconsistency around describing the aspects and dimensions of this process. This inconsistency might be to do with the obstacles around examining several conditions and characteristics interconnected with the behavioral change process. A number of scholars have used quantitative and mixed methods to investigate this process by focusing on the effectiveness of domestic violence perpetrator program (Cannon et al., 2016; Kelly & Westmarland, 2015; Lilley-Walker et al., 2016; Walker et al., 2010). However, they do not generate a new theory. The grounded theory approach is recommended for examining this complicated process in interventions.

This approach is often identified as an appropriate and useful qualitative means of inquiry for providing a comprehensive understanding of a difficult and sensitive topic. Despite limited studies on developing a new theory about the behavioral change process in interventions, a few researchers have attempted to improve a theory about offenders in groups offending interventions. For instance, Holdsworth et al. (2016) examined the conditions around perpetrators' engagement in group programs for offending behavior and utilized a constructivist grounded theory method. This work is unique in that it examines the engagement process among not only offenders but also facilitators through offending programs. Holdsworth et al. (2016) did not focus solely on the rate of participation and the outcomes of the program in interventions but also examined the change as a process. Likewise, Casey and Smith (2010) used grounded theory to explore the factors around the men's participation in anti-violence movements. With the help of the grounded theory approach, participants were able to describe their engagement as "a process that occurs over time, that happens largely through existing social networks, and that is influenced by exposure to sensitizing experiences, tangible involvement opportunities and specific types of meaning-making related to violence" (Casey & Smith, 2010, p. 953). As seen in this quote, different meaning-making under conditions needs to be examined to achieve a better understanding of the research phenomenon.

In conceptualizing the data, participants' classification of ways of taking responsibility in certain objects, actions and events can be examined (Strauss & Corbin, 1998). For instance, participants will have various ways of describing their responsibility in particular settings, and the open coding process defines these classified concepts. However, in the context of perpetrators' identification of the concepts in their relationships with family members, Tilley and Brackley (2005) noted that many perpetrators often avoid accepting that there is conflict in their relationship with their partners. For instance, Tilley and Brackley (2005) highlighted that the dimensions of social and family dynamics were often reconciled by perpetrators when they received disapproval from family members about their violent behavior. This potential mediation or manipulation of the concepts of social and family dynamics may also impact on potential participants' behavioral change in interventions. This potential mediation of the concepts could influence coding by making participants aware of these issues in a critical way.

Gray et al. (2016) examined the motivational dynamics among men who use violence in their intimate relationships and attended an 18-week intervention program. Grounded theory approach was used as the main means of analyzing the data set. The factors related to motivational issues around the process of the men taking responsibility included different levels of motivations based on their relationship status with their current or previous partners. Motivational dynamics are the important variable related to the men taking responsibility. However, these dynamics should not be excluded in the examination around the perpetrators' behavioral change process in order to have a better understanding of the process of taking responsibility. Therefore, grounded theory approach appeared to be beneficial for the research conducted by Gray et al. (2016).

METHODOLOGICAL APPROACHES AND CONCERNS

There are methodological concerns about examining perpetrators' taking of responsibility for their violence in intervention processes. The process of taking responsibility for violent behavior can be examined through recognizing complex lived experiences by considering different aspects and conditions around this process. The ways in which the conceptual framework plays a significant role in building in-depth knowledge is by implementing the grounded theory approach. For instance, Strauss & Corbin (1998) argued that conducting an initial literature review is very beneficial to the research following grounded theory principles. These strengths include theoretical orientation, a leading overarching research question, building a theoretical sampling and sufficient credibility (McGhee et al., 2007). Likewise, McCallin (2003) stated that the researchers should apply for a robust literature review before data collection to provide the rationale of the study and make a contribution to the field.

While it emphasizes the importance of applying an initial literature review and building a conceptual framework, the discussion on specific obstacles in researching men's behavioral change steps in intervention programs is critical to consider their complex relationships. Methodological concerns are likely to focus on two areas: the timeframe of the men's taking of responsibility and participants' unique perceptions and experiences about interventions. On the first of these, there are obstacles to exploring the men building a healthy intimate relationship throughout long-term intervention processes. In the context of domestic violence perpetrator programs, there are unique challenges as many men drop out of programs (Gondolf, 2012) or they might be unwilling to share their experiences during stages that they go through. Some researchers might have limited time during which they can examine this process of behavioral change. As for the second, the constructing theories about intervention processes among perpetrators are challenging as there is limited and inconsistent evidence on this particular research topic. One of the biggest obstacles could be the recruitment of participants as perpetrators. Studies on the processes of men taking responsibility indicate that researchers may need more time to develop an effective relationship with participants (Cohen et al., 2013). In order to recruit individuals who have just attended interventions, it is essential to build trust with them.

Studies show that methodological challenges exist when addressing the process of men taking responsibility in the context of domestic violence interventions. To reduce these challenges, the researchers need to consider participants' unique positions in their socio-cultural contexts during data collection and analysis (Yoshihama, 2000), even though being aware of participants' cultural background or other unique traditional expectations might be an advantage for balancing power dynamics in the field work. For instance, recognizing insider and outsider positions and assumptions might improve building a trustful relationship with participants. Although generalizability, credibility, validity, reliability and building rapport and trust with participants seem to be the key limitations, strategies for providing reflexivity may decrease these potential weaknesses. These strategies include peer review, debriefing sessions for participants, triangulation and memo-writing. The following section gives an overview of certain strategies in the grounded theory approach that might be helpful for examining the behavioral change processes in the interventions.

Grounded Theory Approach

Grounded theory is an approach that examines or tests an existing theory (Glaser & Straus, 1967), and the verification of any theory generated through research is a key aspect of grounded theory (Gibson & Brown, 2009). However, there is no sharp line between theory generation and theory verification (Punch & Oancea, 2014). While these perspectives are contested in the literature, this article highlights the importance of using the existing theoretical frameworks to produce a theory from data about taking new actions in interventions. For instance, key theories can emerge from the literature review which will contribute to understanding the existing knowledge about potential issues of perpetrators' behavioral change steps and engagement in interventions.

Grounded theory approach enables the gathering of the everyday life experiences of individuals; their perspectives; the interactive process between researcher and participants; and descriptive knowledge based on participants' words (Marshall & Rossman, 2014). Moreover, theoretical coding provides an understanding of how participants' racial and cultural identities may influence their meanings. It is important to present an overview of the link between participants' "racialized identifications and their interrelations with other social differences" to analyze their meanings of interview data (Gunaratnam, 2003, p. 109). A grounded theory approach helps to illuminate this process by providing theoretical codes that develop a theory from the major categories (Ng & Hase, 2008).

The strengths of the grounded theory approach are that it can address research questions about perpetrators' issues around taking responsibility for their violence. The biggest strength of the grounded theory approach is the open-minded, creative and iterative process of analysis which can be used to explore new categories and dimensions and develop a new theory (Cho & Lee, 2014; Gibson & Hartman, 2014; Goldkuhl & Cronholm, 2010). On the other hand, Goldkuhl & Cronholm (2010) highlight that existing theories and evidence can support the data and enable a new theory to emerge. The grounded theory approach offers the unique opportunity to create an in-depth literature review and background knowledge around the topic before data begins to be collected. This can be done to reveal the values and experiences of perpetrators and program providers on engagement in interventions. Grounded theory approach also presents a broad way of generating theory by addressing theoretically remarkable features of the developing analysis (Carey, 2012; Cho & Lee, 2014). According to Wuest (2000), theory development which is modifiable with new data is the most significant strength of the grounded theory approach because in this way "the theory can be expanded, revised, and adjusted to maintain its usefulness in explaining relationships with helping systems" (p. 68).

The main steps of this approach include an initial piece of analytical writing whilst building the data and categorizing segments of an interview transcript through a coding process (Charmaz, 2006). Such a coding process develops the theory by comparing and contrasting data to gain an in-depth understanding of participants' worldviews. This article emphasizes the importance of applying memo-writing which is an essential approach in grounded theory as it provides analysis of the data and codes at the beginning of the research and elaborates processes and expectations (Carey, 2012). Grounded theory can explain the key issues around the process of perpetrators taking responsibility and implementing non-violent behavior during interventions.

The grounded theory approach can analyze perpetrators' experiences that have had an effect on their complex interpretations and meanings. Perpetrators of domestic violence might experience complicated feelings and thoughts during interventions. These individuals are more likely to justify or deny their violent behavior (Bancroft, 2003; Dobash et al., 2000; Dutton & Hemphill, 1992; Ehrensaft & Vivian, 1999; Harne & Radford, 2008; Hearn, 1998; Lewis, 2004; Pence & Paymar, 1993; Wood, 2004; Stokoe, 2010). Grounded theory allows the researchers to investigate participants' experiences by applying codes, memos and comparative methods to reveal conditions and situations that influence the men's taking of responsibility for their violence (Jones & Alony, 2011).

TECHNIQUES IN DATA ANALYSIS AND SYNTHESIS

Theoretical sampling, memo writing, open coding, axial coding, selective coding and comparative analysis will be clarified in this section to explain how to interpret participants' meanings by developing a theory. In order to examine perpetrators' hidden perspectives and feelings about their engagement in interventions, the grounded theory approach offers an analytical process to achieve successful data collection and to allow theoretical themes to emerge (Charmaz, 2014). The core characteristics of grounded theory are the continual and critical analysis of data; consideration of preconceptions; flexibility; and focus on the process in the analysis (Strauss & Corbin, 1998). The main principles of data analysis in the grounded theory approach are to develop codes, categories, memos and engage in constant comparisons (Ng & Hase, 2008).

Theoretical Sampling

The examination of perpetrators' behavioral change steps is an unexplored area and it focuses on particular characteristics of participants. Theoretical sampling builds on a framework and develops the concepts and categories in the grounded theory approach (Denscombe, 2014; Harris, 2003). As a result, theoretical sampling needs to be conducted in order to explore relevant data and elaborate categories for the emerging theory (Charmaz, 2006). Research participants' unique perspectives and experiences contribute to the evolving theory by investigating how various concepts exist under different circumstances. Theoretical sampling is based on concepts derived from data during data collection and analysis, which allows the study to develop relevant concepts (Corbin & Strauss, 2008). Theoretical sampling provides flexibility in the research as the research starts with collecting and analyzing data simultaneously at the beginning of the research rather than making any hypothesis (Baker, Wuest & Noerager, 1992; Coyne, 1997).

The experiences of participants and the researcher as a conceptualization of emic and etic perspectives in grounded theory methodology should be taken into account (Schwandt, 2014). Codes include sensitizing concepts from the literature review and emergent codes from the data. Coding includes unique experiences of events, expectations and practices of individuals; these help to identify core issues and connections between and among activities and structures (Charmaz, 1990). This research highlights how grounded theory approach can identify the relationship between perpetrators' major issues or structures and unique experiences of engagement processes.

Memo Writing

Memos are key exercises which can be undertaken throughout the data analysis in order to improve theoretical and methodological ideas. Memo writing elaborates the processes, biases, experiences (Charmaz, 2006), the researcher's thoughts, hunches, related questions and ideas (Bowen, 2008; Strauss & Corbin, 1998) in the interview transcripts to develop analytical categories. It also helps to set up a conceptual framework in a theoretical saturation process by identifying participants' activities and interactions (Goulding, 2002; Roulston, 2014; Thornberg & Charmaz, 2014). In memo writing, comparing and contrasting a great number of concepts and patterns allows the researchers to develop classifications (Heath & Cowley, 2004; Strauss & Corbin, 1998). Further, this provides the classification for a large amount of coding (Charmaz, 2006).

Memo writing provides not only observations on the interviews but also theoretical and methodological notes to achieve an analytical perspective and improve the researcher's examination of the participants' experiences and perspectives. The grounded theory approach increases the degree of participants' voices; much research has employed this approach to understand the sensitive and invisible issues among perpetrators (Edin, Lalos, Högberg, & Dahlgren, 2008; Frost, 2004; Frost & Connolly, 2004; Stanley, Miller, Foster, & Thomson, 2010; Tilley & Brackley, 2005).

Phases of Coding

Given how memo writing is useful for recognizing such sensitive issues, coding phases should be examined in order to illustrate the steps of data analysis in grounded theory approach. For example, an open coding phase uses common concepts and categories that emerge based on their properties (Strauss & Corbin, 1998). In the examination of the men's taking of responsibility in interventions, this coding helps the research to identify large concepts. For instance, Tilley and Brackley (2005) described the concepts that have been listed based on their frequency in the data. At the initial process of open coding, Alkhatib (2012) used actual words that were used in certain circumstances by participants in the examination of attitudes and emotional effects of honor killings on victims' family members. In both examples, the open coding process played a significant role in conceptualizing and classifying raw data.

In describing the process of conceptualizations, the potential meanings of concepts as a microanalysis process need to be explored and this includes comparative analysis by applying line by line coding and looking at incidence to incidence to capture similarities and differences among incidences (Strauss & Corbin, 1998). The researchers should focus on participants' experiences and attitudes in various conditions including locations, timeframes and interactions. They also need to critically compare and contrast their interactions with their partners, program providers, community members and family members in relation to their engagement in interventions.

The researchers should identify categories in specific properties and dimensions, but not all dimensions fit every object. These dimensions include different timeframes within which perpetrators might apply such intervention strategies in terms of their relationship with partners, family members or friends. These dimensions will be about their relationship with family members, relatives, community members and program facilitators. The properties are about the characteristics of a category. Certain properties, including frequency, duration, degree, type and purpose, should be focused on. According to the suggestion by Glaser (1992), the researchers need to continually ask the data: "What category or property of a category does this incidence indicate?" in order to examine similarities and differences (p. 39). In terms of subcategories, the next section gives an overview of how axial coding develops the relationships among subcategories.

In the second phase of coding, axial coding provides the connections among and between subcategories and dimensions related to categories (Charmaz, 2006; Strauss & Corbin, 1998). The axial coding provides grouped codes based on conceptual categories that highlight similarities in codes (Fielding et al., 1998). Axial coding specifies the ways in which categories stem from the data in an open coding process (Scott & Howell, 2008; Strauss & Corbin, 1998). The main goal is to describe "how categories crosscut and link" in their properties and dimensions (Strauss & Corbin, 1998, p. 124). Axial coding illuminates subcategories in-depth by examining the questions of "when, where, why, who, how, and with what consequences" (Strauss & Corbin, 1998, p. 125). This level of examination provides a conceptual level of connections among categories by applying various tasks. These tasks include setting up various properties and dimensions of a major category; identifying taking responsibility for their violence in individuals' specific circumstances, interactions and consequences, and relating their subcategories (Strauss & Corbin, 1998).

In the third phase, selective coding is described as a "process of integrating and refining the theory" (Strauss & Corbin, 1998, p. 143). Selective coding contributes to building more abstract level data analysis and identifies a major pattern in the data. For instance, the research about perpetrators' experience in interventions can integrate categories, revise and organize memos to build and refine the theory (Strauss & Corbin, 1998). More than one central theme might emerge and the researchers should choose these themes as central categories. More than one central category helps to understand the relationship between them and allows for a subsequent theory to emerge (Jones & Alony, 2011). Even though achieving two fully-developed categories is difficult in the explanatory research (Glaser, 1992), more than one category would also describe the central concern of the research phenomenon. Therefore, the researcher should not force the data to achieve one central category if two core categories emerge in the data. These categories will also clarify how participants make sense of their experiences.

The research moves from description to conceptualization by describing "dimensional extremes" (Strauss & Corbin, 1998, p. 152). For instance, men's reasons for implementing non-violent acts or taking responsibility in interventions might be related to their own motivations, the influences of family, community members and so on. This research can use diagrams for the integration by describing concepts and their relationships in a logical and clear way as these integrative diagrams present the data in an abstract way. In order to refine the theory, the research can review the internal consistency and logic of the theoretical scheme and develop poor categories (Strauss & Corbin, 1998). Therefore, the researchers can develop a category sufficiently by demonstrating properties and dimensions as well as variations in a concept in order to avoid any potential bias on their part (Strauss & Corbin, 1998).

Theoretical saturation of a theory requires that no more new properties and dimensions emerge from the data (Strauss & Corbin, 1998). This refinement process may include the following questions: "in the participants' perspective, how does this phenomenon proceed, with what variability and what conditions, in both macro and micro environments?" (Scott & Howell, 2008, p. 13). However, this process is not straightforward because some ideas might not fit the theory, and this needs to be realized. It is important to illuminate the ways in which the abstract conceptualization fits the raw data to understand participants' actions in certain circumstances (Heath & Cowley, 2004; Strauss & Corbin, 1998). Theoretical codes reflect the broad range of perspectives on data at the conceptual level (Dey, 1999).

The major approach for a validating scheme seems to be to "go back and compare the scheme against raw data" (Strauss & Corbin, 1998; p. 159). Another way is to ask participants to read it and comment on it. However, this approach is inappropriate for sensitive topics because some participants may not be able to recognize and describe their hidden perspectives in relation to the theoretical explanations. The application of a reflective coding matrix allows the central phenomenon to be defined holistically (Scott & Howell, 2008). While theoretical coding is beneficial in analyzing the conceptual level of categorizations, memos provide key identifications and observations of the research (Gilgun, 2015).

Constant and Comparative Method

Constant comparative analysis helps in examining the experiences in certain circumstances and events as suggested by Strauss and Corbin (1998). In applying the constant and comparative method, the researchers can attempt to simultaneously work on the data analysis between transcripts and codes (Strauss & Corbin, 1998). Theoretical comparisons allow the research to understand the meanings of acts, events and experiences; to explore potential properties and dimensions; to gain abstractions; and to consider potential biases and viewpoints (Strauss & Corbin, 1998).

The constant comparative method allows the researchers to compare codes with other codes, incidents with other incidents and to identify the similarities and differences in order to enable the improvement of concepts (Strauss & Corbin, 1998). The main critical point is that the researchers should consider the key issues within existing theories as guidance in the data collection and sensitization of unique issues and angles of the research topic, and the researchers should avoid trying to force the data into the theory (Ezzy, 2002; Heath & Cowley, 2004).

This method plays a significant role in identifying participants' unique feelings and thoughts in certain circumstances in relation to their taking responsibility in interventions. For instance, the research by Wuest and Merritt-Gray (1999) applied a constant comparative method by providing dense descriptions of the process in data and its relationship with pre-existing conceptions in the literature. This application helped Wuest & Merritt-Gray (1999) to create a theory. Similarly, Bowen (2006) highlights the importance of examining the relationship between data from interviews and an initial literature review in order to sensitize concepts to theory. Thus, constant comparative analysis enables the researchers to conceptualize the data by comparing data against other data and itself (Duchscher & Morgan, 2004; Ezzy, 2002) and improve the theoretical saturation (Bowen, 2008). Applying codes, writing memos and the constant comparative method will help to reach central categories in order to examine how participants make sense of issues of their implementation of new and non-violent behavior in their intimate relationship. While providing beneficial tools and strategies in the data analysis process, ethical issues need to be clarified in this sensitive topic and hard to reach population. For instance, the researchers' identities and positionalities are more likely to influence the research process (Edwards, 1998; Gadd, 2004; Gunaratnam, 2003; Marcus, 1994; Opie, 1992).

Reflexivity has enabled the researchers to recognize their potential biases, roles and influences on the research process and minimize them by focusing on participants' viewpoints and experiences (Finlay, 2002). Reflexivity helps reduce their potential bias by making them aware of their relationship with participants during the research (Holloway & Biley, 2011; Kolb, 2012; Woodby et al., 2011). Dimensions of reflexivity include the meanings of race, gender, culture, class, socio-economic status, religion, emotional strength and physical appearance (Mertens & Ginsberg, 2008; Towson & Pulla, 2015). Bourke (2014) highlights that these dimensions may influence the research process. This reflexivity allows the researchers to examine their awareness of various dimensions that are related to the relationship with research participants (Pillow, 2003). In addition to these recognitions, the researchers should be aware of the influences that arise from their positions and subjective ideas without trying to reduce them in order to achieve objective results (Ezzy, 2002; Hammersley & Atkinson, 1995). Reflexivity in the grounded theory approach appears to reduce potential obstacles in examining the participants' lived experiences in domestic violence perpetrator programs. The main reason for highlighting reflexivity is to show how the qualitative researcher needs to be aware of the potential power relations between them and participants that often influence data collection and the interpretation of the data.

LIMITATIONS

Whilst the grounded theory is the most appropriate approach to examine the processes of behavioral change, there are some weaknesses in this approach. For instance, insufficient clarity about the period of time in which interviews will be conducted might be called a limitation. Also, there are questions about definitions of theoretical saturation in terms of how and when the researcher achieves the saturation exactly. Issues arise from the process of the research. Further weaknesses are the requirement for a researcher to have sufficient knowledge about the forms of the grounded theory approach and the difficulties of estimating the research timetable (Cho & Lee, 2014). Despite these limitations, this article acknowledges the significance of implementing the grounded theory approach as a suitable methodology for social processes present within complex and sensitive topics (Harry, Sturges & Klingner, 2005; José, Osinski & Hermann, 2017; Parry, 1998).

This article solely focused on the data analysis process in the grounded theory approach in order to illustrate the usefulness of a new theory emerging for examining the process of men taking responsibility in interventions. The lack of discussion on the issues around validity, reliability and objectivity is the limitation of this article. While ethical considerations in examining the men's taking of responsibility in interventions are vital, this article did not concentrate on ethical considerations. The major goal was to explore how the data analysis process based on the grounded theory approach helps to achieve an in-depth understanding of perpetrators' behavioral change process by focusing on taking responsibility for their violence. At the same time, it acknowledges the importance of taking into account issues of ethics and trustworthiness.

CONCLUSION

This article has discussed how the grounded theory approach and reflexivity might play a significant role in examining the processes of the men taking responsibility, in particular focusing on the implementation of non-violent behavior, and a healthy and respectful intimate relationship in domestic violence intervention programs. However, many obstacles can be faced in this sensitive research including the limited timeframe and some participants' potential denials about their violent behavior. Furthermore, their high rate of drop-out from these programs is the major obstacle in accessing them. As a result, this article clarifies the benefits of the grounded theory approach in analyzing sensitive and complex lived experiences in intervention processes. This indicates the strategies in grounded theory approach for these types of sensitive areas of research in terms of understanding participants' perspectives and experiences in the process of intervention. The approaches of theoretical sampling, memo-writing, phases of coding and comparative analysis are important strategies in the grounded theory approach for examining invisible perspectives in sensitive research. Thus, the grounded theory approach can investigate such sensitive and difficult research topics in a qualitative inquiry.

REFERENCES

Alkhatib, S. (2012). *Social death: A grounded theory study of the emotional and social effects of honour killing on victims' family members- A Palestinian family perspective*. University of Manchester.

Baker, C., Wuest, J., & Noerager, P. (1992). Method slurring: The grounded theory/ phenomenology example. *Journal of Advanced Nursing*, *17*(11), 1355–1360. doi:10.1111/j.1365-2648.1992.tb01859.x PMID:1430643

Bancroft, L. (2003). *Why does he do that? Inside the minds of angry and controlling men*. Penguin Publishing Group.

Bourke, B. (2014). Positionality: Reflecting on the research process. *Qualitative Report*, *19*(33), 1–9.

Bowen, G. A. (2006). Grounded theory and sensitizing concepts. *International Journal of Qualitative Methods*, *5*(3), 12–23. doi:10.1177/160940690600500304

Bowen, G. A. (2008). Naturalistic inquiry and the saturation concept: A research note. *Qualitative Research*, *8*(1), 137–152. doi:10.1177/1468794107085301

Cannon, C., Hamel, J., Buttell, F., & Ferreira, R. J. (2016). A Survey of Domestic Violence Perpetrator Programs in the United States and Canada: Findings and Implications for Policy and Intervention. *Partner Abuse*, *7*(3), 226–276. doi:10.1891/1946-6560.7.3.226

Carbajosa, P., Catalá-Miñana, A., Lila, M., Gracia, E., & Boira, S. (2017). Responsive versus treatment-resistant perpetrators in batterer intervention programs: Personal characteristics and stages of change. *Psychiatry, Psychology and Law*, *24*(6), 1–15. doi:10.1080/13218719.2017.1347933 PMID:31984001

Carey, D. M. (2012). *Qualitative research skills for social work: Theory and practice*. Ashgate Publishing, Ltd.

Casey, E., & Smith, T. (2010). "How can I not?": Men's pathways to involvement in anti-violence against women work. *Violence Against Women*, *16*(8), 953–973. doi:10.1177/1077801210376749 PMID:20679189

Charmaz, K. (1990). "Discovering" chronic illness: Using grounded theory. *Social Science & Medicine*, *30*(11), 1161–1172. doi:10.1016/0277-9536(90)90256-R PMID:2360052

Charmaz, K. (2006). *Constructing grounded theory: A practical guide through qualitative analysis*. SAGE Publications.

Cho, J. Y., & Lee, E. (2014). Reducing confusion about grounded theory and qualitative content analysis: Similarities and differences. *The Qualitative Report, 19*(2003), 1–20.

Chovanec, M. G. (2009). Facilitating change in group work with abusive men: Examining stages of change. *Social Work with Groups*, *32*(1), 125–142. doi:10.1080/01609510802413055

Cohen, L., Manion, L., & Morrison, K. (2013). *Research methods in education* (7th ed.). Routledge. doi:10.4324/9780203720967

Corbin, J., & Strauss, A. (2008). *Basics of qualitative research: Techniques and procedures for developing grounded theory* (Vol. 25). SAGE Publications. doi:10.4135/9781452230153

Coyne, I. (1997). Sampling in qualitative research. Purposeful and theoretical sampling; merging or clear boundaries? *Journal of Advanced Nursing*, *26*(3), 623–630. doi:10.1046/j.1365-2648.1997.t01-25-00999.x PMID:9378886

Cuevas, D. A., & Bui, N. H. (2016). Social factors affecting the completion of a batterer intervention program. *Journal of Family Violence*, *31*(1), 95–107. doi:10.100710896-015-9748-0

Daniels, J. W., & Murphy, C. M. (1997). Stages and processes of change in batterers' treatment. *Cognitive and Behavioral Practice*, *4*(1), 123–145. doi:10.1016/S1077-7229(97)80015-6

Denscombe, M. (2014). *The good research guide: For small-scale social research projects* (5th ed.). McGraw-Hill Education.

Dey, I. (1999). *Grounding grounded theory: Guidelines for qualitative inquiry*. Academic Press.

Dobash, R. P., Dobash, R. E., Cavanagh, K., & Lewis, R. (2000). A research evaluation of British programmes for violent men. *Journal of Social Policy*, *28*(2), 205–233. doi:10.1017/S0047279499005589

Duchscher, J. E. B., & Morgan, D. (2004). Grounded theory: Reflections on the emergence vs. forcing debate. *Journal of Advanced Nursing*, *48*(6), 605–612. doi:10.1111/j.1365-2648.2004.03249.x PMID:15548251

Dutton, D., & Hemphill, K. (1992). Patterns of socially desirable responding among perpetrators and victims of wife assault. *Violence and Victims*, *7*(29), 29–40. doi:10.1891/0886-6708.7.1.29 PMID:1504031

Edin, K. E., Lalos, A., Högberg, U., & Dahlgren, L. (2008). Violent men: Ordinary and deviant. *Journal of Interpersonal Violence*, *23*(2), 225–244. doi:10.1177/0886260507309342 PMID:18162638

Edwards, R. (1998). A critical examination of the use of interpreters in the qualitative research process. *Journal of Ethnic and Migration Studies*, *24*(1), 197–208. doi:10.1080/1369183X.1998.9976626

Ehrensaft, M. K., & Vivian, D. (1999). Is partner aggression related to appraisals of coercive control by a partner? *Journal of Family Violence*, *14*(3), 251–266. doi:10.1023/A:1022862332595

Ezzy, D. (2002). *Qualitative analysis practice and innovation*. Routledge.

Finlay, L. (2002). "Outing" the researcher: The provenance, process, and practice of reflexivity. *Qualitative Health Research*, *12*(4), 531–545. doi:10.1177/104973202129120052 PMID:11939252

Frost, A. (2004). Therapeutic engagement styles of child sexual offenders in a group treatment program: A grounded theory study. *Sexual Abuse*, *16*(3), 191–208. doi:10.1177/107906320401600302 PMID:15326880

Frost, A., & Connolly, M. (2004). Reflexivity, reflection, and the change process in offender work. *Sexual Abuse*, *16*(4), 365–380. doi:10.1177/107906320401600408 PMID:15560417

Gadd, D. (2004). Making sense of interviewee–interviewer dynamics in narratives about violence in intimate relationships. *International Journal of Social Research Methodology*, *7*(5), 383–401. doi:10.1080/1364557092000055077

Gibson, B., & Hartman, J. (2014). *Rediscovering grounded theory*. SAGE Publications.

Gibson, W., & Brown, A. (2009). *Working with qualitative data*. SAGE Publications. doi:10.4135/9780857029041

Gilgun, J. F. (2015). Beyond description to interpretation and theory in qualitative social work research. *Qualitative Social Work: Research and Practice*, *14*(6), 741–752. doi:10.1177/1473325015606513

Glaser, B., & Straus, A. (1967). *Discovery of grounded theory: strategies for qualitative research*. Transaction Publishers.

Glaser, B. G. (1992). *Emergence vs forcing: Basics of grounded theory analysis*. Sociology Press.

Goldkuhl, G., & Cronholm, S. (2010). Adding theoretical grounding to grounded theory: Toward multi-grounded theory. *International Journal of Qualitative Methods*, *9*(2), 187–206. doi:10.1177/160940691000900205

Gondolf, E. W. (2012). *The future of batterer programs: Reassessing evidence-based practice*. UPNE.

Goulding, C. (2002). *Grounded theory: A practical guide for management, business and market researchers*. SAGE Publications. doi:10.4135/9781849209236

Gray, R., Broady, T., Gaffney, I., Lewis, P., Mokany, T., & O'Neil, B. (2016). 'I'm working towards getting back together': Client accounts of motivation related to relationship status in men's behaviour change programmes in New South Wales, Australia. *Child Abuse Review*, *25*(3), 171–182. doi:10.1002/car.2318

Gunaratnam, Y. (2003). *Researching "race" and ethnicity: Methods, knowledge and power*. SAGE Publications. doi:10.4135/9780857024626

Guru, S. (2006). Working with Asian perpetrators of domestic violence— The British experience. *Practice*, *18*(3), 153–166. doi:10.1080/09503150600904037

Hammersley, M., & Atkinson, P. (1995). *Ethnography: Principles in practice*. Tavistock.

Hancock, T. U., & Siu, K. (2009). A culturally sensitive intervention with domestically violent Latino immigrant men. *Journal of Family Violence*, *24*(2), 123–132. doi:10.100710896-008-9217-0

Harne, L., & Radford, J. (2008). Tackling domestic violence: Theories, policies and practice. McGraw-Hill International.

Harris, I. (2003). What does "the discovery of grounded theory" have to say to medical education? *Advances in Health Sciences Education: Theory and Practice*, *8*(1), 49–61. doi:10.1023/A:1022657406037 PMID:12652168

Harry, B., Sturges, K. M., & Klingner, J. K. (2005). Mapping the process: An exemplar of process and challenge in grounded theory analysis. *Educational Researcher*, *34*(2), 3–13. doi:10.3102/0013189X034002003

Hearn, J. (1998). *The violences of men: How men talk about and how agencies respond to men's violence to women* (1st ed.). SAGE Publications. doi:10.4135/9781446279069

Heath, H., & Cowley, S. (2004). Developing a grounded theory approach: A comparison of Glaser and Strauss. *International Journal of Nursing Studies*, *41*(2), 141–150. doi:10.1016/S0020-7489(03)00113-5 PMID:14725778

Holdsworth, E., Bowen, E., Brown, S., & Howat, D. (2016). The development of a program engagement theory for group offending behavior programs. *International Journal of Offender Therapy and Comparative Criminology*, 1–21. doi:10.1177/0306624X15624177 PMID:26769679

Holloway, I., & Biley, F. C. (2011). Being a qualitative researcher. *Qualitative Health Research*, *21*(7), 968–975. doi:10.1177/1049732310395607 PMID:21266705

Jones, M., & Alony, I. (2011). Guiding the use of grounded theory. *International Journal of Doctoral Studies*, *6*, 95–114. doi:10.28945/1429

José, D., Osinski, M., & Hermann, R. (2017). The construction process of grounded theory in administration. *Contaduría y Administración*, *62*(3), 985–1000. doi:10.1016/j.cya.2016.06.012

Kalogo, C. (2015). *Factors influencing treatment completion of involuntary groups*. University of St. Thomas.

Kelly, L., & Westmarland, N. (2015). *Domestic violence perpetrator programmes: steps towards change.* Project Mirabal Final Report.

Kolb, S. M. (2012). Grounded theory and the constant comparative method: Valid research strategies for educators. *Journal of Emerging Trends in Educational Research and Policy Studies*, *3*(1), 83–86.

Lewis, R. (2004). Making justice work: Effective legal interventions for domestic violence. *British Journal of Criminology*, *44*(2), 204–224. doi:10.1093/bjc/44.2.204

Lilley-Walker, S.-J., Hester, M., & Turner, W. (2016). Evaluation of European domestic violence perpetrator programmes: Toward a model for designing and reporting evaluations related to perpetrator treatment interventions. *International Journal of Offender Therapy and Comparative Criminology*, 1–17. doi:10.1177/0306624X16673853 PMID:27884945

Marcus, G. E. (1994). What comes (just) after 'post'? the case of ethnography. In N. K. Denzin & Y. S. Lincoln (Eds.), *Handbook of Qualitative Research* (pp. 563–574). Sage Publications.

Marshall, C., & Rossman, G. B. (2014). *Designing qualitative research.* Sage Publications.

Mccallin, A. (2003). Grappling with the literature in a grounded theory study. *Contemporary Nurse*, *15*(1-2), 61–69. doi:10.5172/conu.15.1-2.61 PMID:14649509

McCarry, M. (2007). Masculinity studies and male violence: Critique or collusion? *Women's Studies International Forum*, *30*(5), 631–638. doi:10.1016/j.wsif.2007.07.006

McConnaughy, E. A., Prochaska, J. O., & Velicer, W. F. (1983). Stages of change in psychotherapy: Measurement and sample profiles. *Psychotherapy (Chicago, Ill.)*, *20*(3), 368–375. doi:10.1037/h0090198

McGhee, G., Marland, G. R., & Atkinson, J. M. (2007). Grounded theory research: Literature reviewing and reflexivity. *Journal of Advanced Nursing*, *60*(3), 334–342. doi:10.1111/j.1365-2648.2007.04436.x PMID:17908129

Mertens, D. M., & Ginsberg, P. E. (2008). Deep in ethical waters. *Qualitative Social Work: Research and Practice*, *7*(4), 484–503. doi:10.1177/1473325008097142

Ng, K., & Hase, S. (2008). Grounded suggestions for doing a grounded theory business research. *Electronic Journal of Business Research Methods*, *6*(2), 155–170.

Opie, A. (1992). Qualitative research, appropriation of the "other" and empowerment. *Feminist Review*, *40*(1), 52–69. doi:10.1057/fr.1992.5

Parry, K. W. (1998). Grounded theory and social process : A new direction for leadership research. *The Leadership Quarterly*, *9*(1), 85–105. doi:10.1016/S1048-9843(98)90043-1

Pence, E., & Paymar, M. (1993). *Education groups for men who batter: The Duluth model*. Springer Publishing Company. doi:10.1891/9780826179913

Pillow, W. (2003). Confession, catharsis, or cure? Rethinking the uses of reflexivity as methodological power in qualitative research. *International Journal of Qualitative Studies in Education: QSE*, *16*(2), 175–196. doi:10.1080/0951839032000060635

Prochaska, J. O., & DiClemente, C. C. (1982). Transtheoretical therapy: Toward a more integrative model of change. *Psychotherapy (Chicago, Ill.)*, *19*(3), 276–288. doi:10.1037/h0088437

Punch, K. F., & Oancea, A. (2014). *Introduction to research methods in education* (Vol. 10). SAGE Publications.

Rondeau, G., Brodeur, N., Brochu, S., & Lemire, G. (2001). Dropout and completion of treatment among spouse abusers. *Violence and Victims*, *16*(2), 127–143. doi:10.1891/0886-6708.16.2.127 PMID:11345474

Rosenbaum, A., Gearan, P. J., & Ondovic, C. (2001). Completion and recidivism among court-and self-referred batterers in a psychoeducational group treatment program: Implications for intervention and public policy. *Journal of Aggression, Maltreatment & Trauma*, *5*(2), 199–220. doi:10.1300/J146v05n02_12

Rothman, E. F., & Coutinho, P. (2007). Batterer intervention program enrollment and completion Massachusetts. *Violence Against Women*, *13*(5), 527–543. doi:10.1177/1077801207300720 PMID:17478677

Roulston, K. (2014). Analysing Interviews. In U. Flick (Ed.), *The SAGE handbook of Qualitative Data Analysis - Analysing Interviews* (pp. 297–313). SAGE Publications. doi:10.4135/9781446282243

Sayem, A. M., & Nury, A. T. M. S. (2013). An assessment of attitude towards equitable gender norms among Muslim women in Bangladesh. *Women's Studies International Forum*, *40*, 102–110. doi:10.1016/j.wsif.2013.04.007

Schwandt, T. A. (2014). *The SAGE Dictionary of Qualitative Inquiry*. SAGE Publications.

Scott, K. W., & Howell, D. (2008). Clarifying analysis and interpretation in grounded theory: Using a conditional relationship guide and reflective coding matrix. *International Journal of Qualitative Methods*, *7*(2), 1–15. doi:10.1177/160940690800700201

Stanley, N., Miller, P., Foster, H. R., & Thomson, G. (2010). *Children and families experiencing domestic violence: Police and children's social services' responses*. Issue January.

Stokoe, E. (2010). "I'm not gonna hit a lady": Conversation analysis, membership categorization and men's denials of violence towards women. *Discourse & Society*, *21*(1), 59–82. doi:10.1177/0957926509345072

Strauss, A., & Corbin, J. M. (1998). *Basics of qualitative research: techniques and procedures for developing grounded theory*. Sage Publications.

Thornberg, R., & Charmaz, K. (2014). Grounded theory and theoretical coding. In *The SAGE Handbook of Qualitative Data Analysis*. SAGE Publications. doi:10.4135/9781446282243.n11

Tilley, D. S., & Brackley, M. (2005). Men who batter intimate partners: A grounded theory study of the development of male violence in intimate partner relationships. *Issues in Mental Health Nursing*, *26*(3), 281–297. doi:10.1080/01612840590915676 PMID:16020047

Towson, L., & Pulla, V. (2015). Preparing social workers for empowerment: The place of positionality and its continual intricacies. In V. Pulla & B. B. Mamid (Eds.), *Some aspects of community empowerment and resilience* (pp. 31–49). Allied Publishers.

Walker, D. D., Neighbors, C., Mbilinyi, L. F., O'Rourke, A., Zegree, J., Roffman, R. A., & Edleson, J. L. (2010). Evaluating the impact of intimate partner violence on the perpetrator: The perceived consequences of domestic violence questionnaire. *Journal of Interpersonal Violence*, *25*(9), 1684–1698. doi:10.1177/0886260509354592 PMID:20139346

Waller, B. (2016). Aggression and violent behavior broken fixes : A systematic analysis of the effectiveness of modern and postmodern interventions utilized to decrease IPV perpetration among Black males remanded to treatment. *Aggression and Violent Behavior*, *27*, 42–49. doi:10.1016/j.avb.2016.02.003

Williams, O. J. (1994). Group work with African American men who batter: Toward more ethnically sensitive practice. *Journal of Comparative Family Studies*, *25*(1), 91–104. doi:10.3138/jcfs.25.1.91

Wood, J. T. (2004). Monsters and victims: Male felons' accounts of intimate partner violence. *Journal of Social and Personal Relationships*, *21*(5), 555–576. doi:10.1177/0265407504045887

Woodby, L. L., Williams, B. R., Wittich, A. R., & Burgio, K. L. (2011). Expanding the notion of researcher distress: The cumulative effects of coding. *Qualitative Health Research*, *21*(6), 830–838. doi:10.1177/1049732311402095 PMID:21393618

Wuest, J. (2000). Negotiating with helping systems: An example of grounded theory evolving through emergent fit. *Qualitative Health Research*, *10*(1), 51–70. doi:10.1177/104973200129118246 PMID:10724752

Wuest, J., & Merritt-Gray, M. (1999). Not going back: Sustaining the separation in the process of leaving abusive relationship. *Violence Against Women*, *5*(2), 110–133. doi:10.1177/10778012990005002002

Yoshihama, M. (2000). Reinterpreting strength and safety in a socio-cultural context: Dynamics of domestic violence and experiences of women of Japanese descent. *Children and Youth Services Review*, *22*(3), 207–229. doi:10.1016/S0190-7409(00)00076-1

KEY TERMS AND DEFINITIONS

Grounded: Well established and balanced to the organization or work.

Interventions: Making efforts to reduce or stop the individuals' violent behavior or wrong attitudes that damage people by providing strategies for them to intervene in these attitudes.

Perpetrators of Domestic Violence: Individuals who act violently towards their family members such as their partners and children.

Process: It is the steps to achieve successful outcomes.

Sensitive Research: Researching the difficult topics are related to the vulnerable groups, marginalized population, or unvoiced people by carefully considering ethical considerations.

Taking Responsibility: Being aware of the actions that the individuals have taken and getting knowledgeable about the core issues and their positions on these issues.

Theory: Clarifying the things by providing key concepts, principles, and structures.

Chapter 15
The Importance of Qualitative Methods in Mixed Methods Studies

Jacquelynne A. Boivin
https://orcid.org/0000-0002-5763-5707
Bridgewater State University, USA

ABSTRACT

This chapter presents a thorough review of the literature dedicated to learning more about mixed methods research design. Explanatory sequential mixed methods studies and exploratory sequential mixed methods studies are the two types of mixed methods research design models that this chapter presents in detail. To contextualize different ideas related to research design, the author provides examples of research studies that exemplify different research designs falling within mixed methods. The main objective of this chapter is to highlight the important role that qualitative research design plays in mixed methods research. Such a presentation of the literature aims to argue that qualitative data substantiates quantitative data as a means to heighten the regard which the qualitative methodology receives.

INTRODUCTION

Historically, as qualitative research methods have struggled to gain the level of respect and prowess as quantitative research methods, mixed methodology lacks a substantial amount of guidance in the literature. As a Ph.D. candidate at Lesley University in Cambridge, MA, I desired to conduct a phenomenology for my dissertation, but struggled to structure a strictly-qualitative study in a way that would appease my

DOI: 10.4018/978-1-7998-7600-7.ch015

dissertation committee. I soon learned that many students earning doctorates in the field of Education, and other disciplines within the humanities and social sciences, faced the same struggle. The only way their qualitative data could be justified would be if they had quantitative data to support their qualitative data. This was when I started to investigate mixed methods approaches to research design.

In this chapter, two options of mixed methods studies are be explored. This first is explanatory sequential mixed methods study and the second is exploratory sequential mixed methods study. These two types of mixed methods research can be described a truly symbiotic relationship. The two different portions of each of these research methods provide deeper insight to the other. The qualitative portion informs the quantitative portion, and vice versa. The main objective for supplying such an exploration of mixed methods research is to understand the vastly important role of qualitative methodology within mixed methods. In turn, this discussion dispels the fallacy that the benefit of mixed methods is that quantitative data substantiates qualitative data. This chapter argues that the opposite is true. Mixed methods research is powerful and insightful because it allows qualitative data to substantiate quantitative data. Qualitative methods, in summary, have more value in mixed methods research than quantitative methods.

BACKGROUND

Russell and Greggory (2003) articulated that qualitative research struggles to agree upon consistent expectations in regard to reliability and validity. The inconsistencies within qualitative research methods has contributed to the idea that qualitative research is not as accurate, valid, or reliable as quantitative research methods. This chapter aims to disrupt this notion by placing heavy emphasis on the value and importance of quantitative research methodology. While some academic scholars may argue that mixed methods research is the only version of qualitative research that is valid and reliable because it has quantitative research methods to substantiate the qualitative data, this chapter takes the opposite viewpoint. What if qualitative data substantiates quantitative data regardless of the mixed methods approach. While the claims in this chapter are broad in nature, the generalized perspective allows for researchers of all specializations to reimagine qualitative methodology.

MAIN FOCUS OF THE ARTICLE

This chapter addresses the importance of qualitative data within mixed methods research and is divided into three main sections: (a) explanatory sequential mixed methods study, (b) exploratory sequential mixed methods study, and (c) the necessity of qualitative methods in mixed methods design. While this chapter provides insights and recommendations, it evokes questions that warrant explanation.

MIXED METHODS RESEARCH

Mixed methods research is, in summary, a combination of both qualitative and quantitative research design. As Maxwell (2016) explained in his article, mixed methods research design occurred in academia well before it had the label of "mixed methods." Studies that utilized a combination of qualitative and quantitative research design have been documented since the 1950s (Maxwell, 2016). Onwuegbuzie, Johnson, and Turner (2007) contented that integrative research, multimethod research, multiple methods, triangulated studies, mixed research, and blended research are a few labels given to mixed methods research. Regardless of the label, mixed methods studies share many commonalities.

Onwuegbuzie et al. (2007) summarized mixed methods research design as consisting of four central domains. The first domain is philosophical perspectives and the second domain is logistics surrounding inquiry. The third domain is procedural guidelines and the fourth is ideas surrounding socio-politics. These four domains were used to create a board definition of mixed methods research. Onwuegbuzie et al. (2007) wrote that mixed methods research is, "an intellectual and practical synthesis based on qualitative and quantitative research; it is the third methodological or research paradigm (along with qualitative and quantitative research" (p. 129). This definition properly reflects ideas by Cameron (2011).

Cameron (2011) wrote that when defining mixed methods research, there are five essential tenets that warrant careful consideration. The Five P's recognize the controversies and challenges that mixed methods research faces (Cameron, 2011). For each of the P's, Cameron (2011) cited Bazeley's (2004) learning objectives for the researcher. The first P stands for "paradigms." The associated learning objective is that the research must obtain a solid understanding of the paradigms associated with the guiding research questions at hand, while considering the theoretical framework. The second P stands for "pragmatism." Bazeley's (2003) learning objective that corresponds to the second P requests that the research commit to a thorough review of the literature. Such a review leads to the researcher being able to understand all perspectives surrounding the topic under investigation and to learn more about the past experiences of other researchers who have completed the same or similar research methodology.

The third P stands for "praxis." This corresponds to the learning objective that the research should be able to justify his or her choice of methodology for learning more about the guiding research questions. This process also involves reaching out to other researchers to learn more about coping with errors and the unknown. Collaborative efforts are imperative. The fourth P stands for "proficiency," which refers to the learning objective of the researcher being skilled at executing the research methodology that he or she selects. The fifth, and final, P stands for "publishing." The researcher objective for publishing is to represent the study in a way that highlights what is new and novel for this research topic. What sets this study apart from other published works that share similarities? The Five P's are useful to consider when evaluating different types of mixed methods research and critically reviewing the type of qualitative method employed.

Two types of mixed methods studies are explored in this chapter with emphasis on the role of qualitative research methodology. The central objective in doing so is to encourage the reader to rethink qualitative research equally, or more, valid and reliable as quantitative data. Everest (2014) highlighted that it is yet to be proven via research which type of research methodology, qualitative or quantitative, evokes the most impact on society. Everest (2014) argued that the debate is not close to being over because research questions mainly lack "homogeneous originality" (p. 13). So, until research questions become much more simplistic, there may not be a clear depiction of one research method being deemed the best. However, there is still plenty of data and literature to convince even the most quantitative-purist that there is a high degree of importance within qualitative research design and execution. Maxwell (2010) added that mixed methods research design grew in popularity due to the ambition of many qualitative researchers in wanting to provide a more precise definition of "some," or "most" when explaining findings, to appease those who favor quantitative data. Researchers should consider the benefits and disadvantages of including quantitative data with qualitative data, as Maxwell (2010) explained.

Maxwell (2010) identified several advantages and several disadvantages of including quantitative data alongside qualitative data. The first advantage is that quantitative data contributes to the generalizability of the study within its own setting and participants. The second advantage that Maxwell (2010) argued was that quantitative data helps diminish biases and increases emphasis on diversity. The third advantage that Maxwell (2010) emphasized was that quantitative data can clarify similarities and patterns that may go unnoticed by the researcher if qualitative data is only collected. The balance can widen the researcher's perspective and illuminate aspects of the population being studied that could have otherwise gone unnoticed.

Maxwell (2010) also shared that there are four distinct disadvantages of using quantitative data alongside qualitative data. The first disadvantage is that quantitative data promotes a generalization of findings that often times disregard a group that is not represented in the study. Qualitative data, in contrast, tends to be more local in nature and accounts for populations not included. This may be tied to the vastly smaller sample size associated with qualitative research methodology with lower expectations for what is deemed necessary for reliability and validity. quantitative research methodology, in contrast, relies greatly on statistical significance and requires adequate participant numbers for reliability and validity claims.

The second disadvantage of mixing quantitative research methods with qualitative research methods is that quantitative research promotes the use of a variance model, which undermines what makes qualitative research unique from quantitative research (Maxwell, 2010). Such varied perspectives can lead to implicit or explicit assumptions and findings. The third disadvantage that Maxell (2010) included was that researches can be overly dependent upon quantitative data. He wrote, "Numbers can't *replace* the actual description of evidence but can provide a supplementary type of support for the conclusions when it's impossible to present all the evidence" (Maxwell, 2010, p. 480). Once again, the emphasis is put on qualitative data. The fourth and final disadvantage of including quantitative data with qualitative data is that while statistics can add a façade of precision and rigor, numbers can be presented in a study and not add any true value to the study itself (Maxwell, 2010). The quantitative representations may be included as a way to increase standing as more "real" to critics of qualitative research.

Clearly, mixed methods research is not as distinctly defined and reviewed as some scholars may have previously assumed. Rather, mixed methods research is filled with nuances and opinions. What is apparent from the preceding section is that similar to qualitative research and quantitative research, it is a subset of mixed methods research methodologies with many advantages and disadvantages. Researchers need to take as much, if not more, care in crafting a mixed methods study as other methods of research (Everest, 2014; Maxwell, 2010; Onwuegbuzie et al., 2007). With an enhanced perspective of the history, definition, controversies, and opinions surrounding mixed methods research, it is possible to learn more about two popular types of mixed methods research. While the next two sections of this chapter explore explanatory sequential mixed methods research design and exploratory sequential mixed methods research design, there are other types of mixed methods research that are worth further exploration for scholars who are curious as to which type fits their future research questions.

While this chapter looks at explanatory and exploratory, two other design types within mixed methods are triangulation and embedded (Cameron, 2009). Triangulation and embedded are involve two types of data collection techniques employed concurrently. Embedded can also have sequential data collection. Triangulation merges together the data during analysis. Embedded refers to the researcher embedding either qualitative or quantitative data within a larger design using the other type of data. Parallel mixed methods studies, where qualitative and quantitative data collection happen at the same time, have their own advantages and challenges and researchers who think that parallel data collection better fits his/her study should pursue learning more about that type of mixed methods approach.

When choosing what type of mixed methods design best fits the research questions under investigation, Cameron (2009) advocated for determining where the researcher stands in terms of research paradigm perspective. Those who deem themselves "purist," believe that research methodologies should not be mixed. "Situationists" believe it all depends upon the situation. "Pragmatists" are those who support mixed methods research and are the direct contrast to "purists." While the two central types of mixed methods research that are explained for the usage of this chapter are qualitative and quantitative, they can also be described as positivist (quantitative) and "postpositivist" (qualitative). Cameron (2009) also promoted the idea that qualitative research, within itself, is multidisciplinary.

This chapter explores explanatory sequential mixed methods study next.

EXPLANATORY SEQUENTIAL MIXED METHODS STUDY

There are many variations of explanatory sequential mixed methods studies, but for the purposes of this chapter, Creswell and Creswell's (2018) definition is used. Creswell and Creswell's (2018) definition of explanatory sequential mixed methods aligns with other variations and provides the greatest degree of overlap.

An explanatory sequential mixed methods study begins with quantitative data collection from a larger group of participants (Creswell & Creswell, 2018). This is typically via survey data for statistical analysis. From that sample, the researcher selects a representative pool of participants for qualitative interviewing. The interview questions serve the main purpose of exploring preliminary findings represented in the quantitative data. For example, if I supplied a survey to 25 participants about their approaches to teaching math to elementary students, I may have most participants respond very negatively about their experiences teaching math and a couple who were very positive about the experience. It would be in the best interest of the integrity of the study to select a smaller group, maybe ten participants, who represent the various perspectives to learn more about why they responded the way they did on the quantitative survey. This portion of the research methodology is incredibly important because it can either affirm or challenge any preliminary findings that were evident in the quantitative portion. Whether the preliminary findings affirm or challenge the quantitative data's preliminary findings are equally important because even if the data does not provide answers to a guiding research question, it may make clear what guiding research questions need to be addressed with future research. Creswell and Poth (2018) contended that there are benefits of using different types of qualitative research methodology when executing explanatory sequential mixed methods studies.

The subsequent subsections look more closely at two types of qualitative research design and particular attention is placed on their role in balancing with quantitative research methodology in the same study.

Phenomenology

The first qualitative research methodology that this section explores is phenomenology. A phenomenology is a qualitative research method that explores the shared lived experiences of a group of people to learn more about a particular phenomenon. This is typically done through participants answering surveys or completing interviews. The data is analyzed with the intent of narrowing down the "essence of the experience" (Creswell & Creswell, 2018, p. 13). The essence is a summarizing sentiment shared by participants.

According to Creswell and Poth (2018), there are a variety of types of phenomenological research designs. These include, but are not limited to, transcendental phenomenology and hermeneutical phenomenology. Transcendental phenomenology is most similar to the overarching definition of phenomenology that was previously described. Through identifying individuals who have a shared lived experience and gathering data about those experiences, the researcher then, "analyses the data by reducing the information to significant statements or quotes and combines the statements into themes" (Creswell & Poth, 2018, p. 78).

In contrast, hermeneutical phenomenology is greatly depended upon "the texts of life (Creswell & Poth, 2018). The specific procedures have been left to interpretation of researchers, but overall, there are a series of rules that take place at the same time throughout the study. These include the researcher (a) being interested by a phenomenon out of concern, (b) reflecting of the essential themes that contribute to the lived experience, (c) writing a description of the phenomenon that involves interpretation. Regardless of the type of phenomenology that a research implements, phenomenological research design can most certainly play an integral role in an explanatory sequential mixed methods study.

An example of a mixed methods study that used phenomenology as the qualitative portion of the mixed methods approach was Tarver's (2006) study titled, "An explanatory sequential mixed method study of well-being, religious coping, and service utilization patterns of African American caregivers of ADRD elders." ADRD is an abbreviation for Alzheimer's Disease and Related Disorders. In the first phase, that was quantitative in nature, the researcher collected descriptive statistics and conducted correlations. Tarver found the preliminary finding that the "neither concern for ADRD elder's safety nor religious coping style (positive or negative) has a significant influence on service utilization" (p. iv). In addition Tarver (2006) wrote about the preliminary finding that there was no correlation between age and service usage. Higher levels of well-being were associated with the presence of family's support. The final preliminary finding from the quantitative data was that the older the person was, the less likely the person was to have family support and "Purpose in Life" (Taver, 2006, p. iv). Phenomenology played an integral role in the second phase of this explanatory sequential mixed methods study.

In the second half of Tarver's (2006) mixed methods study, he explained that qualitative methodology was employed with the intentionality of gaining a deeper understanding of the preliminary finding he outlined from the quantitative portion. He interviewed ten caregivers to learn about their shared lived experiences in relation to the guiding research questions. The interview protocol was created with the intentionality of learning more about the preliminary findings established in the quantitative portion. Tarver (2006) discovered themes around prayer, thankfulness, trust, and more. The commonalities between the quantitative preliminary findings and the themes that were extracted from the interview transcripts after coding led to implication "for culturally responsive service delivery to African American caregivers" (p. iv).

As Maxwell (2010) previous discussed in the disadvantages of including quantitative data with qualitative data, numbers can pose as a façade of rigor and precision. However, when an explanatory sequential study purposefully uses the numerical data to better guide the qualitative methodology, then this pitfall is avoided and upholds the quality that scholars expect and strive to achieve. In Tarver's (2006) explanatory sequential mixed methods study, Maxwell's (2010) advantages of combining the two methodologies were achieved, and the disadvantages were avoided. From Tarver's (2006) presentation of results, the quantitative data would not have been sufficient in crafting strong findings in response to the guiding research questions, but rather, they were only strong enough to present them as preliminary findings that warranted further investigation. Rather than pursue a larger scale of numerical investigation, Tarver (2006) chose to use phenomenological interviewing as a means to learn more about the preliminary findings.

The next subsection presents a definition of a case study to serve as another example of the importance of qualitative methodology as part of explanatory sequential mixed methods studies.

Case Study

According to Creswell and Poth (2018) a case study begins with identifying a case, which is a culture-sharing group. The case is found by the researcher in a "real-life, contemporary context, or setting" (Yin, 2014, as cited in Creswell & Poth, 2018, p. 96). In order to study a case, the researcher gathers multiple means of data. These can include, but are not limited to, interviews, observational notes, audiovisual data, documentation, and other types of reports (Creswell & Poth, 2018). The key to this data collection is that a high level of detail about the case must be represented. This leads to a case description and case themes. A case study may focus on a single case, or multiple cases. It depends on the guiding research questions and the availability of participants.

There are several types of case studies. For the purpose of this chapter, a brief description of three of them will suffice. The first is instrumental case study (Creswell & Poth, 2018). In this type of case study, the researcher chooses an issue or area of concern. The researcher then identifies a single bounded case that exemplifies the issue or area of concern that was first identified.

In contrast, a second type of case study is an intrinsic case study. This is incredibly similar to a narrative study because it looks closely at a case that is unique or unusual (Creswell & Poth, 2018). What sets it apart from a narrative study is that the case is set in its own context or setting and the analysis takes place within those confines.

The third type of case study is the collective case study. In summary, a collective case study looks at one issue, but by looking at multiple cases. By exploring several cases that resemble the same issue, the research is able to analyze multiple perspectives on the same issue. The same procedure should be followed for each case, in order to retain consistency. Creswell and Poth (2018) added, "As a general rule, qualitative researchers are reluctant to generalize from one case to another because the contexts of cases differ" (p. 99). If the cases are representative and inclusive, then generalizations are more appropriate.

Regardless of the type of case study that a researcher implements, it can most certainly play an integral role in an explanatory sequential mixed methods study. An example of an explanatory sequential mixed methods study that utilized the case study approach in the qualitative portion of the methodology is Makgakga and Makwakwa's (2016) explanatory sequential mixed methods study titled, "Exploring learners' difficulties in solving grade 12 differential calculus: a case study of one secondary school in Polokwane district." This particular study highlights the important role that case study methodology can serve in the qualitative portion of an explanatory sequential mixed methods study. The focus of this study was on twelfth grader students facing challenging when solving calculus questions the relate to differentiation. Not uncommon to case study methodology, the researchers utilized convenience sampling. They used a high school in South Africa and first began with administering a diagnostic test to serve as the quantitative portion of the data collection and analysis. The quantitative data led to some preliminary findings. One of these preliminary findings was that "52% of the learners performed better in finding the derivative of functions by using the rules of differentiation, compared with 23% of the learners who performed better on the questions by using first principles" (Makgakga & Makwakwa, 2016, p. 13). Such statistical data helped guide qualitative data collection.

The qualitative portion of the study consisted of a series of semi-structured interviews with a portion of the students and the teacher (Makgakga & Makwakwa, 2016). The researchers noted that the qualitative interview protocol was created after considering the students' responses to the quantitative diagnostic assessment. The qualitative data reaffirmed the quantitative preliminary findings, but increased the level of detail in the findings. The researchers added, "Using the rules of differential, learners experienced difficulties when given functions involving functions" (Makgakga & Makwakwa, 2016, p. 13). This then led to the recommendation that students learn how to differentiate functions using the roles of differentiation and first principles. This is done most effectively by students correcting errors in others' work. Makgakga & Makwakwa (2016) argue that the case study methodology executed in the qualitative portion of the study was integral to formulating strong findings and imperative recommendations. They added that even though the case study was limited in terms of the number of participants, the location, and other constraints, the case study methodology allowed for in-depth data collection that would not have been possible on a larger scale.

Grounded Theory

A third type of qualitative research method is grounded theory. In this methodology, the central intention is to discover or create a theory, which goes a step further than looking closely at a shared experience amongst a group of participants, like in a phenomenology (Creswell & Poth, 2018). Grounded theory, in summary, is the process by which the researcher uses a large number of participants to learn about their viewpoints. Their perspectives are analyzed and cumulated to form a generalized explanation of the action, process, or interaction that is under investigation.

An important characteristic of grounded theory is that is investigates a particular process or action that has specific steps or phases over a given duration of time (Creswell & Poth, 2018). The central goal of this research is to create a theory about the process. An important aspect of grounded theory research methodology is the step of memoing. To create a memo is to record ideas while collecting and analyzing data. This makes the record-keeping an ongoing process that reflects the flow of the process that the study is investigating. Data is typically collected through interviewing participants. The researcher compares interview data as more interviews are collected. This ongoing analysis beings to shape theory development.

Other data analysis approaches in grounded theory research methodology depend upon the type of grounded theory study the researcher is conducting (Creswell & Poth, 2018). One option is categorizing data and then focusing on one category as the basis of the theory through axial coding and selective coding. Another option is to depict the theory as a diagram, a series of hypotheses/propositions, or frame it as a discussion (Strauss & Corbin, 1998, as cited in Creswell & Poth, 2018).

There are two main types of grounded theory research (Creswell & Poth, 2018). The first type is systematic and procedural and the second type is constructivist. In the systematic and procedural approach, the researcher conducts 20-30 interviews and selects a category that represents an event or development. The researcher institutes theoretical sampling and implements a constant comparative method of data analysis. Open coding then leads to categorization involving casual conditions, strategies, contextual and intervening conditions, and consequences. This leads to the researcher developing theoretical hypotheses.

The second type of grounded theory, designated as constructivist, consists of focusing on the diversity of different settings and variations amongst perspectives (Creswell & Poth, 2018). The framework is much more flexible because it is reliant upon the researcher's viewpoint. Great emphasis is still put on consistent coding, memoing, and sampling. This constructivist version, however, is less dependent upon diagrams, systematic approaches, and jargon.

Some common challenges inherent within grounded theory research methodology include the researcher struggling to decide if a category is saturated or if the theory has enough detail (Creswell & Poth, 2018). Discriminant sampling can help increase saturation. While grounded theory research has been increasing in popularity regardless of subject-area, ranging from education to medicine, there is still room for improvement in provided more consistent guidelines.

Now that the main ideas of grounded theory research methodology have been delineated, this subsection will describe an explanatory sequential mixed methods study that implements grounded theory as the qualitative portion of the study. De Smet, Meganck, Van Nieuwenhove, Desmet, and Truijens' (2019) article titled, "No change? A grounded theory analysis of depressed patients' perspectives on non-improvement in psychotherapy" is a prime example of an explanatory sequential mixed method study that emphasizes importance of qualitative data. The central objective of this study was to understand the effects of psychotherapy as an important area of concern for research and clinical practices. The issue was that the outcomes of such therapies tend to be negative and these negative outcomes are reported as the result of pre-post quantitative questionnaires that lack patient perspective on their day-to-day struggles and successes (De Smet et al., 2019). To better understand these quantitative results, the researchers wanted to combine such findings with qualitative research methodology.

This study first began with quantitative data analysis through administering pre-post inventories (De Smet et al., 2019). All participants were in a time-limited psychodynamic and cognitive behavioral therapy for major depression. Participants were then invited to participate in semi-structured interviews. This was the qualitative portion that served as the grounded theory part of the study. The researchers used axial coding of the interview transcript, which led to selective coding, to better refine the theory. The study found that mixing qualitative, first-person analysis into quantitative analysis, especially when understanding the phenomenon of non-improvement. Furthermore, the study illuminated that combining qualitative and quantitative data collection aided the researchers in obtaining an enhanced understanding of treatment effects. The researchers credited grounded theory as the research methodology that enabled them to learn more about the quantitative results for better data collection of similar patients in the future.

Ethnography

A fourth type of qualitative study is called ethnography. Creswell and Poth (2018) made connections between ethnographies and grounded theory studies. They state that while a grounded theory study creates a theory after the researchers analyses many participants who share in the same process, interaction, or action, the individuals tend to not be the same location. This means that they typically do not interact frequently to have a shared language, belief system, or pattern of behaviors. Ethnographies differ from grounded theory studies because this type of research focuses on such patterning of language, beliefs and behaviors. Ethnographies focus on a group of people who share a culture, making it also similar to phenomenology. This type of qualitative research is a process of day-to-day participant observations and interviews through total immersion of the researcher in participants' lives. An important aspect of ethnography that is worth mention is that the analysis of data is through the researcher's etic scientific perspective when creating a generalized cultural interpretation.

The cultural interpretation is dependent upon the researcher's description of the group's themes that relate to the theoretical concepts under investigation. Creswell and Poth (2018) added that the most important ethnographies begin with looking at how a group functions because there is not much known about the group. The study type, in turn, provides new understanding or clarity about the group. There are many different types of ethnographies, but the two most popular are realist ethnography and critical ethnography. This type of ethnography is known for its ability to diminish researcher-bias. Realist ethnography are the most traditional in nature and a popular choice by cultural anthropologists, especially. Information is presented objectively, written in the third-person. Observations and verbatim quotes are provided and while the research is physically present in the location of data-collection, he or she does not provide thoughts or interpretations during data collection. Even the most mundane details are collected and reported as part of this process.

In contrast, critical ethnographies involve a much more passionate researcher, who is typically an advocate for the topic that is under investigation. The study is conducted as a response to a current event that effects systems of power/authority or marginalizes a given group. Creswell and Poth (2018) summarized critical ethnography succinctly by writing that this type of study is typically intertwined with values, peoples' empowerment, status-quo upheaval, and changes in power/authority. An example of an ethnography is described next.

The study that is described next is an example of an ethnographic research study as a result of there being so much quantitative research having already been conducted by other researchers (Goonan, Mirosa, & Spence, 2013). The researchers thought that there needed to be qualitative research added to the research canon. This connects to explanatory sequential research because the quantitative research that others have conducted helped to shape the qualitative research that Goonan (2013) conducted. Their study was titled, "Getting a taste for food waste: A mixed methods ethnographic study into hospital food waste before patient consumption conducted at three New Zealand foodservice facilities." A series of fourth different methods of collecting qualitative data were completed. These included document analysis, observations, focus group sessions, and semistructured interviews. Such a variety can be deemed an ethnography, since ethnographic research often times involves multiple sources of data (Creswell & Poth, 2018). Goonan et al. (2013), however, deemed this study a mixed methods ethnography. The study aimed to learn more about the creation of waste from hospital kitchens. Previous quantitative research focused on the waste that patients generated, not the kitchens. Similar to an explanatory sequential mixed methods study that aims to utilize quantitative data to guide the collection of qualitative data, Goonan's study did that, but it did not qualify as an explanatory sequential mixed methods study because Goonan et al. (2013) did not collect and analysis quantitative data. This example is an opportunity to highlight the dearth of explanatory sequential mixed methods studies that utilize ethnography in the qualitative portion.

Another study worth discussing was an explanatory sequential mixed methods study that was about a type of ethnographic research, but the study itself utilized a phenomenological approach in the qualitative portion. Addissie, Davey, Newport, Addissie, MacGregor, Feleke, and Farsides' (2014) study titled, "A mixed-methods study on perceptions towards use of Rapid Ethical Assessment to improve informed consent processes for health research in a low-income setting" serves as a powerful example of an explanatory sequential mixed methods study that implements ethnography as the qualitative portion. This study used Rapid Ethical Assessment (REA), which is a type of fast-paced ethnographic assessment that the researcher administers at the start of the study. The REA guides the consent process and is used to support universal ethical guidelines. Addissie et al.'s (2014) study intended to learn if REA could be a practical and useful tool in Ethiopia. Quantitative data was first collected through an online, self-administered questionnaire. These were closed-response questions, except for a final open-ended question. The questions directly connected to the variables the researchers outlined in the review the literature. The second phase of data collection was qualitative in nature (Addissie et al., 2014). Through in-depth interviews, each question was created after thorough analysis of the quantitative data.

Phenomenology, case study, grounded theory, and ethnography are just four examples of the important role that qualitative methodology plays in explanatory sequential mixed methods study. This chapter will now define and provide examples of exploratory sequential mixed methods study.

EXPLORATORY SEQUENTIAL MIXED METHODS STUDY

Similar to explanatory sequential mixed methods studies, are many variations of exploratory sequential mixed methods studies, but for the purposes of this chapter, Creswell and Creswell's (2018) definition is used. Creswell and Creswell's (2018) definition of exploratory sequential mixed methods aligns with other variations and provides the greatest degree of overlap.

Schoonenboom and Johnson (2017) furthered Creswell and Creswell's (2018) ideas regarding exploratory sequential mixed methods design, but placed necessary emphasis on the idea that the initial qualitative results and tested and generalized by the subsequent quantitative data. This definition highlights that quantitative data does not necessarily add depth. The second half of an exploratory sequential mixed methods study is a confirmation tool on a larger scale typically. This is also called an "explanatory-confirmatory design" (Schoonenboom & Johnson, 2017, p. 127). Similar to explanatory sequential mixed methods studies, exploratory sequential mixed methods studies can integrate a variety of qualitative methodologies. The difference is that the qualitative methodology takes place before the quantitative methodology. This section will now explore how phenomenology and case study approaches function within exploratory sequential mixed methods studies with examples.

Phenomenology

To reiterate, a phenomenology, in summary, is a qualitative study that focuses on the shared lived experiences of a group of individuals. The typical form of data collection in a phenomenology is via interviews and the analysis is done through coding procedures to extract themes that contribute to the findings. When a phenomenology is part of an exploratory sequential mixed methods study, however, the findings that are formulated from the themes are preliminary findings that are used to customize the instrumentation used in the quantitative portion of the study.

An example for an exploratory sequential mixed methods study that implemented phenomenology as the qualitative portion is Máquez and Delgado's (2017) study titled, "An exploratory sequential design to validate measures of moral emotions." The main purpose of the study was to validate, "measures of knowledge of the moral emotions of contempt, anger, and disgust" (Máquez & Delgado, 2017, p. 261). The study began with implementing descriptive phenomenological research design because the researchers explained that they wanted to analyze the subjective experiences of participants. The researchers recruited 60 participants for the qualitative phase. Similar to how the qualitative portion typically has fewer participants than the quantitative portion of an explanatory sequential mixed methods study, Máquez and Delgado's (2017) study did the same approach. In the quantitative portion, they had many more participants: 102 participants in total. Interestingly, however, is that none of the participants who participated in the qualitative portion of the study were invited to participate in the quantitative portion.

The researchers provided detailed transparency as to their justification for implementing an exploratory sequential mixed methods approach. Máquez and Delgado (2017) explained that the qualitative portion was a necessary component that allowed them to design their quantitative survey instrument. They added that by first executing in-depth interviews, the questions that they included on the quantitative survey reflected personal meanings that they extracted from the interview transcripts. This diminished participant confusion as to how terminology was being used, thus increasing clarity and sharing understanding amongst participants and researchers.

Case Study

As a reminder, a case study is a qualitative research methodology that looks closely at a culture-sharing group (Creswell & Poth, 2018). Multiple means of data collection are used in this qualitative portion of the study. This subsection explores an example of an exploratory sequential mixed methods study that implemented a case study approach in the qualitative portion.

In West's (2011) study titled, "Teaching middle school Jazz: An exploratory sequential mixed methods study," he was interested in learning more about the middle school music teachers' past experiences and current ways of thinking and actions related to middle school jazz education. West (2011) strived to execute this study through "the lens of perceived ability to teach middle school jazz" (p. xiv). Similar to the exploratory sequential mixed methods study described in the previous subsection that involved phenomenology as the qualitative portion of the study, West's study used the qualitative portion (which was a case study) as a tool in creating the quantitative instrumentation.

The case study part of the research, which served as the qualitative portion of this exploratory mixed methods study, involved only two middle school music teachers (West, 2011). West (2011) conducted interviews with each teacher and then observed each of their classes three times. West (2011) emphasized that following Creswell's (2007) recommendation for case study participants, that the limited participants should exemplify maximum variation. In other words, they should be as different as possible, but yet still sharing the same culture-sharing variable. West (2011) explained that the two teachers who were interviewed and observed had various differences in their teaching approach.

The quantitative portion of the study came after West (2011) was able to analyze the qualitative data. The coding led to preliminary findings that led to four categories. These included professional background, current actions, previous experiences, and current thoughts. These four categories served as the four central sections of the quantitative survey that was completed by 264 middle school music teachers. The qualitative and quantitative results were combined during the analysis to formulate a list of effective teaching practices that middle school music teachers execute. These included teachers requiring students to play "traditional" jazz instruments when joining a jazz ensemble, students using major scales when learning improvisation, teachers executing modeling, teachers asking students to listen to and watch jazz performances, teachers bringing in jazz musicians to work with students, and teachers preforming with students in call-and-response activities. Such findings would not have been possible without such rich qualitative data collection through case study methodology.

West's (2011) study is a prime example of how a case study promotes stronger quantitative research methodology and enhances the depth and meaning of findings.

Grounded Theory

In brief, grounded theory is the qualitative research methodology in which the researcher investigates a process or development with the means of creating a theory. Dugar's (2017) study titled, "Consumers' perceptions of voluntary and involuntary deconsumption: An exploratory sequential scale development study" serves as an example of an exploratory sequential mixed methods study that integrated grounded theory as the initial qualitative portion. Interestingly, when researching an example to include in this chapter, it was a challenging task to find an exploratory sequential mixed methods study that uses grounded theory. This may be because grounded theory is a process within itself and to say that the theory is established before the quantitative data has been collected and analyzed, in true sequential form, is not always possible or useful.

The main objective of this study was to learn more about the decrease or lack of consumption (also known as deconsumption) amongst the "baby boomer" generation (Dugar, 2017). The study aimed to learn more about the concept of voluntary and involuntary deconsumption, to better conceptualize the role of motivation within deconsumption, and to analyze the components, structures, ways to measure, and utilizations of deconsumption. The study began with grounded theory combined with critical incident in a relationship context (CIRC). Dugar (2017) first explored the theoretical explanation and meaning of the deconsumption process. Participants' views and perspectives were the central source of qualitative data. Dugar (2017) added, "common experiences were analyzed using a constant-comparative method to identify the conditions, contexts, motivations, strategies, and consequences of deconsumption (p. 5). This led to theoretical explanations that were unified. The qualitative, preliminary findings that were produced by this portion of the study were used to create the instrumentation utilized in the quantitative portion. The questions were created with the theoretical framework of voluntary and involuntary deconsumption.

Interestingly, Dugar (2017) explained that the main purpose of the quantitative portion of the study was to further refine the theoretical framework that the qualitative portion had created. The quantitative survey also allowed Dugar (2017) to confirm the integrity of the theoretical framework to ensure the hypotheses consistently matched more data, only quantitative in nature this time. In sum, this study found that the factors that determine a person's experience with deconsumption was based on whether the deconsumption was voluntary or involuntary, the gender of the respondent, and where on the continuum of being a "baby boomer" the respondent landed. In addition, Dugar (2017) added that voluntary deconsumption entailed four main areas, which included social agency/activism, lack of materialism, life acceptance, and a high level of life purpose. Involuntary deconsumption consisted of three central areas that included a sense of feeling like a victim, materialism, and a lack of life acceptance. Such findings were made possible due to the rigorous and vivid grounded theory work.

Ethnography

As previously explained, ethnography is a qualitative research methodology that involves the researcher fully immersing his/herself into the lives of the individuals who are being studies (Creswell & Poth, 2018). An example of an exploratory sequential mixed methods study that utilizes ethnography as the qualitative methodology is Crede and Borrego's (2013) study titled, "From ethnography to items: A mixed methods approach to developing a survey to examine graduate engineering student retention." The central objective of this study is to learn about what promotes and hinders the retention of undergraduate engineering students at their institutions of higher education.

Crede and Borrego (2013) began their exploratory sequential mixed methods study with first conducting a series of observations of and interviews with undergraduate engineering students. They clearly articulated that ethnography guided their observations and interviews, which lasted a total of nine months. As Creswell and Poth (2018) emphasized, a long duration of time increases the richness of the qualitative data collected. The qualitative data led Crede and Borrego (2013) to four important themes that were later utilized in creating the quantitative survey questions. The themes included the school's climate and organization of research groupings, the international diversity present on campus, the students' sense of value at the institution, and the balance of opportunities for students to learn as individuals and in groups.

Crede and Borrego (2013) summarized the important role that ethnography played by writing that they gathered a deeper understanding of the qualitative data while also keeping in mind the broad application of such understandings. The broad applications were useful later when considering the quantitative data collection methodology. The ethnographic data "gave voice to the students, which provided rich details on participant language, cultural considerations, and helped formulate hypotheses for the survey" (Crede & Borrego, 2013, p. 67). The emphasis put on student voice matches the idea that Creswell and Poth (2018) presented in that ethnographies often leave the researcher's voice out of it, unless it is a critical ethnography. Crede and Borrego (2013) study seemed to have been more of a realist ethnography.

The quantitative portion of the study began with creating a survey based on the themes extracted from the ethnographic half of the study (Crede & Borrego, 2013). They shared that the qualitative data contributed to the actual survey questions. In addition, qualitative data informed the researchers as to the participant group's diversity. Their diversity affected their values, cultural norms, and expectations. The differences in perspectives helped increase the validity of the survey because the researchers considered it greatly when developing the survey. The quantitative surveys found that they had varying internal consistency values, based upon the country or region the participant is from. There was a notable difference in the responses from the participants who came from the Middle East. The researchers claimed that they believe the Middle Eastern students stood out as vastly different because there were no Middle Eastern students in the ethnographic portion of the study, thus their differences were not accounted for in the creation of the survey. Such an observation is useful insight into exploratory sequential mixed methods studies that use ethnography as the qualitative portion. Future researchers need to thoroughly consider the representation of participants in the qualitative portion versus the quantitative portion. This could take place to either ensure validity when discussing limitations, or as a means for purposeful participant sampling to ensure consistency of cultural background or to take into account other differences.

By exploring how qualitative methods enhance mixed methods studies through analysis of example studies, this chapter has contextualized the importance of qualitative methodology. This chapter now provides a review of literature that furthers this message, which is greatly directed toward quantitative-purists.

THE NECESSITY OF QUALITATIVE METHODS IN MIXED METHODS DESIGN

This third and final major section of this chapter addresses the "nay-sayers" who may argue that the only benefit of mixed methods research is that quantitative data makes the qualitative data more valid and reliable. On the contrary, it is the qualitative portion that strengthens the ideas represented by the quantitative data. This section gives the voice and power to the qualitative portion of the study. Whether it comes first or second in order of execution (explanatory sequential or exploratory sequential), the qualitative portion is integral to a mixed methods design. Creswell, Shope, Plano Clark, and Green (2006) wrote that qualitative methodology is the main force responsible for the success of mixed methods research.

Hesse-Biber (2010) wrote about the importance of qualitative approaches in mixed methods practice. She described the existence of a "methodological orthodoxy" in how scholars practice mixed methods research. Hesse-Biber (2010) claimed that in the current view of mixed methods practice, quantitative methods are viewed "as primary with an overall mixed methods design that is in the service of testing out quantitatively generated theories about the social world" (p. 455). Her article aimed to challenge this school of thought by detracting from the focus on positivism and replacing it with qualitative approaches in mixed methods practice. In summary, her central argument was that qualitative research intends to give individuals power by telling their stories. She added that it is the researcher who takes these stories and makes meaning of the social world. Hesse-Biber (2010) even furthered the argument made earlier in this chapter about the importance of case study design in mixed methods study. She conducted a series of intensive case studies to further exemplify the powerful balance that results when combining research methods when emphasis is put on the qualitative approach. In her mixed methods studies, the case study portion was the driving force of the studies' creation and analysis.

Hesse-Biber (2010) supported the claim that qualitative research methodology deserves increased respect and emphasis in mixed methods due to its core assumption that reality has multiple variations and is continuously constructed by society.

Hammarberg, Kirkman, and de Lacey (2016) defined quantitative research as the research methodology used when answering a research question that requires factual data. They added that quantitative research is used when variables can be defined and isolated and also when seeking generalizations about attitudes, opinions, thoughts, beliefs, or preferences. In quantitative research, according to Hammarberg, Kirkman, and de Lacey (2016), a hypothesis can be formed before data is collected when variables can be linked.

Alternatively, qualitative research methods are starkly opposite in nature and in a favorable way. Hammarberg et al. (2016) explained that qualitative methods answer questions pertaining to meaning and experience. The participants' perspectives play a vital role in this process. They added that this is not the type of data researchers can count on or measure easily to further explain how this method contrasts with quantitative methods. After clearly defining the differentiating characteristics between qualitative research methods and quantitative research methods, Hammarberg et al. (2016) discussed how to judge qualitative research. They explained that to use the terms "reliable" and "valid" have been disputed by many scholars, thus leaving the research community more recently preferring the term "verification." Hammarberg et al. (2016) define this as the level of integrity and robustness. They furthered their elaboration by sharing that qualitative research cannot possible adopt an objective stance due to the fact that the research serves as an instrument. Hammarberg et al. (2016) referred to Dezin and Lincoln (1998) when explaining that the subjects contribute to the data interpretation and synthesis. Furthermore, Hammarberg, et al. (2016) cited Leninger (1994) when stating that qualitative researchers use different means by which they defend the integrity of their research. These include "trustworthiness, credibility, applicability and consistency are the evaluative criteria" (Hammarberg, 2016, p. 499). This section will now describe what each of these areas of integrity mean for qualitative research to highlight it's integrity as a research methodology when compared to quantitative research methodology.

The first area that marks a qualitative study's integrity, is trustworthiness. By this, Hammarberg et al. (2016) explained the study's procedure needs to be as thoroughly and meticulously delineated as any quantitative study. The necessary components that need explicit detail include the procedural steps, decisions that guided the methodology's execution, and data management and generation. The second area of qualitative research integrity is credibility. At first glance, the term "credibility" seems nuanced and ambiguous. According to Hammarberg et al. (2016), credibility refers to the study's sufficient descriptions of the context with detail in how the researcher or researchers cared for the results. Truth value or internal validity are two other designations that Hammarberg et al. (2016) use for discussing qualitative methodology creditability. They added that there are three ways in which a research can defend the study's credibility: (a.) reflexivity, which is the researcher's influence on the study, (b.) triangulation, which is the multiple ways in which the guiding research questions can be answered through different forms of data collection, and (c.) clear and concise delineation of the interpretation process of data. When the quotes taken verbatim from the data collection do not align with the interpretation specifications that the researcher thoroughly describes, the credibility of the study is then decreased.

The next area of integrity is applicability (Hammarberg et al., 2016). This refers to the ability to transfer research findings, thus making it the criteria that relates most to the study's external validity. A study is considered applicable when it fits into contexts that are beyond the study's specific setting and participants. Researchers read the study and deem it applicable if they are able to apply it to their personal experiences. One common misconception is that applicability increases as the sample sizes increases. This is not the case, as a larger sample size can mean that the researchers do not investigate the research questions with the same amount of detail and depth that is possible with smaller sample sizes. Applicability is increased when sampling strategies are clearly "purposive, convenient, theoretical or snowballed. Maximum variation sampling may be used to seek representation of diverse perspectives on the topic" (Hammarberg et al., 2016, p. 500). The participants should have many similarities only if that is a necessary criterion for investigating a shared perspective, understanding, experience, or other criteria in order to answer the guiding research questions. While many quantitative-purists argue that biases are an inherent part of qualitative research methodology, Hammarberg et al. (2016) contended that since participants who are selected to participate in a qualitative study were selected because they have a quality that enables them to share insights regarding the phenomenon under investigation, researcher biases are inconsequential. Methods like bracketing help researchers avoid the pitfalls of subjectivity and biases (Creswell & Poth, 2018).

The final area of integrity for qualitative research is consistency (Hammarberg, 2016). Consistency refers to the reliability and dependability of the results. This means that if another researcher were given the same data, he or she would notice the same trends and themes. It does not mean that the if another researcher conducted the same study again, he or she would necessarily obtain the same results. Often times, researchers strive to obtain as much variation in participants' experiences with the phenomenon under investigation. This variation limits researcher expectations.

Hammarberg et al. (2016) end their article with a conclusion that highlights the misconceptions that quantitative-purists and qualitative-doubters often acknowledge within the research communities. Hammarberg et al. (2016) stated that qualitative research is not the easiest option and not merely a collection of stories or anecdotes. Rather, it incorporates philosophical or theoretic frameworks that can be quite complex in nature. Analysis of qualitative studies can be rigorous and meet the criteria of integrity without including statistical analysis. Numbers are not the qualifying factor to make a study legitimate. Research areas like health intervention and education cannot always be quantified and sometimes individuals' stories and perspectives carry more weight than any massive, statistical study could. Hammarberg et al. (2016) highlight an idea that may go unnoticed by those unfamiliar with qualitative research. As many scholars know, the word-limit for journal articles can be frustratingly small. Qualitative researchers, in participle, struggle with this because of ten times their data representations need more room than is allowed in order to fully depict their study's integrity.

CONCLUSION

In sum, this chapter described the importance of qualitative research methods within the context of mixed methods research. By exploring explanatory sequential mixed methods studies and exploratory sequential mixed methods study, this chapter emphasized the qualitative portion of each type. To better characterize this analysis, examples of these two types of studies showed how the qualitative portions of the studies made the findings possible. Phenomenology, case study, grounded theory, and ethnography were four examples of qualitative portions of sequential mixed methods studies that were shared. The delineation of the types of studies with examples led to the third section that focused on the necessity of qualitative research methods. The final section focused on the importance and integrity of such research to help lessen the volume of quantitative research purists. The objective of this chapter was to encourage researchers to think about how qualitative research methodology could enhance quantitative research by turning a quantitative study into a mixed methods study.

ACKNOWLEDGMENT

This research received no specific grant from any funding agency in the public, commercial, or not-for-profit sectors.

REFERENCES

Addissie, A., Davey, G., Newport, M. J., Addissie, T., MacGregor, H., Feleke, Y., & Farsides, B. (2014). A mixed-methods study on perceptions towards use of Rapid Ethical Assessment to improve informed consent processes for health research in a low-income setting. *BMC Medical Ethics*, *15*(1), 1–12. doi:10.1186/1472-6939-15-35 PMID:24885049

Cameron, R. (2009). A sequential mixed model research design: Design, analytical and display issues. *International Journal of Multiple Research Approaches*, *3*(2), 140–152. doi:10.5172/mra.3.2.140

Cameron, R. (2011). Mixed methods research: The five Ps framework. *Electronic Journal of Business Research Methods*, *9*(2), 96–108.

Crede, E., & Borrego, M. (2013). From ethnography to items: A mixed methods approach to developing a survey to examine graduate engineering student retention. *Journal of Mixed Methods Research*, *7*(1), 62–80. doi:10.1177/1558689812451792

Creswell, J. W., & Creswell, J. D. (2018). *Research Design: Qualitative, Quantitative, and Mixed Methods Approaches* (5th ed.). Sage.

Creswell, J. W., & Poth, C. N. (2018). *Qualitative Inquiry & Research Design* (4th ed.). Sage.

Creswell, J. W., Shope, R., Plano Clark, V. L., & Green, D. O. (2006). How interpretive qualitative research extends mixed methods research. *Research in the Schools*, *13*(1), 1–11.

De Smet, M. M., Meganck, R., Van Nieuwenhove, K., Desmet, M., & Truijens, F. L. (2019). No Change? A Grounded Theory Analysis of Depressed Patients' Perspectives on Non-improvement in Psychotherapy. *Frontiers in Psychology*, *10*(588), 1–17. doi:10.3389/fpsyg.2019.00588 PMID:30971973

Dugar, K. K. (2017). *Consumers' perceptions of voluntary and involuntary deconsumption: An exploratory sequential scale development study* [Unpublished doctoral dissertation]. The University of Denver.

Everest, T. (2014). Resolving the qualitative-quantitative debate in healthcare research. *Medical Practice and Reviews*, *5*(1), 6–15. doi:10.5897/MPR.2013.0107

Goonan, S., Mirosa, M., & Spence, H. (2014). Getting a taste for food waste: A mixed methods ethnographic study into hospital food waste before patient consumption conducted at three New Zealand foodservice facilities. *Journal of the Academy of Nutrition and Dietetics*, *114*(1), 63–71. doi:10.1016/j.jand.2013.09.022 PMID:24231365

Hammarberg, K., Kirkman, M., & de Lacey, S. (2016). Qualitative research methods: When to use them and how to judge them. *Human Reproduction (Oxford, England)*, *31*(3), 498–501. doi:10.1093/humrep/dev334 PMID:26759142

Hesse-Biber, S. (2010). Qualitative approaches to mixed methods practice. *Qualitative Inquiry -. Qualitative Inquiry*, *16*(6), 455–468. doi:10.1177/1077800410364611

Makgakga, S., & Makwakwa, E. G. (2016). *Exploring learners' difficulties in solving grade 12 differential calculus: A case study of one secondary school in Polokwane district* [Unpublished doctoral dissertation]. The University of South Africa.

Márquez, M. G., & Delgado, A. R. (2017). An exploratory sequential design to validate measures of moral emotions. *Psicothema*, *29*(2), 261–267. PMID:28438252

Maxwell, J. A. (2010). Using numbers in qualitative research. *Qualitative Inquiry*, *16*(6), 475–482. doi:10.1177/1077800410364740

Maxwell, J. A. (2016). Expanding the history and range of mixed methods research. *Journal of Mixed Methods Research*, *10*(1), 12–27. doi:10.1177/1558689815571132

Onwuegbuzie, A. J., Johnson, B. R., & Turner, L. A. (2007). Toward a definition of mixed methods research. *Journal of Mixed Methods Research*, *1*(2), 112–133. doi:10.1177/1558689806298224

Russell, C. K., & Gregory, D. M. (2003). Evaluation of qualitative research studies. *Evidence-Based Nursing*, *6*(2), 36–40. doi:10.1136/ebn.6.2.36 PMID:12710415

Sage. (2019). Learn to use an exploratory sequential mixed method design for instrument development. *Research Methods Datasets*. Retrieved from https://methods.sagepub.com/base/download/DatasetStudentGuide/exploratory-sequential-mixed-method-instrument-development-united-states

Schoonenboom, J., & Johnson, R. B. (2017). How to Construct a Mixed Methods Research Design. *Kölner Zeitschrift für Soziologie und Sozialpsychologie*, *69*(S2, Suppl 2), 107–131. doi:10.100711577-017-0454-1 PMID:28989188

Tarver, D. D. (2006). *An explanatory sequential mixed method study of well-being, religious coping, and service utilization patterns of African American caregivers of ADRD elders* [Unpublished doctoral dissertation]. The University of Nebraska-Lincoln.

West, C. L. (2011). *Teaching Middle School Jazz: An Exploratory Sequential Mixed Methods Study* [Unpublished doctoral dissertation]. The University of Michigan.

Compilation of References

Aaltonen, A., & Tempini, N. (2014). Everything counts in large amounts: A critical realist case study on data-based production. *Journal of Information Technology*, *29*(1), 97–110. doi:10.1057/jit.2013.29

Ackroyd, S., & Fleetwood, S. (2000). Realism in contemporary organisation and management studies. In S. Ackroyd & S. Fleetwood (Eds.), Realist perspectives on Management and Organisations (pp. 3-25). Psychology Press.

Ackroyd, S. (2012). Contemporary Realism and Organisation Change. In D. M. Boje, B. Burnes, & J. Hassard (Eds.), *The routledge companion to organizational change* (pp. 230–244). Routledge.

Adams, T. E. (2017). Critical autoethnography, education, and a call for forgiveness. *International Journal of Multicultural Education*, *19*(1), 79–88. doi:10.18251/ijme.v19i1.1387

Adams, T. E. (2020). Critical rhetoric, relationality, and temporality: A case for forgiveness. *International Journal of Communication*, *14*, 899–911.

Addissie, A., Davey, G., Newport, M. J., Addissie, T., MacGregor, H., Feleke, Y., & Farsides, B. (2014). A mixed-methods study on perceptions towards use of Rapid Ethical Assessment to improve informed consent processes for health research in a low-income setting. *BMC Medical Ethics*, *15*(1), 1–12. doi:10.1186/1472-6939-15-35 PMID:24885049

Aguerrebere, J. (2009). *Panel discussion at Teachers' use of data to impact teaching and learning*. Alliance for Excellent Education.

Ajil, A., & Blount-Hill, K. L. (2020). "Writing the other as other": Exploring the othered lens in academia using collaborative autoethnography. *Decolonization of Criminology and Justice*, *2*(1), 83–108. doi:10.24135/dcj.v2i1.19

Alderfer, C. P. (1967). An organizational syndrome. *Administrative Science Quarterly*, *12*(3), 440–460. doi:10.2307/2391314

Alharahsheh, H. H., & Pius, A. (2020). A Review of key paradigms: Positivism vs interpretivism. *Global Academic Journal of Humanities and Social Sciences*, *2*(3), 39–43.

Alkhatib, S. (2012). *Social death: A grounded theory study of the emotional and social effects of honour killing on victims' family members- A Palestinian family perspective*. University of Manchester.

Allen, L. (2011). "Picture this": Using photo-methods in research on sexualities and schooling. *Qualitative Research, 11*(5), 487–504. doi:10.1177/1468794111413224

Altrighter, H., & Posch, P. (1989). Does the grounded theory approach offer a guiding paradigm for teacher research? *Cambridge Journal of Education, 19*(1), 21–31. doi:10.1080/0305764890190104

Amsden, J., & VanWynsberghe, R. (2005). Community mapping as a research tool with youth. *Action Research, 3*(4), 357–381. doi:10.1177/1476750305058487

Anderson, G. L. (1989). Critical ethnography in education: Origins, current status, and new directions. *Review of Educational Research, 59*(3), 249–270. doi:10.3102/00346543059003249

Anderson, P. J., Blatt, R., Christianson, M. K., Grant, A. M., Marquis, C., Neuman, E. J., Sonenshein, S., & Sutcliffe, K. M. (2006). Understanding mechanisms in organizational research reflections from a collective journey. *Journal of Management Inquiry, 15*(2), 102–113. doi:10.1177/1056492605280231

Anderson, T., & Shattuck, J. (2012). Design-based research: A decade of progress in education research? *Educational Researcher, 41*(1), 16–25. doi:10.3102/0013189X11428813

Andrade, A. D. (2009). Interpretive research aiming at theory building: Adopting and adapting the case study design. *Qualitative Report, 14*(1), 42.

Ansley, F., & Gaventa, J. (1997). Researching for democracy and democratizing research. *Change: The Magazine of Higher Learning, 29*(1), 46–53. doi:10.1080/00091389709603114

Anyon, J. (1980). Social class and the hidden curriculum of work. *Journal of Education, 162*(1), 67–92. doi:10.1177/002205748016200106

Archer, L., Dewitt, J., & Osborne, J. (2015). Is science for us? Black students' and parents' views of science and science careers. *Science Education, 99*(2), 199–237. doi:10.1002ce.21146 PMID:28579645

Argyris, C. (1968). Conditions for competence acquisition and therapy. *The Journal of Applied Behavioral Science, 4*(2), 147–177. doi:10.1177/002188636800400201

Argyris, C. (1970). *Intervention theory and method*. Addison-Wesley.

Argyris, C. (1980). *Inner contradictions of rigorous research*. Academic Press.

Argyris, C. (1982). *Reasoning, learning, and action: Individual and organizational*. Jossey-Bass.

Argyris, C. (1995). Action science and organizational learning. *Journal of Managerial Psychology, 10*(6), 20–26. doi:10.1108/02683949510093849

Argyris, C. (1996). Actionable knowledge: Design causality in the service of consequential theory. *The Journal of Applied Behavioral Science, 32*(4), 390–406. doi:10.1177/0021886396324004

Argyris, C., Putnam, R., & Smith, D. (1985). *Action science: Concepts, methods, and skills for research and intervention*. Jossey-Bass.

Avgerou, C. (2013). Social mechanisms for causal explanation in social theory based IS research. *Journal of the Association for Information Systems, 14*(8), 399–419. doi:10.17705/1jais.00341

Avgerou, C. (2019). Contextual explanation: Alternative approaches and persistent challenges. *Management Information Systems Quarterly, 43*(3), 977–1006. doi:10.25300/MISQ/2019/13990

Avison, D. E., Lau, F., Myers, M. D., & Nielsen, P. A. (1999). Action research. *Communications of the ACM, 42*(1), 94–97. doi:10.1145/291469.291479

Baber, L., Fletcher, R., & Graham, E. (2015). Black Men Attending Community Colleges: Examining an institutional approach toward equity. *New Directions for Community Colleges, 172*(172), 97–107. doi:10.1002/cc.20167

Bachman, L. F. (1990). *Fundamental considerations in language testing*. OUP.

Bachman, L. F., & Palmer, A. S. (1996). *Language testing in practice: Designing and developing useful language tests*. OUP.

Bailey, P. H., & Tilley, S. (2002). Storytelling and the interpretation of meaning in qualitative research. *Journal of Advanced Nursing, 38*(6), 574–583. doi:10.1046/j.1365-2648.2000.02224.x PMID:12067396

Baker, C., Wuest, J., & Noerager, P. (1992). Method slurring: The grounded theory/phenomenology example. *Journal of Advanced Nursing, 17*(11), 1355–1360. doi:10.1111/j.1365-2648.1992.tb01859.x PMID:1430643

Ballard, H. L., & Belsky, J. M. (2010). Participatory action research and environmental learning: Implications for resilient forests and communities. *Environmental Education Research, 16*(5–6), 611–627. doi:10.1080/13504622.2010.505440

Ball, S. J. (2016). Subjectivity as a site of struggle: Refusing neoliberalism? *British Journal of Sociology of Education, 37*(8), 1129–1146. doi:10.1080/01425692.2015.1044072

Bancroft, L. (2003). *Why does he do that? Inside the minds of angry and controlling men*. Penguin Publishing Group.

Baquedano-López, P., Alexander, R. A., & Hernández, S. J. (2013). Equity issues in parental and community involvement in schools: What teacher educators need to know. *Review of Research in Education, 37*(1), 149–182. doi:10.3102/0091732X12459718

Barnard, A. (2016, December 13). Battle Over Aleppo Is Over, Russia Says, as Evacuation Deal Reached. The New York Times. https://www.nytimes.com/2016/12/13/world/middleeast/syria-aleppo-civilians.html?_r=1

Barnes, M. E. (2017). Encouraging interaction and striving for reciprocity: The challenges of community-engaged projects in teacher education. *Teaching and Teacher Education*, *68*, 220–231. doi:10.1016/j.tate.2017.09.004

Barone, T., & Eisner, E. W. (2012). Arts-based research. *Sage (Atlanta, Ga.)*.

Battey, D. (2013). "Good" mathematics teaching for students of color and those in poverty: The importance of relational interactions within instruction. *Educational Studies in Mathematics*, *82*(1), 125–144. doi:10.100710649-012-9412-z

Battiste, M. (1998). Enabling the autumn seed: Toward a decolonized approach to aboriginal knowledge, language, and education. *Canadian Journal of Native Education*, *22*(1), 16–27.

Baxter, P., & Jack, S. (2008). Qualitative case study methodology: Study design and implementation for novice researchers. *Qualitative Report*, *13*(4), 544–559.

Beal, C. D., & Flynn, J. (2015). Toward the digital water age: Survey and case studies of Australian water utility smart-metering programs. *Utilities Policy*, *32*, 29–37. doi:10.1016/j.jup.2014.12.006

Beilin, R. (2005). Photo-elicitation and the agricultural landscape: 'Seeing' and "telling" about farming, community and place. *Visual Studies*, *20*(1), 56–68. doi:10.1080/14725860500064904

Belanger, S. L. (1998). Knowing through art. In W. Braud & R. Anderson (Eds.), Transpersonal research methods for the social sciences: Honoring human experience (pp. 166-168). SAGE.

Bell, D. (2004). *Silent covenants: Brown v. Board of Education and the unfulfilled hopes for racial reform.* Oxford University Press.

Bell, D. (1980). *Shades of Brown: New perspectives on school desegregation.* Teachers College Press.

Bhaskar, R. (1975). *A realist theory of science*. Leeds Books Ltd.

Bhaskar, R. (1989). *The possibility of naturalism: a philosophical critique of the contemporary human sciences*. Harvester Wheatsheaf.

Bhattacharya, K. (2020). Autohistoria-teoría: Merging self, culture, community, spirit, and theory. *Journal of Autoethnography*, *1*(2), 198–202. doi:10.1525/joae.2020.1.2.198

Bishop, R. (2011). *Freeing ourselves*. Sense Publishers. doi:10.1007/978-94-6091-415-7

Bloomberg, L. D., & Volpe, M. (2016). *Completing your qualitative dissertation: A road map from beginning to end* (3rd ed.). SAGE Publications.

Bolman, L. (2013). Chris Argyris: The iconoclast. In *The Palgrave handbook of organizational change thinkers* (pp. 17–36). doi:10.1007/978-3-319-52878-6_29

Boone, E. J. (1985). *Developing programs in adult education.* Prentice-Hall, Inc.

Boucher, M. L. (2018a). Interrogating whiteness: Using photo-elicitation to empower teachers to talk about race. In M. L. Boucher (Ed.), *Participant empowerment through photo elicitation in ethnographic education research: New research and approaches* (pp. 201–225). Springer.

Boucher, M. L. (Ed.). (2018b). *Participant empowerment through photo elicitation in ethnographic education research: New research and approaches*. Springer., doi:10.1007/978-3-319-64413-4

Boucher, M. L. Jr. (2014). More than an ally: A successful white teacher who builds solidarity with his African American students. *Urban Education*, *51*(1), 82–107. doi:10.1177/0042085914542982

Boucher, M. L. Jr. (2017). The art of observation: Issues and potential of using photo-methods in critical ethnography with adolescents. *International Journal of Adult Vocational Education and Technology*, *8*(2), 1–14. doi:10.4018/IJAVET.2017040101

Boucher, M. L. Jr, & Helfenbein, R. J. (2015). The push and the pull: Deficit models, Ruby Payne, and becoming a "Warm Demander.". *The Urban Review*, *47*(4), 742–758. doi:10.100711256-015-0332-y

Boughton, B., & Durnan, D. (2014). Cuba's "Yes, I Can" mass adult literacy campaign model in Timor-Leste and Aboriginal Australia: A comparative study. *International Review of Education / Internationale. Zeitschrift für Erziehungswissenschaft*, *60*(4), 559–580. doi:10.100711159-014-9421-5

Bourke, B. (2014). Positionality: Reflecting on the research process. *Qualitative Report*, *19*(33), 1–9.

Bourke, B., & State, M. (2014). "Positionality: Reflecting on the research process" by Brian Bourke. *Qualitative Report*, *19*(18), 1–9.

Bowen, G. A. (2006). Grounded theory and sensitizing concepts. *International Journal of Qualitative Methods*, *5*(3), 12–23. doi:10.1177/160940690600500304

Bowen, G. A. (2008). Naturalistic inquiry and the saturation concept: A research note. *Qualitative Research*, *8*(1), 137–152. doi:10.1177/1468794107085301

Bracken, S. (2007). The importance of language, context, and communication as components of successful partnership. *New Directions for Community Colleges*, *2007*(139), 41–47. doi:10.1002/cc.291

Brandt, H. M., Freedman, D. A., Friedman, D. B., Choi, S. K., Seel, J. S., Guest, M. A., & Khang, L. (2016). Planting healthy roots: Using documentary film to evaluate and disseminate community-based participatory research. *Family & Community Health*, *39*(4), 242–250. doi:10.1097/FCH.0000000000000120 PMID:27536929

Braun, V., & Clarke, V. (2012). Thematic analysis. In H. M. Cooper & P. M. Camic (Eds.), APA handbook of research methods in psychology, Vol 2: Research designs: Quantitative, qualitative, neuropsychological, and biological (pp. 57–71). American Psychological Association. doi:10.1037/13620-004

Bray, W. S. (2011). A collective case study of the influence of teachers' beliefs and knowledge on error-handling practices during class discussion of mathematics. *Journal for Research in Mathematics Education*, *42*(1), 2–38. doi:10.5951/jresematheduc.42.1.0002

Briars, D. J., Larson, M., Strutchens, M. E., & Barnes, D. (2015). A Call for Mathematics Education Colleagues and Stakeholders to Collaboratively Engage with NCTM: In Response to Martin's Commentary. *Journal of Urban Mathematics Education*, *8*(2), 23–36.

Bridger, L. (2013). Seeing and Telling Households: A Case for Photo Elicitation and Graphic Elicitation in Qualitative Research. *Graduate Journal of Social Science*, *10*(2), 106–131.

Bringle, R., Clayton, P., & Price, M. (2009). Partnerships in service-learning and civic engagement. *Partnerships: A Journal of Service-Learning & Civic Engagement, 1*(1), 1-20.

Bringle, R., & Hatcher, J. (2002). Campus-community partnerships: The terms of engagement. *The Journal of Social Issues*, *58*(3), 503–516. doi:10.1111/1540-4560.00273

Brookfield, S. (2018). Critical adult education theory: Traditions and influence. In M. Milana, S. Webb, J. Holford, R. Waller, & P. Jarvis (Eds.), *The Palgrave international handbook on adult and lifelong education and learning* (pp. 53–74). Palgrave Macmillan. doi:10.1057/978-1-137-55783-4_4

Brown, B. A., Henderson, J. B., Gray, S., Donovan, B., Sullivan, S., Patterson, A., & Waggstaff, W. (2016). From description to explanation: An empirical exploration of the African-American pipeline problem in STEM. *Journal of Research in Science Teaching*, *53*(1), 146–177. doi:10.1002/tea.21249

Brown, H. D. (2014). *Principles of language learning and teaching: A course in second language acquisition*. Pearson.

Brown, H. D., & Lee, H. K. (2015). *Teaching by principles: An interactive approach to language pedagogy*. Pearson.

Bryant, A. (2009, September). Grounded theory and pragmatism: The curious case of Anselm Straus. *Forum Qualitative, 10*(3). https://www.qualitative-research.net/index.php/fqs/article/view/1358/2850

Bryant, A. (2007). *The SAGE handbook of grounded theory*. Sage. doi:10.4135/9781848607941

Bryant, A. (2019). The varieties of grounded theory. *Sage (Atlanta, Ga.)*.

Burkholder, C., & Frawley, A. (2019). Revisiting and Disrupting Uncritical Diversity Narratives through Autoethnography and Cellphilming. *LEARNing Landscapes*, *12*(1), 63–76. doi:10.36510/learnland.v12i1.979

Burrell, G., & Morgan, G. (1979). *Sociological paradigms and organisational analysis* (Vol. 248). Heinemann.

Buys, N., & Bursnall, S. (2007). Establishing university-community partnerships: Processes and benefits. *Journal of Higher Education Outreach & Engagement*, *29*(1), 73–86.

Bytheway, J. (2018). Using grounded theory to explore learners' perspectives of workplace learning. *International Journal of Work-Integrated Learning*, *19*(3), 249–259.

Cahill, C., Alberto Quijada Cerecer, D., Reyna Rivarola, A. R., Hernández Zamudio, J., & Alvarez Gutiérrez, L. (2019). 'Caution, we have power': Resisting the 'school-to-sweatshop pipeline' through participatory artistic praxes and critical care. *Gender and Education*, *31*(5), 576–589. doi:10.1080/09540253.2019.1582207

Cahnmann, M. (2003). The craft, practice, and possibility of poetry in educational research. *Educational Researcher*, *32*(3), 29–36. doi:10.3102/0013189X032003029

Call-Cummings, M., Hauber-Özer, M., & Dazzo, G. (2019). Examining researcher identity development within the context of a course on PAR: A layered narrative approach. *Qualitative Report*, *24*(9), 2178–2196. https://nsuworks.nova.edu/tqr/vol24/iss9/6

Cameron, R. (2009). A sequential mixed model research design: Design, analytical and display issues. *International Journal of Multiple Research Approaches*, *3*(2), 140–152. doi:10.5172/mra.3.2.140

Cameron, R. (2011). Mixed methods research: The five Ps framework. *Electronic Journal of Business Research Methods*, *9*(2), 96–108.

Canagarajah, S. (2008). The politics of English language teaching. In N. Van Deusen-Scholl & N. H. Hornberger (Eds.), Encyclopedia of language and education (2nd ed., pp. 213-237). Springer Science+Business Media. doi:10.1007/978-0-387-30424-3_16

Canale, M., & Swain, M. (1980). Theoretical bases of communicative approaches to second language teaching and testing. *Applied Linguistics*, *1*(1), 1–47. doi:10.1093/applin/1.1.1

Candy, L. (2011). Research and creative practice. In L. Candy & E. A. Edmonds (Eds.), *Interacting: art, research and the creative practitioner* (pp. 33–59). Libri.

Cannon, C., Hamel, J., Buttell, F., & Ferreira, R. J. (2016). A Survey of Domestic Violence Perpetrator Programs in the United States and Canada: Findings and Implications for Policy and Intervention. *Partner Abuse*, *7*(3), 226–276. doi:10.1891/1946-6560.7.3.226

Carbajosa, P., Catalá-Miñana, A., Lila, M., Gracia, E., & Boira, S. (2017). Responsive versus treatment-resistant perpetrators in batterer intervention programs: Personal characteristics and stages of change. *Psychiatry, Psychology and Law*, *24*(6), 1–15. doi:10.1080/13218719.2017.1347933 PMID:31984001

Carey, D. M. (2012). *Qualitative research skills for social work: Theory and practice*. Ashgate Publishing, Ltd.

Carlin, A., & Kim, Y. (2019). Teaching qualitative research: Versions of grounded theory. *The Grounded Theory Review, 18*(1). http://groundedtheoryreview.com/2019/12/29/teaching-qualitative-research-versions-of-grounded-theory/

Carspecken, P. F. (1996). *Critical ethnography in educational research: A theoretical and practical guide.* Routledge.

Casey, E., & Smith, T. (2010). "How can I not?": Men's pathways to involvement in anti-violence against women work. *Violence Against Women, 16*(8), 953–973. doi:10.1177/1077801210376749 PMID:20679189

Chambers, R. (2002). *Participatory workshops: A sourcebook of 21 sets of ideas and activities.* Earthscan.

Chandler, D., & Torbert, B. (2003). Transforming inquiry and action: Interweaving 27 flavors of action research. *Journal of Action Research, 1*(2), 133–152. doi:10.1177/14767503030012002

Charmaz, K. (2000). Grounded theory: constructivist and objectivist methods. In N. Denzin, & Y. Lincoln (Eds.), Handbook of qualitative research (pp. 509-535). Sage.

Charmaz, K. (1990). "Discovering" chronic illness: Using grounded theory. *Social Science & Medicine, 30*(11), 1161–1172. doi:10.1016/0277-9536(90)90256-R PMID:2360052

Charmaz, K. (2006). *Constructing grounded theory: A practical guide through qualitative analysis.* SAGE Publications.

Charmz, K. (2015). Grounded theory: Methodology and theory construction. In *International Encyclopedia of the Social & Behavioral Sciences* (pp. 402–407). Elsevier. doi:10.1016/B978-0-08-097086-8.44029-8

Checkland, P., & Holwell, S. (1998). Action research: Its nature and validity. *Systemic Practice and Action Research, 11*(1), 9–21. doi:10.1023/A:1022908820784

Chiu, C.-M., Hsu, M.-H., & Wang, E. T. (2006). Understanding knowledge sharing in virtual communities: An integration of cosial capital and social cognitive theories. *Decision Support Systems, 42*(3), 1872–1888. doi:10.1016/j.dss.2006.04.001

Cho, J. Y., & Lee, E. (2014). Reducing confusion about grounded theory and qualitative content analysis: Similarities and differences. *The Qualitative Report, 19*(2003), 1–20.

Chong, D., & Druckman, J. N. (2007). Framing theory. *Annual Review of Political Science, 10*(1), 103–126. doi:10.1146/annurev.polisci.10.072805.103054

Chovanec, M. G. (2009). Facilitating change in group work with abusive men: Examining stages of change. *Social Work with Groups, 32*(1), 125–142. doi:10.1080/01609510802413055

Christ, T. W. (2013). The worldview matrix as a strategy when designing mixed methods research. *International Journal of Multiple Research Approaches, 7*(1), 110–118. doi:10.5172/mra.2013.7.1.110

Chun Tie, Y., Birks, M., & Francis, K. (2019). Grounded theory research: A design framework for novice researchers. *SAGE Open Medicine*, *7*, 1–8. doi:10.1177/2050312118822927 PMID:30637106

Cipollone, K., Zygmunt, E., & Tancock, S. (2018). "A paradigm of possibility": Community mentors and teacher preparation. *Policy Futures in Education*, *16*(6), 709–728. doi:10.1177/1478210317751270

Clandinin, D. J., & Connelly, F. M. (2000). *Narrative inquiry: Experience and story in qualitative research*. Jossey-Bass.

Clandinin, D. J., & Johnson, R. B. (2017). Narrative inquiry and case study research. In R. B. Christensen (Ed.), *Educational research: Qualitative, quantitative, and mixed approaches* (pp. 416–441). Sage.

Clandinin, D. J., & Rosiek, J. (2007). Mapping a landscape of narrative inquiry: Borderland spaces and tensions. In D. J. Clandinin (Ed.), *Handbook of narrative inquiry: Mapping a methodology* (pp. 35–75). Sage. doi:10.4135/9781452226552.n2

Clarke, A. (2003). Situational analysis: Grounded theory mapping after the postmodern turn. *Symbolic Interaction*, *26*(4), 553–576. doi:10.1525i.2003.26.4.553

Clarke, R. A. (1988, May). Information technology and dataveillance. *Communications of the ACM*, *31*(5), 498–512. doi:10.1145/42411.42413

Clark-Ibáñez, M. (2004). Framing the social world with photo-elicitation interviews. *The American Behavioral Scientist*, *47*(12), 1507–1527. doi:10.1177/0002764204266236

Clark-Ibáñez, M. K. (2007). Inner-city children in sharper focus: Sociology of childhood and photo-elicitation Interviews. In G. C. Stanczak (Ed.), *Visual research methods: Image, society, and representation* (pp. 167–196). Sage Publications.

Clark-Ibáñez, M. K. (2008). Gender and being "bad": Inner city students' photographs. In P. Thompson (Ed.), *Doing Visual Research with Children and Young People* (pp. 95–113). Routledge.

Cobb, J. (2017, June 22). Old Questions But No New Answers in the Philando Castile Verdict. The New Yorker. https://www.newyorker.com/news/daily-comment/old-questions-but-no-new-answers-in-the-philando-castile-verdict

Cobern, W. W. (1996). Worldview theory and conceptual change in science education. *Scientific Literacy and Cultural Studies Project*, 15. http://scholarworks.wmich.edu/cgi/viewcontent.cgi?article=1014&context=science_slcsp

Cochran-Smith, M. (2009). "Re-culturing" teacher education: Inquiry, evidence, and action. *Journal of Teacher Education*, *60*(5), 458–468. doi:10.1177/0022487109347206

Cochran-Smith, M., Carney, M., Keefe, E., Burton, S., Chang, W.-C., Fernandez, M., ... Baker, M. (2018). *Reclaiming accountability in teacher education*. Teachers College Press. doi:10.1007/978-981-13-2026-2

Cochran-Smith, M., & Villegas, A. M. (2015). Framing teacher preparation research: An overview of the field, part I. *Journal of Teacher Education*, *66*(1), 7–20. doi:10.1177/0022487114549072

Coghlan, D., & Brannick, T. (2014). Doing action research in your own organization. *Sage (Atlanta, Ga.)*.

Cohen, L., Manion, L., & Morrison, K. (2013). *Research methods in education* (7th ed.). Routledge. doi:10.4324/9780203720967

Cole, M., & Engestrom, Y. (1993). A cultural-historical approach to distributed congition. In G. Salomon (Eds.), Distributed cognitions; Psychological and educational considerations (pp. 1-46). Cambridge University Press.

Coleman, E. G. (2010). Ethnographic Approaches to Digital Media. *Annual Review of Anthropology*, *39*(1), 1–16. doi:10.1146/annurev.anthro.012809.104945

Coleman, J. S. (1964). *Introduction to mathematical sociology*. Free Press.

Colombo, M., & Azzurra, S. (2005). The discursive construction of community identity. *Journal of Community & Applied Social Psychology*, *15*(1), 48–62. doi:10.1002/casp.809

Common Core State Standards Initiative. (2010). *Common Core State Standards for Mathematics*. National Governors Association Center for Best Practices and the Council of Chief State School Officers.

Cooke, M. (2000). Five arguments for deliberative democracy. *Political Studies*, *48*(5), 947–969. doi:10.1111/1467-9248.00289

Corbin, J., & Strauss, A. (2008). *Basics of qualitative research: Techniques and procedures for developing grounded theory* (Vol. 25). SAGE Publications. doi:10.4135/9781452230153

Corti, L., & Fielding, N. (2016, October-December). Opportunities From the Digital Revolution: Implications for Researching, Publishing, and Consuming Qualitative Research. *Digital Representations: Opportunities for re-using and publishing digital qualitative data*, 1-13.

Council for the Accreditation of Educator Preparation. (2013). *2013 CAEP Standards*. http://caepnet.org/standards/introduction

Council of Chief State School Officers Task Force on Teacher Preparation and Entry into the Profession. (2012). *Our responsibility, our promise: Transforming teacher preparation and entry into the profession*. Council of Chief State School Officers. https://ccsso.org/sites/default/files/2017-10/Our%20Responsibility%20Our%20Promise_2012.pdf

Council of Chief State School Officers. (2016). *Accountability in teacher preparation: Policies and data in the 50 states and DC*. https://www.ccsso.org/sites/default/files/2017-10/50StateScan092216.pdf

Council of Chief State School Officers. (2018). *Measuring What Matters*. http://www.ccsso.org/sites/default/files/2018-02/Measuring%20What%20Matters.pdf

Cox, G. (2013). Researching resistance to open education resource contribution: An activity theory approac. *E-Learning and Digital Media, 10*(2), 148–160. doi:10.2304/elea.2013.10.2.148

Coyne, I. (1997). Sampling in qualitative research. Purposeful and theoretical sampling; merging or clear boundaries? *Journal of Advanced Nursing, 26*(3), 623–630. doi:10.1046/j.1365-2648.1997.t01-25-00999.x PMID:9378886

Crede, E., & Borrego, M. (2013). From ethnography to items: A mixed methods approach to developing a survey to examine graduate engineering student retention. *Journal of Mixed Methods Research, 7*(1), 62–80. doi:10.1177/1558689812451792

Crenshaw, K. W. (1988). Race, reform, and retrenchment: Transformation and legitimation in anti-discrimination law. *Harvard Law Review, 101*(7), 1331–1387. doi:10.2307/1341398

Crenshaw, K. W. (1991). Mapping the margins: Intersectionality, identity politics, and violence against women of color. *Stanford Law Review, 43*(6), 1241–1299.

Crenshaw, K. W. (1995). Mapping the margins: Intersectionality, identity politics, and violence against women of color. In K. Crenshaw, N. Gotanda, G. Peller, & K. Thomas (Eds.), *Critical race theory: The key writings that formed the movement* (pp. 357–383). The New Press.

Crenshaw, K. W. (2017). *On intersectionality: Essential writings*. The New Press.

Creswell, J. W. (2007). Qualitative inquiry and research design: Choosing among five approaches. *Sage (Atlanta, Ga.)*.

Creswell, J. W., & Creswell, J. D. (2018). *Research Design: Qualitative, Quantitative, and Mixed Methods Approaches* (5th ed.). Sage.

Creswell, J. W., & Poth, C. N. (2018). *Qualitative Inquiry & Research Design* (4th ed.). Sage.

Creswell, J. W., & Poth, C. N. (2018). *Qualitative inquiry and research design: Choosing among five approaches* (4th ed.). SAGE Publications.

Creswell, J. W., Shope, R., Plano Clark, V. L., & Green, D. O. (2006). How interpretive qualitative research extends mixed methods research. *Research in the Schools, 13*(1), 1–11.

Crilly, N., Blackwell, A. F., & Clarkson, P. J. (2006). Graphic elicitation: Using research diagrams as interview stimuli. *Qualitative Research, 6*(3), 341–366. doi:10.1177/1468794106065007

Croghan, R., Griffin, C., Hunter, J., & Phoenix, A. (2008). Young people's constructions of self: Notes on the use and analysis of the Photo-Elicitation methods. *International Journal of Social Research Methodology, 11*(4), 345–356. doi:10.1080/13645570701605707

Crotty, M. (1998). The foundations of social research: Meaning and perspective in the research process. *Sage (Atlanta, Ga.)*.

Cueva, K., Revels, L., Cueva, M., Lanier, A. P., Dignan, M., Viswanath, K., Fung, T. T., & Geller, A. C. (2018). Culturally-relevant online cancer education modules empower Alaska's community health aides/practitioners to disseminate cancer information and reduce cancer risk. *Journal of Cancer Education, 33*(5), 1102–1109. doi:10.100713187-017-1217-4 PMID:28405897

Cuevas, D. A., & Bui, N. H. (2016). Social factors affecting the completion of a batterer intervention program. *Journal of Family Violence, 31*(1), 95–107. doi:10.100710896-015-9748-0

Cummins, J. (2000). *Language, power and pedagogy: Bilingual children in the crossfire.* Multilingual Matters. doi:10.21832/9781853596773

Cuthrell, K., & Lys, D. (2014). *Using data for program improvement.* American Association of Colleges of Teacher Education. https://secure.aacte.org/apps/rl/res_get.php?fid=1638&ref=rl

Daniels, J. W., & Murphy, C. M. (1997). Stages and processes of change in batterers' treatment. *Cognitive and Behavioral Practice, 4*(1), 123–145. doi:10.1016/S1077-7229(97)80015-6

Dare, E. A., Ellis, J. A., & Roehrig, G. H. (2018). Understanding science teachers' implementations of integrated STEM curricular units through a phenomenological multiple case study. *International Journal of STEM Education, 5*(1), 1–19. doi:10.118640594-018-0101-z PMID:30631694

Darling-Hammond, L., Wilhoit, G., & Pittenger, L. (2014). Accountability for College and Career Readiness: Developing a New Paradigm. *Education Policy Analysis Archives, 22*(86), 1–38.

DataPortal. (2021). *Digital Around the World.* https://datareportal.com/global-digital-overview

Datnow, A., & Hubbard, L. (2015). Teachers' use of assessment data to inform instruction: Lessons from the past and prospects for the future. *Teachers College Record, 117*(4), 1e26.

Daugherty, J. (2009). Engineering professional development design for secondary school teachers: A multiple case study. *Journal of Technology Education, 21*(1), 10–24. doi:10.21061/jte.v21i1.a.1

Davidson, M. N., & Foster-Johnson, L. (2001). Mentoring in the preparation of graduate Researchers of color. *Review of Educational Research, 71*(4), 549–574. doi:10.3102/00346543071004549

Davis, F. (1986). *A technology acceptance model for empirically testing new end-user information systems: Theory and results* (Doctoral dissertation). Sloan School of Management, Massachusetts Institute of Technology. http://hdl.handle.net/1721.1/15192

Davis, D., & Veenstra, C. (2014). Community Involvement in STEM Learning. *Journal for Quality and Participation, 37*(1), 30–33.

Davis, S., & Peck, C. (in press). Using data for program improvement in teacher education: A study of promising practices. *Teacher College Press.*

Davis, S., & Peck, C. A. (2018). Building capacity and commitment for data use in teacher education programs. In E. Mandinach & E. Gummer (Eds.), *Data for continuous programmatic improvement: Steps colleges of education must take to become a data culture* (pp. 68–92). Harvard Education Press.

Dawson, P. (2014). Reflections: On time, temporality and change in organizations. *Journal of Change Management, 14*(3), 285–308. doi:10.1080/14697017.2014.886870

Dawson, P. (2019). *Reshaping change: A processual perspective* (2nd ed.). Routledge. doi:10.4324/9781351264921

Dawson, P., & Buchanan, D. (2005). The way it really happened: Competing narratives in the political process of technological change. *Human Relations, 58*(7), 845–865. doi:10.1177/0018726705057807

De Smet, M. M., Meganck, R., Van Nieuwenhove, K., Desmet, M., & Truijens, F. L. (2019). No Change? A Grounded Theory Analysis of Depressed Patients' Perspectives on Non-improvement in Psychotherapy. *Frontiers in Psychology, 10*(588), 1–17. doi:10.3389/fpsyg.2019.00588 PMID:30971973

Delgado, R. (1989). Storytelling for oppositionists and others: A plea for narrative. *Michigan Law Review, 87*(8), 2411–2441. doi:10.2307/1289308

Delgado, R., & Stefancic, J. (2012). *Critical race theory: An introduction.* NYU Press.

DeLugan, R. M., Roussos, S., & Skram, G. (2014). Linking academic and community guidelines for community-engaged scholarship. *Journal of Higher Education Outreach & Engagement, 18*(1), 155–168.

Dennen, V. P. (2014). Become a blogger: Trajectories, norms, and activites in a community of practice. *Computers in Human Behavior, 36*, 350–358. doi:10.1016/j.chb.2014.03.028

Denner, J., Cooper, C. R., Lopez, E. M., & Dunbar, N. (1999). Beyond "Giving Science Away": How University Community Partnerships Inform Youth Programs, Research, and Policy. *Society for Research in Child Development, 8*(1), 1–20. doi:10.1002/j.2379-3988.1999.tb00010.x

Dennis, R. F. (1983). Evaluation models for assessing effects of media-based nutrition education. In Effective communication in primary health care. Food and nutrition bulletin supplement 13, United Nations University.

Denscombe, M. (2014). *The good research guide: For small-scale social research projects* (5th ed.). McGraw-Hill Education.

Denson, C. D., Avery, Z. K., & Schell, J. W. (2010). Critical inquiry into urban African-American students' perceptions of engineering. *Journal of African American Studies, 14*(1), 61–74. doi:10.100712111-009-9107-4

Denzin, N. (2007). Grounded theory and the politics of interpretation. In A. Bryant & K. Charmaz (Eds.), *The Sage handbook of grounded theory* (pp. 454–471). Sage. doi:10.4135/9781848607941.n21

Denzin, N. K. (2006). The politics and ethics of performance pedagogy: Toward a pedagogy of hope. In D. S. Madison & J. Hamera (Eds.), *The Sage handbook of performance studies* (pp. 325–338). Sage. doi:10.4135/9781412976145.n18

Denzin, N. K. (2017). Critical qualitative inquiry. *Qualitative Inquiry, 23*(1), 8–16. doi:10.1177/1077800416681864

Denzin, N. K., & Lincoln, Y. S. (2008). *The landscape of qualitative research* (Vol. 1). Sage Publications.

Devas, N., & Grant, U. (2003). Local government decision-making—citizen participation and local accountability: Some evidence from Kenya and Uganda. *Public Administration and Development, 23*(4), 307–316. doi:10.1002/pad.281

Dewey, J. (1938). *Education and experience.* Simon and Schuster.

Dey, I. (2004). Grounded theory. In G. G. C. Seale (Ed.), *Qualitative research practice* (pp. 80–93). Sage. doi:10.4135/9781848608191.d9

Dey, L. (1999). *Grounding grounded theory: Guidelines for qualitative inquiry.* Academic Press.

Diaz, G. (2002). Artistic inquiry: On Lighthourse Hill. In C. Bagley & M. B. Cancienne (Eds.), *Dancing with data* (pp. 147–161). Peter Lang.

Dickens, L., & Watkins, K. (1999). Action research: Rethinking Lewin. *Management Learning, 30*(2), 127–140. doi:10.1177/1350507699302002

Dickson, G. (Ed.). (2000a). Participatory action research: Theory and practice. In M.J. Stewart (Ed.), Community nursing: Promoting Canadians' health (2nd ed., pp. 542-563). W.B. Saunders.

Dickson, G. (2000b). Aboriginal grandmothers' experience with health promotion and participatory action research. *Qualitative Health Research, 10*(2), 188–213. doi:10.1177/104973200129118363 PMID:10788283

Digsite. (2020). *Qualitative Research is changing.* https://www.digsite.com/qualitative-research-is-changing

Dipko, S., Billington, C., & Brick, P. (2012). The effect of mode on participant responses to qualitative research in virtual worlds. *The American Association for Public Opinion Research (AAPOR) 67th Annual Conference, 2012.* http://www.websm.org/db/12/15291/Web%20Survey%20Bibliography/The_Effect_of_Mode_on_Participant_Responses_to_Qualitative_Research_in_Virtual_Worlds/

Diversi, M. (1998). Glimpses of street life: Representing lived experience through short stories. *Qualitative Inquiry, 4*(2), 131–147. doi:10.1177/107780049800400201

Dixson, A. D., & Rousseau, C. K. (2005). And we are still not saved: Critical race theory in education ten years later. *Race, Ethnicity and Education, 8*(1), 7–27. doi:10.1080/1361332052000340971

Dobash, R. P., Dobash, R. E., Cavanagh, K., & Lewis, R. (2000). A research evaluation of British programmes for violent men. *Journal of Social Policy, 28*(2), 205–233. doi:10.1017/S0047279499005589

Dolle, J. R., Gomez, L. M., Russell, J. L., & Bryk, A. S. (2013). More than a network: Building professional communities for educational improvement. *National Society for the Study of Education Yearbook*, *112*, 443–463.

Domínguez, M. (2019). Decolonial innovation in teacher development: Praxis beyond the colonial zero-point. *Journal of Education for Teaching*, *45*(1), 47–62. doi:10.1080/02607476.2019.1550605

Donnor, J. (2013). *Education as the property of whites: African Americans continued quest for good schools. In Handbook of critical race theory in education*. Routledge.

Douglas, K., & Carless, D. (2013). An invitation to performative research. *Methodological Innovations Online*, *8*(1), 53–64. doi:10.4256/mio.2013.0004

Dowd, A. (2005). *Data don't drive: Building a practitioner-driven culture of inquiry to assess community college performance*. Lumina Foundation for Education. https://www.luminafoundation.org/files/publications/datadontdrive2005.pdf

Duchscher, J. E. B., & Morgan, D. (2004). Grounded theory: Reflections on the emergence vs. forcing debate. *Journal of Advanced Nursing*, *48*(6), 605–612. doi:10.1111/j.1365-2648.2004.03249.x PMID:15548251

Dugar, K. K. (2017). *Consumers' perceptions of voluntary and involuntary deconsumption: An exploratory sequential scale development study* [Unpublished doctoral dissertation]. The University of Denver.

Dutton, D., & Hemphill, K. (1992). Patterns of socially desirable responding among perpetrators and victims of wife assault. *Violence and Victims*, *7*(29), 29–40. doi:10.1891/0886-6708.7.1.29 PMID:1504031

Dutton, W. H. (2013). The social shaping of digital research. *International Journal of Social Research Methodology*, *16*(3), 177–195. doi:10.1080/13645579.2013.774171

Dwyer, C., Millett, C., & Paye, D. (2006). *A culture of evidence: A secondary assessment and learning outcomes*. Educational Testing Service.

Dym, C. L., Agogino, A. M., Eris, O., Frey, D. D., & Leifer, L. J. (2005). Engineering design thinking, teaching, and learning. *Journal of Engineering Education*, *94*(1), 103–120. doi:10.1002/j.2168-9830.2005.tb00832.x

Eagleman, A. N. (2013). Acceptance, motivations, and usage of social media as a marketing communications tool amongst employees of sport national governing bodies. *Sport Management Review*, *16*(4), 488–497. doi:10.1016/j.smr.2013.03.004

Easton, J. Q. (2009). Using data systems to drive school improvement. Keynote address at the STATS-DC 2009 Conference, Bethesda, MD.

Easton, J. Q. (2010). Helping states and districts swim in an ocean of new data. Keynote address at the Educate with Data: Stats-DC 2010, Bethesda, MD.

Easton, G. (2010). Critical realism in case study research. *Industrial Marketing Management*, *39*(1), 118–128. doi:10.1016/j.indmarman.2008.06.004

Eccles, J. (2009). Who am I and what am I going to do with my life? Personal and collective identities as motivators of action. *Educational Psychologist*, *44*(2), 78–89. doi:10.1080/00461520902832368

Eckert, N. (2021, March 9). Video: The 9-minute encounter that rocked Minnesota and the world. Twin Cities. https://www.twincities.com/2021/03/08/george-floyd-video-derek-chauvin-trial-rocks-world3838220

Edin, K. E., Lalos, A., Högberg, U., & Dahlgren, L. (2008). Violent men: Ordinary and deviant. *Journal of Interpersonal Violence*, *23*(2), 225–244. doi:10.1177/0886260507309342 PMID:18162638

Edwards, C. W. (2019). Overcoming imposter syndrome and stereotype threat: Reconceptualizing the definition of a scholar. *Taboo: The Journal of Culture and Education*, *18*(1), 3. doi:10.31390/taboo.18.1.03

Edwards, R. (1998). A critical examination of the use of interpreters in the qualitative research process. *Journal of Ethnic and Migration Studies*, *24*(1), 197–208. doi:10.1080/1369183X.1998.9976626

Ehrensaft, M. K., & Vivian, D. (1999). Is partner aggression related to appraisals of coercive control by a partner? *Journal of Family Violence*, *14*(3), 251–266. doi:10.1023/A:1022862332595

Eisenhardt, K. M. (1989). Building theories from case study research. *Academy of Management Review*, *14*(4), 532–550. doi:10.5465/amr.1989.4308385

Eisenhardt, K. M., & Graebner, M. E. (2007). Theory building from cases: Opportunities and challenges. *Academy of Management Journal*, *50*(1), 25–32. doi:10.5465/amj.2007.24160888

Eisner, E. W. (1997). The promise and perils of alternative forms of data representation. *Educational Researcher*, *26*(6), 4–10. doi:10.3102/0013189X026006004

Eisner, E. W. (2008). Art and knowledge. In G. J. Knowles & A. L. Cole (Eds.), *Handbook of the arts in qualitative research* (pp. 3–12). Sage.

Elementary and Secondary Education Act of 1965. (n.d.). *The Social Welfare History Project*. Retrieved from https://socialwelfare.library.vcu.edu/programs/education/elementary-and-secondary-education-act-of-1965/

Ellinger, A. D., & Watkins, K. E. (1998). Updating the critical incident technique after forty-four years. In *Academy of Human Resource Development 1998 Conference Proceedings* (pp. 285-292). Academy Human Resource Development.

Ellinger, A. D., & McWhorter, R. (2016). Qualitative case study research as empirical study. *International Journal of Adult Vocational Education and Technology*, *7*(3), 1–13. doi:10.4018/IJAVET.2016070101

Ellis, C. D., & Calafell, B. M. (2020). Toward praxis: Interrogating social justice within autoethnography. *Journal of Autoethnography, 1*(2), 203–207. doi:10.1525/joae.2020.1.2.203

Ellis, C., & Bochner, A. (2014). Merging culture and personal experience in critical autoethnography. In R. M. Boylorn & M. P. Orbe (Eds.), *Critical autoethnography: Intersecting cultural identities in everyday life* (pp. 9–10). Routledge.

Ellis, C., & Bochner, A. P. (2000). Autoethnography, personal narrative, reflexivity: Researcher as subject. In N. K. Denzin & Y. S. Lincoln (Eds.), *Handbook of Qualitative Research* (2nd ed., pp. 733–768). SAGE.

Enos, S., & Morton, K. (2003). Developing a theory and practice of campus-community partnerships. In *Building partnerships for service-learning* (pp. 20–41). Jossey-Bass.

Erel, U., Reynolds, T., & Kaptani, E. (2017). Participatory theatre for transformative social research. *Qualitative Research, 17*(3), 302–312. doi:10.1177/1468794117696029

Everest, T. (2014). Resolving the qualitative-quantitative debate in healthcare research. *Medical Practice and Reviews, 5*(1), 6–15. doi:10.5897/MPR.2013.0107

Ezzy, D. (2002). *Qualitative analysis practice and innovation.* Routledge.

Fals Borda, O. (1979). Investigating reality in order to transform it: The Colombian experience. *Dialectical Anthropology, 4*(1), 33–55. doi:10.1007/BF00417683

Fals Borda, O. (1988). *Knowledge and people's power.* Indian Social Institute.

Fals Borda, O. (2006). Participatory (action) research in social theory: Origins and challenges. In P. Reason & H. Bradbury (Eds.), *Handbook of action research* (pp. 27–37). Sage.

Fals-Borda, O. (1987). The application of participatory action-research in Latin America. *International Sociology, 2*(4), 329–347. doi:10.1177/026858098700200401

Fals-Borda, O., & Rahman, M. A. (1991). *Action and knowledge: Breaking the monopoly with participatory action-research.* Apex Press. doi:10.3362/9781780444239

Fang, M. L., Woolrych, R., Sixsmith, J., Canham, S., Battersby, L., & Sixsmith, A. (2016). Place-making with older persons: Establishing sense-of-place through participatory community mapping workshops. *Social Science & Medicine, 168*, 223–229. doi:10.1016/j.socscimed.2016.07.007 PMID:27492575

Faulkner, S. (2009). *Poetry as method: Reporting research through verse.* Left Coast Press.

Fayard, A. L., & Weeks, J. (2007). Photocopiers and water-coolers: The affordances of informal interaction. *Organization Studies, 28*(5), 605–634. doi:10.1177/0170840606068310

Feldman, A. (2003). Validity and Quality in Self-Study. *Educational Researcher, 32*(3), 26–28. doi:10.3102/0013189X032003026

Ferman, B., & Hill, T. L. (2004). The challenges of agenda conflict in higher-education-community research partnerships: Views from the community side. *Journal of Urban Affairs*, *26*(2), 241–257. doi:10.1111/j.0735-2166.2004.00199.x

Ferrante, J. (1998). *Sociology, a global perspective*. Wadsworth.

Feuer, M., Floden, R., Chudowsky, N., Ahn, J., & National Academy of Education. (2013). *Evaluation of teacher preparation programs: Purposes, methods, and policy options*. National Academy of Education.

Finlay, L. (2002). "Outing" the researcher: The provenance, process, and practice of reflexivity. *Qualitative Health Research*, *12*(4), 531–545. doi:10.1177/104973202129120052 PMID:11939252

Firestone, W. A. (1987). Meaning in method: The rhetoric of quantitative and qualitative research. *American Educational Research Association*, *16*(7), 16–21. doi:10.3102/0013189X016007016

Fishbein, M., & Ajzen, I. (1975). *Belief, Attitude, Intention and Belavior: An Introduiction to Theory and Research*. Addison-Wesley.

Fisher, D., Rooke, D., & Torbert, B. (2003). *Personal and organizational transformations through action inquiry*. Edge/Work Press.

Fishman, B. J., Penuel, W. R., Allen, A. R., Cheng, B. H., & Sabelli, N. O. R. A. (2013). Design-based implementation research: An emerging model for transforming the relationship of research and practice. *National Society for the Study of Education*, *112*, 136–156.

Fishman, J. A. (1996). What do you lose when you lose your language? In G. Cantoni (Ed.), *Stabilizing indigenous languages* (pp. 165–175). Northern Arizona University.

Flanagan, J. C. (1954). The critical incident technique. *Psychological Bulletin*, *51*(4), 327–358. doi:10.1037/h0061470 PMID:13177800

Fletcher, A. J. (2017). Applying critical realism in qualitative research: Methodology meets method. *International Journal of Social Research Methodology*, *20*(2), 181–194. doi:10.1080/13645579.2016.1144401

Flora, C. B., & Flora, J. (2004). *Rural communities: Legacy and change* (2nd ed.). Westview Press.

Forber-Pratt, A. J. (2015). "You're going to do what?" Challenges of autoethnography in the academy. *Qualitative Inquiry*, *21*(9), 821–835. doi:10.1177/1077800415574908

Forbes, J. (1964). Mapping accessibility. *Scottish Geographical Magazine*, *80*(1), 1–12. doi:10.1080/00369226408735915

Foster, P. (2012). Collaborative developmental action inquiry. In D. Coughlan & M. Brydon-Miller (Eds.), *Sage encyclopedia of action research management* (pp. 64–73). Sage.

Foucault, M. (1980). *Power-knowledge: Selected interviews and other writings, 1972-1977* (5th ed.). Pantheon Books.

Frankfurt, H. G. (1958). Peirce's notion of abduction. *The Journal of Philosophy*, *55*(14), 593–597. doi:10.2307/2021966

Frankham, J., & MacRae, C. (2011). Ethnography. In Theory and Methods in Social Research (2nd ed.). SAGE Publications.

Freeman, E. R., Brugge, D., Bennett-Bradley, W. M., Levy, J., & Carrasco, E. R. (2006). Challenges of conducting community-based participatory research in Boston's neighborhoods to reduce disparities in asthma. *Journal of Urban Health*, *83*(6), 1013–1021. doi:10.100711524-006-9111-0 PMID:17103339

Freire, P. (1970). Cultural action and conscientization. *Harvard Educational Review*, *40*(3), 452–477. doi:10.17763/haer.40.3.h76250x720j43175

Freire, P. (1970). *Pedagogy of the oppressed*. Herder & Herder.

Freire, P. (1996). *Pedagogia da autonomia. Saberes necessários à práctica educativa*. Paz e Terra.

Frost, A. (2004). Therapeutic engagement styles of child sexual offenders in a group treatment program: A grounded theory study. *Sexual Abuse*, *16*(3), 191–208. doi:10.1177/107906320401600302 PMID:15326880

Frost, A., & Connolly, M. (2004). Reflexivity, reflection, and the change process in offender work. *Sexual Abuse*, *16*(4), 365–380. doi:10.1177/107906320401600408 PMID:15560417

Gadd, D. (2004). Making sense of interviewee–interviewer dynamics in narratives about violence in intimate relationships. *International Journal of Social Research Methodology*, *7*(5), 383–401. doi:10.1080/1364557092000055077

Gallas, K. (1991). Arts as epistemology: Enabling children to know what they know. *Harvard Educational Review*, *61*(1), 40–51. doi:10.17763/haer.61.1.l7518t5kh06710n4

Gallo, M. L. (2002). Picture this: Immigrant workers use photography for communication and change. *Journal of Workplace Learning*, *14*(2), 49–57. doi:10.1108/13665620210419293

Gamoran, A., & Hannigan, E. C. (2000). Algebra for everyone? Benefits of college-preparatory mathematics for students with diverse abilities in early secondary school. *Educational Evaluation and Policy Analysis*, *22*(3), 241–254. doi:10.3102/01623737022003241

Gándara, P. (2006). Strengthening the academic pipeline leading to careers in math, science, and technology for Latino students. *Journal of Hispanic Higher Education*, *5*(3), 222–237. doi:10.1177/1538192706288820

Gant, V., Cheatham, L., Di Vito, H., Offei, E., Williams, G., & Yatosenge, N. (2019). Social work through collaborative autoethnography. *Social Work Education*, *38*(6), 707–720. doi:10.1080/02615479.2019.1570109

Ganz, M. (2006). *Organizing: People, power and change*. John F Kennedy School of Government, Harvard University.

Gardner, D. P., Larsen, Y. W., Baker, W., Campbell, A., & Crosby, E. A. (1983). *A nation at risk: The imperative for educational reform*. United States Department of Education.

Gasman, M., & Commodore, F. (2014). The state of research on historically Black colleges and universities. *Journal for Multicultural Education*, *8*(2), 89–111. doi:10.1108/JME-01-2014-0004

Gauntlett, D., & Holzwarth, P. (2006). Creative and visual methods for exploring identities. *Visual Studies*, *21*(1), 82–91. doi:10.1080/14725860600613261

Gaventa, J. (1991). Towards a knowledge democracy: Viewpoints on participatory research in North America. In O. Fals-Borda & M. A. Rahman (Eds.), *Action and knowledge: Breaking the monopoly with participatory action-research* (pp. 121–131). Apex Press. doi:10.3362/9781780444239.009

Gay, G. (2010). *Culturally responsive teaching: Theory, research, and practice*. Teachers College Press.

Geertz, C. (1973). The interpretation of cultures: Selected essays (22nd ed.). Basic Books.

George, A. L., & Bennett, A. (2005). *Case studies and theory development in the social sciences*. MIT Press.

Ge, X. (2010, September). Information-seeking behavior in the Digital Age: A multidisciplinary study of academic researchers. *College & Research Libraries*, *71*(5), 435–455. doi:10.5860/crl-34r2

Gibson, B., & Hartman, J. (2014). *Rediscovering grounded theory*. SAGE Publications.

Gibson, J. J. (1966). *The senses considered as perceptual systems*. Houghton Mifflin.

Gibson, J. J. (1979). *The ecological approach to visual perception of experimental psychology*. Lawrence Erlbaum.

Gibson, W., & Brown, A. (2009). *Working with qualitative data*. SAGE Publications. doi:10.4135/9780857029041

Gilgun, J. F. (2015). Beyond description to interpretation and theory in qualitative social work research. *Qualitative Social Work: Research and Practice*, *14*(6), 741–752. doi:10.1177/1473325015606513

Gillborn, D. (2013). The policy of inequity. Using CRT to unmask White supremacy in education policy. Handbook of critical race theory in education, 129-139.

Glaser, B. G. (1978). *Theoretical sensitivity*. Sociology Press.

Glaser, B. G. (1992). *Emergence vs forcing: Basics of grounded theory analysis*. Sociology Press.

Glaser, B., & Straus, A. (1967). *Discovery of grounded theory: strategies for qualitative research*. Transaction Publishers.

Glassman, M., & Erdem, G. (2014). Participatory action research and its meanings: Vivencia, praxis, conscientization. *Adult Education Quarterly*, *64*(3), 206–221. doi:10.1177/0741713614523667

Glassman, M., Erdem, G., & Bartholomew, M. (2013). Action research and its history as an adult education movement for social change. *Adult Education Quarterly*, *63*(3), 272–288. doi:10.1177/0741713612471418

Glesne, C. (1997). That rare feeling: Re-presenting research through poetic transcription. *Qualitative Inquiry*, *3*(2), 202–221. doi:10.1177/107780049700300204

Goldkuhl, G., & Cronholm, S. (2018). Reflection/commentary on a past article: "Multi-grounded theory: An update. *International Journal of Qualitative Methods*, *17*, 1–5.

Goldkuhl, G., & Cronholm, S. (2010). Adding theoretical grounding to grounded theory: Toward multi-grounded theory. *International Journal of Qualitative Methods*, *9*(2), 187–205. doi:10.1177/160940691000900205

Goldkuhl, G., & Lind, M. (2010). *A multi-grounded design research process*. Springer. doi:10.1007/978-3-642-13335-0_4

Gondolf, E. W. (2012). *The future of batterer programs: Reassessing evidence-based practice*. UPNE.

González, N., Moll, L. C., & Amanti, C. (Eds.). (2005). *Funds of knowledge: Theorizing practices in households, communities, and classrooms*. Routledge.

Goonan, S., Mirosa, M., & Spence, H. (2014). Getting a taste for food waste: A mixed methods ethnographic study into hospital food waste before patient consumption conducted at three New Zealand foodservice facilities. *Journal of the Academy of Nutrition and Dietetics*, *114*(1), 63–71. doi:10.1016/j.jand.2013.09.022 PMID:24231365

Gormley, K. J. (2001). Research as a democratic process: Educational community development in Brazil through participatory research. *Convergence*, 34.

Goulding, C. (2002). *Grounded theory: A practical guide for management, business and market researchers*. SAGE Publications. doi:10.4135/9781849209236

Graham-Cagney, A., & Watkins, K. (2014) *Insider Action Research Doctorates: Creating Transformative Learning Spaces in Professional Development in Higher Education.* An unpublished paper prepared for International Conference on Doctoral Education, Orlando, FL.

Gray, R. E. (2003). Performing on and off the stage: The place(s) of performance in arts-based approaches to qualitative inquiry. *Qualitative Inquiry*, *9*(2), 254–267. doi:10.1177/1077800402250957

Gray, R., Broady, T., Gaffney, I., Lewis, P., Mokany, T., & O'Neil, B. (2016). 'I'm working towards getting back together': Client accounts of motivation related to relationship status in men's behaviour change programmes in New South Wales, Australia. *Child Abuse Review*, *25*(3), 171–182. doi:10.1002/car.2318

Greene, D., & David, J. L. (1984). A research design for generalizing from multiple case studies. *Evaluation and Program Planning*, *7*(1), 73–85. doi:10.1016/0149-7189(84)90027-2

Gries, L. E. (2013). Iconographic tracking: A digital research method for visual rhetoric and circulation studies. *Computers and Composition*, *30*(4), 332–348. doi:10.1016/j.compcom.2013.10.006

Grimmet, K. (2016). *Examining the quality of life of young adults with Intellectual Disability: Using photo elicitation to empower an unknown voice* (Doctoral thesis). ProQuest. (10195731)

Guba, E. G., & Lincoln, Y. S. (1994). Competing paradigms in qualitative research. Handbook of Qualitative Research, 2, 163-194.

Guba, E. G. (1990). The paradigm dialog. *Sage (Atlanta, Ga.)*.

Guba, E. G., & Lincoln, Y. S. (1981). *Effective evaluation*. Jossey-Bass.

Guillen, L., & Zeichner, K. (2018). A university-community partnership in teacher education from the perspective of community-based teacher educators. *Journal of Teacher Education*, *69*(2), 140–153. doi:10.1177/0022487117751133

Guishard, M. A., Halkovic, A., Galletta, A., & Li, P. (2018). Toward epistemological ethics: Centering communities and social justice in qualitative research. *Forum Qualitative Sozialforschung/Forum: Qualitative. Social Research*, *19*(3). Advance online publication. doi:10.17169/fqs-19.3.3145

Gunaratnam, Y. (2003). *Researching "race" and ethnicity: Methods, knowledge and power*. SAGE Publications. doi:10.4135/9780857024626

Guru, S. (2006). Working with Asian perpetrators of domestic violence— The British experience. *Practice*, *18*(3), 153–166. doi:10.1080/09503150600904037

Gutmann, A., & Thompson, D. F. (2004). *Why deliberative democracy?* Princeton University Press. doi:10.1515/9781400826339

Gutterman, T., Babchuk, W., Howell Smith, M., & Stevens, J. (2019). Contemporary approaches to mixed methods-grounded theory research: A field-based analysis. *Journal of Mixed Methods Research*, *13*(2), 179–195. doi:10.1177/1558689817710877

Gwon, S. H., & Thongpriwan, V. (2019). PhotoVoice in a prenursing health care systems course. *The Journal of Nursing Education*, *58*(1), 64–64. doi:10.3928/01484834-20190103-14 PMID:30673098

Hackman, J. R. (1985). Doing research that makes a difference. *Doing research that is useful for theory and practice, 126*, 148.

Hamilton, M. L. (Ed.). (1998). Reconceptualizing teaching practice: Self-study in teacher education. Falmer.

Hammarberg, K., Kirkman, M., & de Lacey, S. (2016). Qualitative research methods: When to use them and how to judge them. *Human Reproduction (Oxford, England)*, *31*(3), 498–501. doi:10.1093/humrep/dev334 PMID:26759142

Hammersley, M., & Atkinson, P. (1995). *Ethnography: Principles in practice*. Tavistock.

Hanauer, D. I. (2010). *Poetry as research: Exploring second language poetry writing*. John Benjamins. doi:10.1075/lal.9

Hancock, T. U., & Siu, K. (2009). A culturally sensitive intervention with domestically violent Latino immigrant men. *Journal of Family Violence, 24*(2), 123–132. doi:10.100710896-008-9217-0

Hanson, C. L. (2014). "I learned I am a feminist": Lessons for adult learning from participatory action research with union women. *Canadian Journal for the Study of Adult Education, 27*(1), 1–16.

Harding, S. (1991). *Whose science? Whose knowledge? Thinking from women's lives*. Cornell University Press.

Hardy, S., Gregory, S., & Ramjeet, J. (2009). An exploration of intent for narrative methods of inquiry. *Nurse Researcher, 16*(4), 7–19. doi:10.7748/nr2009.07.16.4.7.c7157 PMID:19653542

Harne, L., & Radford, J. (2008). Tackling domestic violence: Theories, policies and practice. McGraw-Hill International.

Harper, D. (2002). Talking about pictures: A case for photo elicitation. *Visual Studies, 17*(1), 13–26. doi:10.1080/14725860220137345

Harper, S. R. (2010). An anti-deficit achievement framework for research on students of color in STEM. *New Directions for Institutional Research, 148*(148), 63–74. doi:10.1002/ir.362

Harris, I. (2003). What does "the discovery of grounded theory" have to say to medical education? *Advances in Health Sciences Education: Theory and Practice, 8*(1), 49–61. doi:10.1023/A:1022657406037 PMID:12652168

Harris, T. (2015). Grounded theory. *Nursing Standard, 29*(35), 32–39. doi:10.7748/ns.29.35.32.e9568 PMID:25922026

Harry, B., Sturges, K. M., & Klingner, J. K. (2005). Mapping the process: An exemplar of process and challenge in grounded theory analysis. *Educational Researcher, 34*(2), 3–13. doi:10.3102/0013189X034002003

Hearn, J. (1998). *The violences of men: How men talk about and how agencies respond to men's violence to women* (1st ed.). SAGE Publications. doi:10.4135/9781446279069

Heath, H., & Cowley, S. (2004). Developing a grounded theory approach: A comparison of Glaser and Strauss. *International Journal of Nursing Studies, 41*(2), 141–150. doi:10.1016/S0020-7489(03)00113-5 PMID:14725778

Hergenrather, K. C., Rhodes, S. D., Cowen, C. A., Bardhoshi, G., & Pula, S. (2009). Photovoice as community-based participatory research: A qualitative review. *American Journal of Health Behavior, 33*(6), 686–698. doi:10.5993/AJHB.33.6.6 PMID:19320617

Heron, J. (1996). Co-operative Inquiry: Research into the human condition. *Sage (Atlanta, Ga.)*.

Herrington, D., & Daubenmire, P. L. (2016). No Teacher Is an Island: Bridging the Gap between Teachers' Professional Practice and Research Findings. *Journal of Chemical Education*, *93*(8), 1371–1376. doi:10.1021/acs.jchemed.5b00700

Hesse-Biber, S. (2010). Qualitative approaches to mixed methods practice. *Qualitative Inquiry -. Qualitative Inquiry*, *16*(6), 455–468. doi:10.1177/1077800410364611

Hicks, A. (2018). Developing the methodological toolbox for information literacy research: Grounded theory and visual research methods. *Library & Information Science Research*, *40*(3-4), 40. doi:10.1016/j.lisr.2018.09.001

Hinton, L. (2011). Language revitalization and language pedagogy: New teaching and learning strategies. *Language and Education*, *25*(4), 307–318. doi:10.1080/09500782.2011.577220

Hoddy, E. T. (2019). Critical realism in empirical research: Employing techniques from grounded theory methodology. *International Journal of Social Research Methodology*, *22*(1), 111–124. doi:10.1080/13645579.2018.1503400

Holdsworth, E., Bowen, E., Brown, S., & Howat, D. (2016). The development of a program engagement theory for group offending behavior programs. *International Journal of Offender Therapy and Comparative Criminology*, 1–21. doi:10.1177/0306624X15624177 PMID:26769679

Holloway, I., & Biley, F. C. (2011). Being a qualitative researcher. *Qualitative Health Research*, *21*(7), 968–975. doi:10.1177/1049732310395607 PMID:21266705

Honneth, A. (1998). Democracy as reflexive cooperation: John Dewey and the theory of democracy today. *Political Theory*, *26*(6), 763–783. doi:10.1177/0090591798026006001

Horsford, S. D. (2010). Mixed feelings about mixed schools: Superintendents on the complex legacy of school desegregation. *Educational Administration Quarterly*, *46*(3), 287–321. doi:10.1177/0013161X10365825

Houghten, J., Steele, C., & Henty, M. (2003). Changing Research Practices in the Digital Information and Communication Environment. Department of Education, Science and Training. Canberra: Commonwealth of Australia.

Hou, H. (2015). What makes an online community of practice work? A situated study of Chinese student teachers' perceptions of online professional learning. *Teaching and Teacher Education*, *46*, 6–16. doi:10.1016/j.tate.2014.10.005

Howard-Payne, L. (2016). Glasser or Strauss? Considerations for selecting a grounded theory study. *South African Journal of Psychology. Suid-Afrikaanse Tydskrif vir Sielkunde*, *46*(1), 50–62. doi:10.1177/0081246315593071

Howard, T. C. (2003). Culturally relevant pedagogy: Ingredients for critical teacher reflection. *Theory into Practice*, *42*(3), 195–202. doi:10.120715430421tip4203_5

Howard, T. C. (2015). *Why race and culture matter in schools: Closing the achievement gap in America's classrooms*. Teachers College Press.

Howarth, J., Currie, M. A., Morrell, E., Sorensen, J., & Bengle, T. (2017). Challenges of building community-university partnerships in new poverty landscapes. *Community Development (Columbus, Ohio)*, *48*(1), 48–66. doi:10.1080/15575330.2016.1244696

Huff, A. S. (2009). Ontology and epistemology. In *Designing research for publication* (pp. 108–126). Sage.

Hultin, L. (2019). On becoming a sociomaterial researcher: Exploring epistemological practices grounded in a relational, performative ontology. *Information and Organization*, *29*(2), 91–104. doi:10.1016/j.infoandorg.2019.04.004

Hurst, D. (2015). Neoliberalism, markets and accountability: Transforming education and undermining democracy in the United States and England. *Policy Futures in Education*, *3*(1), 3–15. doi:10.2304/pfie.2005.3.1.6

Huxham, C., & Vangen, S. (2005). *Managing to collaborate: The theory and practice of collaborative advantage*. Routledge.

Hyett, N., Kenny, A., & Dickson-Swift, V. (2014). Methodology or method? A critical review of qualitative case study reports. *International Journal of Qualitative Studies on Health and Well-being*, *9*(1), 1–12. doi:10.3402/qhw.v9.23606 PMID:24809980

Hymes, D. (1972). On communicative competence. In J. B. Pride & J. Holmes (Eds.), *Sociolinguistics* (pp. 1–21). Penguin.

Ingham, J., Cadieux, J., & Berrada, A. (2015). e-shopping acceptance: A qualitative and meta-analystic review. *Information & Management*, *52*(1), 44–60. doi:10.1016/j.im.2014.10.002

Iseke, J. (2013). Indigenous storytelling as research. *International Review of Qualitative Research*, *6*(4), 559-577.

Jacoby, B. (2003). *Building partnerships for service-learning*. Jossey-Bass.

Janssen, J., Hale, L., Mirfin-Veitch, B., & Harland, T. (2013). Building the research capacity of clinical physical therapists using a participatory action research approach. *Physical Therapy*, *93*(7), 923–934. doi:10.2522/ptj.20120030 PMID:23559527

Johnson, R., & Turner, W. (2010). *Grounded theory in practice: Is it inherently a mixed method?* Academic Press.

Johnson, P., & Duberley, J. (2000). Understanding management research: An introduction to epistemology. *Sage (Atlanta, Ga.)*.

Jones, K. (2003). The turn to a narrative knowing of persons: One method explored. *Narrative Studies*, *8*(1), 60–71. doi:10.1177/136140960300800109

Jones, M., & Alony, I. (2011). Guiding the use of grounded theory. *International Journal of Doctoral Studies*, *6*, 95–114. doi:10.28945/1429

Jong, W., & Dückers, M. L. (2016). Self-correcting mechanisms and echo-effects in social media: An analysis of the "gunman in the newsroom" crisis. *Computers in Human Behavior*, *59*, 334–341. doi:10.1016/j.chb.2016.02.032

Jorgenson, J., & Sullivan, T. (2010). Accessing children's perspectives through participatory photo interviews. *Forum Qualitative Social Research*, *11*(1), 8.

José, D., Osinski, M., & Hermann, R. (2017). The construction process of grounded theory in administration. *Contaduría y Administración*, *62*(3), 985–1000. doi:10.1016/j.cya.2016.06.012

Jurow, S., Horn, I. S., & Philip, T. M. (2019). Re-mediating knowledge infrastrctures: A site for innovation in teacher education. *Journal of Education for Teaching*, *45*(1), 82–96. doi:10.1080/02607476.2019.1550607

Kacen, L., & Chaitin, J. (2006). "The Times They are a Changing": Undertaking Qualitative Research in Ambiguous, Conflictual, and Changing Contexts. *Qualitative Report*, *11*(2), 209–228.

Kalogo, C. (2015). *Factors influencing treatment completion of involuntary groups*. University of St. Thomas.

Kelly, L., & Westmarland, N. (2015). *Domestic violence perpetrator programmes: steps towards change*. Project Mirabal Final Report.

Kemmis, S., McTaggart, R., & Nixon, R. (2014). *The action research planner: Doing critical participatory action research*. Springer Science+Business Media.

Kemmis, S. (2009). Action research as a practice-based practice. *Educational Action Research*, *17*(3), 463–474. doi:10.1080/09650790903093284

Kennedy, L. M. (2020). Confessions of a novice researcher: An autoethnography of inherent vulnerabilities. *Qualitative Report*, *25*(6), 1526–1539. https://nsuworks.nova.edu/tqr/vol25/iss6/7/

Kennedy, M. M. (1979). Generalizing from single case studies. *Evaluation Quarterly*, *3*(4), 661–678. doi:10.1177/0193841X7900300409

Khachan, V. A. (2009). Diglossic needs of illiterate adult women in Egypt: A needs assessment. *International Journal of Lifelong Education*, *28*(5), 649–660. doi:10.1080/02601370903190086

Khiat, H. (2010). A grounded theory approach: Conceptions of understanding in engineering mathematics learning. *Qualitative Report*, *15*(6), 1459–1488.

Kier, M. W. (2013). *Examining the effects of a STEM career video intervention on the interests and STEM professional identities of rural, minority middle school students*. Dissertation Study. North Carolina State University.

Kietzmann, J., Plangger, K., Eaton, B., Heilgenberg, K., Pitt, L., & Berthon, P. (2013). Mobility at work: A typology of mobile communities of practice and contextual ambidexterity. *The Journal of Strategic Information Systems*, *22*(4), 282–297. doi:10.1016/j.jsis.2013.03.003

Kim, J. (2006). For whom the school bell tolls: Conflicting voices inside an alternative high school. *International Journal of Education & the Arts*, *7*(6), 1–19.

Kimmerle, J., Thiel, A., Gerbing, K., Bientzle, M., & Halatchliski, I. (2013). Knowledge construction in an outsider community: Extending the communities of practice concept. *Computers in Human Behavior*, *29*(3), 1078–1090. doi:10.1016/j.chb.2012.09.010

Kindon, S., Pain, R., & Kesby, M. (2007). Participatory action research: Origins, approaches and methods. In *Participatory action research approaches and methods* (pp. 35–44). Routledge. doi:10.4324/9780203933671-13

Kolb, S. M. (2012). Grounded theory and the constant comparative method: Valid research strategies for educators. *Journal of Emerging Trends in Educational Research and Policy Studies*, *3*(1), 83–86.

Koro-Ljungberg, M., & Cannella, G. S. (2017). Critical qualitative inquiry: Histories, methodologies, and possibilities. *International Review of Qualitative Research*, *10*(4), 327–339. doi:10.1525/irqr.2017.10.4.327

Kosnik, C., Cleovoulou, Y., & Fletcher, T. (2009). The use of interviews in self-study research. In C. A. Lassonde, S. Galman, & C. Kosnik (Eds.), *Self-study research methodologies for teacher educators* (pp. 51–69). Sense Publishers. doi:10.1163/9789087906900_005

Kostiainen, E., Ukskoski, T., Ruohotie-Lyhty, M., Kauppinen, M., Kainulainen, J., & Mäkinen, T. (2018). Meaningful learning in teacher education. *Teaching and Teacher Education*, *71*, 66–77. doi:.tate.2017.12.009 doi:10.1016/j

Kouritzin, S. G. (1999). Face[t]s of first language loss. Routledge.

Kovach, M. (2005). Emerging from the margins: Indigenous methodologies. In L. Brown & S. Strega (Eds.), *Research as resistance: Critical, Indigenous, & anti-oppressive approaches* (pp. 19–36). Canadian Scholars' Press.

Kovach, M. (2009). *Indigenous methodologies: Characteristics, conversations, and contexts*. University of Toronto Press.

Kwon, C. K., & Nicolaides, A. (2017). Managing diversity through triple-loop learning: A call for paradigm shift. *Human Resource Development Review*, 1–15.

LaBoskey, V. K. (2004). The methodology of self study and it's theoretical underpinnings. In J. J. Loughran, M. L. Hamilton, V. K. LaBoskey, & T. Russell (Eds.), *International handbook of self-study of teaching and teacher education practices* (pp. 817–870). Kluwer. doi:10.1007/978-1-4020-6545-3_21

Ladhani, S., & Sitter, K. C. (2020). The revival of anti-racism. *Critical Social Work*, *21*(1), 54–65. doi:10.22329/csw.v21i1.6227

Ladson-Billings, G. (2005). What's the matter with the team? Diversity in teacher education. *Journal of Teacher Education*, *56*(3), 229–234. doi:10.1177/0022487105275917

Ladson-Billings, G., & Tate, I. V. W. (1995). Toward a critical race theory of education. *Teachers College Record*, *97*(1), 47–68.

Langellier, K. M., & Peterson, E. E. (2006). Shifting contexts in personal narrative performance. In D. S. Madison & J. Hamera (Eds.), *The Sage handbook of performance studies* (pp. 151–168). Sage. doi:10.4135/9781412976145.n9

Langley, A. (1999). Strategies for theorizing from process data. *Academy of Management Review*, *24*(4), 691–710. doi:10.5465/amr.1999.2553248

LaRossa, R. (2005, November). Grounded theory methods and qualitative family research. *Journal of Marriage and Family*, *67*(4), 837–857. doi:10.1111/j.1741-3737.2005.00179.x

Lather, P. (1986). Issues of validity in openly ideological research: Between a rock and a soft place. *Interchange*, *17*(4), 63–84. doi:10.1007/BF01807017

Law, J. (1992). Notes on the theory of the Actor-Network: Ordering, strategy, and heterogenity. *Systems Practice*, *5*(4), 379–393. doi:10.1007/BF01059830

Lawler, E. E. (1985). *Doing research that is useful for theory and practice*. Lexington Books.

Leavy, P. (2009). *Method meets art: Arts-based research practice*. Guilford.

Leavy, P., & Harris, A. (2018). *Contemporary feminist research from theory to practice*.

Lee, O. (2002). An action research report on the Korean national digital library. *Information & Management*, *39*(4), 255–260. doi:10.1016/S0378-7206(01)00094-5

Lee, O., & Krajcik, J. (2012). Large-scale interventions in science education for diverse student groups in varied educational settings. *Journal of Research in Science Teaching*, *49*(3), 271–280. doi:10.1002/tea.21009

Lee, W., Xiong, L., & Hu, C. (2012). The effect of Facebook users' arousal and valence on intention to go to the festival: Applying an extension of the technology acceptance model. *International Journal of Hospitality Management*, *31*(3), 819–827. doi:10.1016/j.ijhm.2011.09.018

Leonard, J. (2008). *Culturally specific pedagogy in the mathematics classroom*. Routledge.

Leonard, M., & McKnight, M. (2014). Look and tell: Using photo-elicitation methods with teenagers. *Children's Geographies*, *13*(6), 629–642. doi:10.1080/14733285.2014.887812

Leonardo, Z. (2002). The souls of white folk: Critical pedagogy, whiteness studies, and globalization discourse. *Race, Ethnicity and Education*, *5*(1), 29–50. doi:10.1080/13613320120117180

Lewin, K. (1936). *Principles of topological psychology*. McGraw-Hill., 10. doi:10.1037/10019-000

Lewin, K. (1944). The dynamics of group action. *Educational Leadership*, *1*(4), 195–200.

Lewin, K. (1946). Action research and minority problems. *The Journal of Social Issues*, *2*(1), 34–46. doi:10.1111/j.1540-4560.1946.tb02295.x

Lewin, K. (1947). Frontiers in group dynamics II. Channels of group life; social planning and action research. *Human Relations*, *1*(2), 143–153. doi:10.1177/001872674700100201

Lewin, K. (1997). *Self-hatred among Jews (1941).* American Psychological Association. doi:10.1037/10269-012

Lewis, C., & Throne, R. (2021). Autoethnography and other self-inquiry methods for practice-based doctoral research. In R. Throne (Ed.), *Practice-based and practice-led research for dissertation development*. IGI Global. doi:10.4018/978-1-7998-6664-0

Lewis, J. S. (2018). Releasing a tradition: Diasporic epistemology and the decolonized curriculum. *The Cambridge Journal of Anthropology*, *36*(2), 21–33. doi:10.3167/cja.2018.360204

Lewis, R. (2004). Making justice work: Effective legal interventions for domestic violence. *British Journal of Criminology*, *44*(2), 204–224. doi:10.1093/bjc/44.2.204

Liepins, R. (2000). New energies for an old idea: Reworking approaches to 'community' in contemporary rural studies. *Journal of Rural Studies*, *16*(1), 23–35. doi:10.1016/S0743-0167(99)00042-X

Lilley-Walker, S.-J., Hester, M., & Turner, W. (2016). Evaluation of European domestic violence perpetrator programmes: Toward a model for designing and reporting evaluations related to perpetrator treatment interventions. *International Journal of Offender Therapy and Comparative Criminology*, 1–17. doi:10.1177/0306624X16673853 PMID:27884945

Lima, L. C. (2018). Adult and permanent education in times of crisis: A critical perspective based on Freire and Gelpi. *Studies in the Education of Adults*, *50*(2), 219–238. doi:10.1080/02660830.2018.1523087

Lincon, Y. S., & Guba, E. G. (1985). Naturalistic inquiry. *Sage (Atlanta, Ga.).*

Lin, X., Schwartz, D. L., & Hatano, G. (2005). Toward teachers' adaptive metacognition. *Educational Psychologist*, *40*(4), 245–255. doi:10.120715326985ep4004_6

Lomax, H., Fink, J., Singh, N., & High, C. (2011). The politics of performance: Methodological challenges of researching children's experiences of childhood through the lens of participatory video. *International Journal of Social Research Methodology*, *14*(3), 231–243. doi:10.1080/13645579.2011.563622

Lorde, A. (1984). *Sister outsider: Essays and speeches*. Crossing Press.

Loughran, J. (2005). Researching teaching about teaching: Self-study of teacher education practices. *Studying Teacher Education*, *1*(1), 5–16. doi:10.1080/17425960500039777

Loxley, A., & Seery, A. (2008). *Some philosophical and other related issues of insider research. In Researching education from the inside*. Routledge.

Lucas, S. (1999). *Tracking inequality: Stratification and mobility in American schools*. Teachers College Press.

Lucio-Villegas, E. (2018). Revisiting Paulo Freire: Adult education for emancipation. In The Palgrave international handbook on adult and lifelong education and learning (pp. 151-168). Palgrave Macmillan.

Luttrell, W. (2010). "A camera is a big responsibility": A lens for analysing children's visual voices. *Visual Studies*, *25*(3), 224–237. doi:10.1080/1472586X.2010.523274

Ly, L. (2014). Can cell phones stop police brutality? *CNN*. https://www.cnn.com/2014/11/18/us/police-cell-phone-videos/

Lynch, S., & Kuntz, A. (2019). A critical autoethnography of a doctoral students' research journey: Learning to take risks in the academy. *Curriculum Studies in Health and Physical Education*, *10*(2), 156–171. doi:10.1080/25742981.2019.1588762

Lynn, M. (2004). Inserting the 'Race' into Critical Pedagogy: An analysis of 'race-based epistemologies'. *Educational Philosophy and Theory*, *36*(2), 153–165. doi:10.1111/j.1469-5812.2004.00058.x

Lynn, M., & Parker, L. (2006). Critical race studies in education: Examining a decade of research on U.S. schools. *The Urban Review*, *38*(4), 257–290. doi:10.100711256-006-0035-5

Mace, D. P., & Lueblke, P. (2014). *Creating organizational support for collaborative inquiry*. American Association of Colleges of Teacher Education. https://secure.aacte.org/apps/rl/res_get.php?fid=1638&ref=rl

Madison, D. S. (2003). Performance, personal narrative, and the politics of possibility. In Y. S. Lincoln & N. K. Denzin (Eds.), *Turning points in qualitative research* (pp. 469–486). AltaMira Press.

Madison, S. D. (2012). *Critical ethnography: Method, ethics, and performance* (2nd ed.). Sage Publications.

Mahruf, M., Howes, A. J., & Shohel, C. (2007). Transition from nonformal schools: Learning through photo elicitation in educational fieldwork in Bangladesh. *Visual Studies*, *22*(1), 53–61. doi:10.1080/14725860601167200

Makgakga, S., & Makwakwa, E. G. (2016). *Exploring learners' difficulties in solving grade 12 differential calculus: A case study of one secondary school in Polokwane district* [Unpublished doctoral dissertation]. The University of South Africa.

Malewski, E., Sharma, S., & Phillion, J. (2012). How international field experiences promote cross-cultural awareness in preservice teachers through experiential learning: Findings from a six-year collective case study. *Teachers College Record*, *114*(8), 1–44. PMID:24013958

Mandinach, E. B. (2012). A perfect time for data use: Using data-driven decision making to inform practice. *Educational Psychologist*, *47*(2), 71–85. doi:10.1080/00461520.2012.667064

Mandinach, E., & Gummer, E. (2015). Data-driven decision making: Components of the enculturation of data use in education. *Teachers College Record*, *117*(4), 1–8.

Mandleco, B. (2013). Research with children as participants: Photo elicitation. *Journal for Specialists in Pediatric Nursing*, *18*(1), 78–82. doi:10.1111/jspn.12012 PMID:23289457

Marcus, G. E. (1994). What comes (just) after 'post'? the case of ethnography. In N. K. Denzin & Y. S. Lincoln (Eds.), *Handbook of Qualitative Research* (pp. 563–574). Sage Publications.

Márquez, M. G., & Delgado, A. R. (2017). An exploratory sequential design to validate measures of moral emotions. *Psicothema*, *29*(2), 261–267. PMID:28438252

Marshall, C., & Rossman, G. B. (2014). *Designing qualitative research.* Sage Publications.

Martinovic, D., & Manizade, A. (2017). First-page settings open access article using grounded theory to extend existing PCK framework at the secondary level. *Education Science, 7*(60).

Marx, S., Pennington, J. L., & Chang, H. (2017). Critical autoethnography in pursuit of educational equity: Introduction to the IJME special issue. *International Journal of Multicultural Education*, *19*(1), 1–6. doi:10.18251/ijme.v19i1.1393

Mason, J. (2002). *Qualitative Researching* (2nd ed.). Sage.

Masters, J. (1995). The history of action research. *Action Research Electronic Reader*, *22*, 2005.

Matusov, E. (1996). Intersubjectivty without agreement. *Mind, Culture, and Activity*, *3*(1), 29–45. doi:10.120715327884mca0301_4

Maxwell, J. A. (2010). Using numbers in qualitative research. *Qualitative Inquiry*, *16*(6), 475–482. doi:10.1177/1077800410364740

Maxwell, J. A. (2012). A realist approach for qualitative research. *Sage (Atlanta, Ga.).*

Maxwell, J. A. (2016). Expanding the history and range of mixed methods research. *Journal of Mixed Methods Research*, *10*(1), 12–27. doi:10.1177/1558689815571132

Maxwell, J. A. (2017). The validity and reliability of research: A realist perspective. In D. Wyse, N. Selwyn, E. Smith, & L. E. Suter (Eds.), *The BERA/SAGE handbook of educational research* (Vol. 1, pp. 116–140). Sage. doi:10.4135/9781473983953.n6

Maxwell, J. A. (2018). Collecting qualitative data: a realist approach. In U. Flick (Ed.), *The SAGE handbook of qualitative data collection* (pp. 19–32). Sage. doi:10.4135/9781526416070.n2

May, S. (2013). Indigenous immersion education: International developments. *Journal of Immersion and Content-Based Language Education, 1*(1), 34-69.

Maybee, C., Carlson, J., Selbodnik, M., & Chapman, B. (2015). "It's in the syllabus": Identifying information literacy and data information literacy opportunities using a grounded theory approach. *Libraries Faculty and Staff Scholarship and Research.*

Mccallin, A. (2003). Grappling with the literature in a grounded theory study. *Contemporary Nurse*, *15*(1-2), 61–69. doi:10.5172/conu.15.1-2.61 PMID:14649509

McCarry, M. (2007). Masculinity studies and male violence: Critique or collusion? *Women's Studies International Forum*, *30*(5), 631–638. doi:10.1016/j.wsif.2007.07.006

McCarty, T.L., Romero, M. E., & Zepeda, O. (2006). Reclaiming the gift: Indigenous youth counter-narratives on Native language loss and revitalization. *American Indian Quarterly*, *30*(1-2), 28-48.

McConnaughy, E. A., Prochaska, J. O., & Velicer, W. F. (1983). Stages of change in psychotherapy: Measurement and sample profiles. *Psychotherapy (Chicago, Ill.)*, *20*(3), 368–375. doi:10.1037/h0090198

McEneaney, J., Gillette, R., Farkas, W., & Guziniczak, L. (2016). Online commerce as a digital literacy: A grounded theory approach. *The Journal of Literacy and Technology*, *17*(3), 219–255.

McGhee, G., Marland, G. R., & Atkinson, J. M. (2007). Grounded theory research: Literature reviewing and reflexivity. *Journal of Advanced Nursing*, *60*(3), 334–342. doi:10.1111/j.1365-2648.2007.04436.x PMID:17908129

McIntyre, A. (2008). Participatory action research. *Sage (Atlanta, Ga.)*.

McLaren, P. L., Ryoo, J. J., Crawford, J., & Moreno, D. (2010). Critical praxis. In C. Kridel (Ed.), *Encyclopedia of curriculum studies*. SAGE Publications, Inc., doi:10.4135/9781412958806.n90

McLure, M., & Faraj, S. (2000). "it is what one does": Why people particpate and help others in electronic communities of practice. *The Journal of Strategic Information Systems*, *9*(2-3), 9. doi:10.1016/S0963-8687(00)00045-7

McTaggart, R. (1994). Participatory Action Research: Issues in theory and practice. *Educational Action Research*, *2*(3), 313–337. doi:10.1080/0965079940020302

Mead, M. (1995). Visual anthropology in a discipline of words. In P. Hockings (Ed.), *Principles of visual anthropology* (pp. 3–10). De Gruyter Mouton. doi:10.1515/9783110290691.3

Means, B., & Penuel, W. R. (2005). Research to support scaling up technology-based educational innovations. *Scaling up success: Lessons from technology-based educational improvement*, 176-197.

Means, B., & Harris, C. J. (2013). Towards an Evidence Framework for Design-Based Implementation Research. *Yearbook of the National Society for the Study of Education*, *112*, 350–371.

Meredith, J., Giles, D., & Stommel, W. (2021). *Analysing Digital Interaction*. Palgrave Macmilan.

Merriam, S. B. (1985). The case study in educational research: A review of selected literature. *The Journal of Educational Thought*, *19*(3), 204–217.

Merriam, S. B. (1998). *Qualitative research and case study applications in education*. Jossey-Bass Publishers.

Merriam, S. B. (2009). *Qualitative research: A guide to design and implementation* (3rd ed.). Jossey-Bass.

Merrick, K., Gu, N., & Wang, X. (2011). Case studies using multiuser virtual worlds as an innovative platform for collaborative design. *Journal of Information Technology in Construction*, *16*, 165–188.

Mertens, D. M., & Ginsberg, P. E. (2008). Deep in ethical waters. *Qualitative Social Work: Research and Practice*, *7*(4), 484–503. doi:10.1177/1473325008097142

Mignolo, W. D. (2009). Epistemic disobedience, independent thought and decolonial freedom. *Theory, Culture & Society*, *26*(7-8), 159–181. doi:10.1177/0263276409349275

Miklikowska, M. (2012). Psychological underpinnings of democracy: Empathy, authoritarianism, self-esteem, interpersonal trust, normative identity style, and openness to experience as predictors of support for democratic values. *Personality and Individual Differences*, *53*(5), 603–608. doi:10.1016/j.paid.2012.04.032

Miles, M. B., & Huberman, A. M. (1994). Qualitative data analysis: An expanded sourcebook. *Sage (Atlanta, Ga.)*.

Miles, M. B., Huberman, A. M., & Saldana, J. (2013). Qualitative data analysis. *Sage (Atlanta, Ga.)*.

Miles, M. B., Huberman, A. M., & Saldana, J. (2014). Qualitative data analysis: A methods sourcebook. *Sage (Atlanta, Ga.)*.

Miller, K. (2015). Dear Critics: Addressing concerns and justifying the benefits of photography as a research method. *Forum Qualitative Social Research*, *16*(27), 27.

Mills, J., Bonner, A., & Francis, K. (2006). The development of constructivist grounded theory. *International Journal of Qualitative Methods*, *5*(1), 25–35. doi:10.1177/160940690600500103

Milner, H. R. IV. (2008). Critical race theory and interest convergence as analytic tools in teacher education policies and practices. *Journal of Teacher Education*, *59*(4), 332–346. doi:10.1177/0022487108321884

Mingers, J. (2014). *Systems thinking, critical realism and philosophy: A confluence of ideas*. Routledge. doi:10.4324/9781315774503

Mingers, J. (2015). Helping business schools engage with real problems: The contribution of critical realism and systems thinking. *European Journal of Operational Research*, *242*(1), 316–331. doi:10.1016/j.ejor.2014.10.058

Mingers, J., & Standing, C. (2017). Why things happen–Developing the critical realist view of causal mechanisms. *Information and Organization*, *27*(3), 171–189. doi:10.1016/j.infoandorg.2017.07.001

Morrell, M. E. (2007). Empathy and democratic education. *Public Affairs Quarterly*, *21*(4), 381–403.

Morse, J. (2020). The Changing Face of Qualitative Inquiry. *International Journal of Qualitative Methods*, *19*, 1–7. doi:10.1177/1609406920909938

Moseley, M. J. (1979). *Accessibility, the rural challenge*. Methuen.

Moss, P. (2007). Evidence and decision making, the 106th yearbook of the National Society for the Study of Education. National Society for the Study of Education.

Muller, M. (2014). Curiosity, creativity, and surprise as analytic tools: Grounded theory method. In J. Olson, & W. Kellogg (Eds.), Ways of knowing in HCI (pp. 49-68). Springer.

Murphy, E., & Rodriguez-Mnzanares, M. A. (2014). *Activity theory perspectives on technology in higher education*. IGI Global. doi:10.4018/978-1-4666-4590-5

Nam Mak, B., & Chui, H. (2013). Colleagues' talk and power after work hours: A community of practice in Facebook Status Updates? *Discourse. Context and Media*, *2*(2), 94–102. doi:10.1016/j.dcm.2013.04.002

National Geographic Virtual Library. (2015). *The history of the national geographic society*. http://gale.cengage.co.uk/national-geographic-virtual-library/history.aspx

Navarro, O., Shah, J. K., Valdez, C., Dover, A. G., & Henning, N. (2020). Fighting on all fronts: The push, pull, and persistence of social justice educators and the move to reimagine teacher preparation. *Teacher Education Quarterly*, *47*(3), 9–31.

Ngai, E., Tao, S. C., & Moon, K. L. (2015). Social media research: Theories, constructs, and conceptual frameworks. *International Journal of Information Management*, *35*(1), 33–44. doi:10.1016/j.ijinfomgt.2014.09.004

Ng, K., & Hase, S. (2008). Grounded suggestions for doing a grounded theory business research. *Electronic Journal of Business Research Methods*, *6*(2), 155–170.

Nicolaides, A. (2015). Generative learning: Adults learning within ambiguity. *Adult Education Quarterly*, 1–17.

Niemimaa, M. (2014). *Sociomaterial ethnography: Taking the Matter seriously. In MCIS 2014 Proceedings*. AIS.

Norris, J. (2000). Drama as research: Realizing the potential of drama in education as a research methodology. *Youth Theatre Journal*, *14*(1), 40–51. doi:10.1080/08929092.2000.10012516

O'Brien, D. (2013). Visual research with young d/Deaf people: An investigation ofthe transitional experiences of d/Deaf young people from mainstream schools using auto-driven photo elicitation interviews. *Graduate Journal of Social Science*, *10*(2), 152–175.

Ohito, E. O. (2019). Thinking through the flesh: A critical autoethnography of racial body politics in urban teacher education. *Race, Ethnicity and Education*, *22*(2), 250–268. doi:10.1080/13613324.2017.1294568

Olmos-López, P., & Tusting, K. (2020). Autoethnography and the study of academic literacies: Exploring space, team research and mentoring. *Trabalhos em Lingüística Aplicada*, *59*(1), 264–295. doi:10.1590/010318136565715912020

Onwuegbuzie, A. J., Johnson, B. R., & Turner, L. A. (2007). Toward a definition of mixed methods research. *Journal of Mixed Methods Research*, *1*(2), 112–133. doi:10.1177/1558689806298224

Opie, A. (1992). Qualitative research, appropriation of the "other" and empowerment. *Feminist Review*, *40*(1), 52–69. doi:10.1057/fr.1992.5

Österlind, E. (2008). Acting out of habits–can theatre of the oppressed promote change? Boal's theatre methods in relation to Bourdieu's concept of habitus. *Research in Drama Education*, *13*(1), 71–82. doi:10.1080/13569780701825328

Padgett, D. K., Smith, B. T., Derejko, K., Henwood, B. F., & Tiderington, E. (2013). A picture is worth . . .? Photo Elicitation interviewing with formerly homeless adults. *Qualitative Health Research*, *23*(11), 1435–1444. doi:10.1177/1049732313507752 PMID:24122520

Parry, K. W. (1998). Grounded theory and social process : A new direction for leadership research. *The Leadership Quarterly*, *9*(l), 85–105. doi:10.1016/S1048-9843(98)90043-1

Parsons, E. C. (2014). Unpacking and critically synthesizing the literature on race and ethnicity in science education. Handbook of research in science education, 167-183.

Patton, M. Q. (2002). *Qualitative research & evaluation methods*. Sage Publication, Inc.

Patton, M. Q. (2002). *Qualitative research & evaluation methods: Integrating theory and practice* (3rd ed.). SAGE Publications.

Patton, M. Q. (2002). Qualitative research and evaluation methods. *Sage (Atlanta, Ga.)*.

Paul, F. (2005). Grouping within algebra 1: A structural sieve with powerful effects for low-income, minority, and immigrant students. *Educational Policy*, *19*(2), 262–282. doi:10.1177/0895904804274056

Peck, C. A., McDonald, M., & Davis, S. (2015). *Using data for program improvement in teacher education: A study of promising practices*. American Association of Colleges of Teacher Education. https://secure.aacte.org/apps/rl/resource.php

Peck, C. A., Gallucci, C., & Sloan, T. (2010). Negotiating implementation of high-stakes performance assessment policies in teacher education: From compliance to inquiry. *Journal of Teacher Education*, *61*(5), 451–463. doi:10.1177/0022487109354520

Peck, C. A., & McDonald, M. (2013). Creating "cultures of evidence" in teacher education: Context, policy, and practice in three high-data-use programs. *New Educator*, *9*(1), 1228. doi:10.1080/1547688X.2013.751312

Peck, C. A., & McDonald, M. A. (2014). What is a culture of evidence? How do you get one? And...should you want one? *Teachers College Record*, *116*, 1–27.

Pelias, R. J. (2003). The academic tourist: An autoethnography. *Qualitative Inquiry*, *9*(3), 369–373. doi:10.1177/1077800403009003003

Pelias, R. J. (2004). *A methodology of the heart: Evoking academic and daily life*. AltaMira Press.

Pence, E., & Paymar, M. (1993). *Education groups for men who batter: The Duluth model.* Springer Publishing Company. doi:10.1891/9780826179913

Pennant, A. L. M. (2020). My journey into the 'heart of whiteness' whilst remaining my authentic (Black) self. *Educational Philosophy and Theory*, 1–12. doi:10.1080/00131857.2020.1769602

Pennycook, A. (1989). The concept of method, interested knowledge, and the politics of language teaching. *TESOL Quarterly*, *23*(4), 589–618. doi:10.2307/3587534

Pentland, B., Recker, J., Ryan Wolf, J., & Wyner, G. (2020). Bringing context inside process research with digital trace data. *Journal of the Association for Information Systems*, 21.

Penuel, W. R., Coburn, C. E., & Gallagher, D. J. (2013). Negotiating problems of practice in research–practice design partnerships. *Yearbook of the National Society for the Study of Education*, *112*, 237–255.

Penuel, W. R., & Fishman, B. J. (2013). Large-scale intervention research we can use. *Journal of Research in Science Teaching*, *49*, 281–304. doi:10.1002/tea.21001

Penuel, W. R., Fishman, B. J., Cheng, B. H., & Sabelli, N. (2011). Organizing research and development at the intersection of learning, implementation, and design. *Educational Researcher*, *40*(7), 331–337. doi:10.3102/0013189X11421826

Pettigrew, A. M. (2012). Context and action in the transformation of the firm: A reprise. *Journal of Management Studies*, *49*(7), 1304–1328. doi:10.1111/j.1467-6486.2012.01054.x

Phan, N. T. T., & Locke, T. (2015). Sources of self-efficacy in Vietnamese EFL teachers: A qualitative study. *Teaching and Teacher Education*, *52*, 73–82. doi:10.1016/j.tate.2015.09.006

Phillips, D. C., & Burbules, N. C. (2000). *Postpositivism and educational research.* Rowman & Littlefield.

Pilaar Birch, S. P. (2013). Using social media for research dissemination: The digital research video project. *Internet Archaeology*, *35*(35). Advance online publication. doi:10.11141/ia.35.4

Pillow, W. (2003). Confession, catharsis, or cure? Rethinking the uses of reflexivity as methodological power in qualitative research. *International Journal of Qualitative Studies in Education: QSE*, *16*(2), 175–196. doi:10.1080/0951839032000060635

Pinar, W. F. (2005). The problem with curriculum and pedagogy. *Journal of Curriculum and Pedagogy*, *2*(1), 67–82. doi:10.1080/15505170.2005.10411529

Pinnegar, S., & Daynes, J. G. (2007). Locating narrative inquiry historically: Thematics in the turn to narrative. In D. J. Clandinin (Ed.), *Handbook of narrative inquiry: Mapping a methodology* (pp. 3–34). Sage. doi:10.4135/9781452226552.n1

Pinnegar, S., & Erickson, L. B. (2009). Uncovering self-studies in accreditation review processes. In C. A. Lassonde, S. Galman, & C. Kosnik (Eds.), *Self-study research methodologies for teacher educators* (pp. 151–168). Sense Publishers.

Plutynski, A. (2011). Four problems of abduction: A brief history. *HOPOS The Journal of the International Society for the History of Philosophy of Science, 1*(2), 227–249. doi:10.1086/660746

Pratt, M. L. (1992). *Imperial eyes: Travel writing and transculturation.* Routledge. doi:10.4324/9780203163672

Prigol, I. E., & Behrens, M. (2019). Grounded theory: Methodology applied in education research. *Educação e Realidade, 44*(3), 1–20.

Prins, E. (2005). Framing a conflict in a campus-community partnership. *Journal of Planning Education and Research, 25*(1), 57–74. doi:10.1177/0739456X04270370

Prins, E. (2007). 'Aquí no somos unidos'/'We're not united here': Adult literacy and the obstacles to solidarity in postwar El Salvador. *International Journal of Qualitative Studies in Education: QSE, 20*(4), 401–431. doi:10.1080/09518390601176598

Prochaska, J. O., & DiClemente, C. C. (1982). Transtheoretical therapy: Toward a more integrative model of change. *Psychotherapy (Chicago, Ill.), 19*(3), 276–288. doi:10.1037/h0088437

Prosser, J., & Loxley, A. (2007). Enhancing the contribution of visual methods to inclusive education. *Journal of Research in Special Educational Needs, 7*(1), 55–68. doi:.2007.00081.x doi:10.1111/j.1471-3802

Punch, K. F., & Oancea, A. (2014). *Introduction to research methods in education* (Vol. 10). SAGE Publications.

Pyrch, T. (2007). Participatory action research and the culture of fear: Resistance, community, hope and courage. *Action Research, 5*(2), 199–216. doi:10.1177/1476750307077324

Pyyry, N. (2013). 'Sensing with "photography and 'thinking with' photographs in research into teenage girls" hanging out. *Children's Geographies, 13*(2), 149–163. doi:10.1080/14733285.2013.828453

Quantz, R. A., & O'Connor, T. W. (1988). Writing critical ethnography: Dialogue, multivoicedness, and carnival in cultural texts. *Educational Theory, 38*(1), 95–109. doi:10.1111/j.1741-5446.1988.00095.x

Rambe, P. (2012). Activity theory and technology mediated interaction: Cognitive scaffolding using question-based consultation on Facebook. *Australasian Journal of Educational Technology, 28*(8), 1333–1361. doi:10.14742/ajet.775

Rampal, A., & Mander, H. (2013). Lessons on food and hunger: Pedagogy of empathy for democracy. *Economic and Political Weekly, 48*(28), 50–57.

Rao, V., & Woolcock, M. (2003). Integrating qualitative and quantitative approaches in program evaluation. In F. J. Bourguignon & L. P. da Silva (Eds.), *The impact of economic policies on poverty and income distribution: Evaluation techniques and tools* (pp. 165–190). Oxford University Press.

Rasberry, G. W. (2002). Imagine, inventing a data-dancer. In C. Bagley & M. B. Cancienne (Eds.), *Dancing the data* (pp. 106–120). Peter Lang.

Rasmussen, B. (2010). The "good enough" drama: Reinterpreting constructivist aesthetics and epistemology in drama education. *Research in Drama Education, 15*(4), 529–546. doi:10.1080/13569783.2010.512187

Ravitch, D. (2004). *The language police: How pressure groups restrict what students learn*. Vintage.

Ravitch, D. (2010). *The death and life of the great American school system: How testing and choice are undermining education*. Basic Books.

Reason, P., & Bradbury, H. (Eds.). (2001). *Handbook of action research*. Sage.

Reason, P., & Torbert, W. (2001). The action turn: Towards a transformational social science. *Concepts and Transformation, 6*(1), 1–37.

Reed-Danahay, D. (2017). Bourdieu and critical autoethnography: Implications for research, writing, and teaching. *International Journal of Multicultural Education, 19*(1), 144–154. doi:10.18251/ijme.v19i1.1368

Reeves, T. C., Herrington, J., & Oliver, R. (2005). Design research: A socially responsible approach to instructional technology research in higher education. *Journal of Computing in Higher Education, 16*(2), 96–115. doi:10.1007/BF02961476

Regan, P. M., & Jesse, J. (2019). Ethical challenges of edtech, big data and personalized learning: Twenty-first century student sorting and tracking. *Ethics and Information Technology, 21*(3), 167–179. doi:10.100710676-018-9492-2

Renaud, M. (2001). Serendipity and policy relevance. In *Social Science for Knowledge and Decision Making*. OECD Publications.

Richardson, M. (1998). Poetics in the field and on the page. *Qualitative Inquiry, 4*(4), 451–462. doi:10.1177/107780049800400401

Richard, V. M., & Lahman, M. K. E. (2014). Photo-elicitation: Reflexivity on method, analysis, and graphic portraits. *International Journal of Research & Method in Education, 38*(1), 3–22. doi:10.1080/1743727X.2013.843073

Ritterbusch, A. E. (2019). Empathy at knifepoint: The dangers of research and lite pedagogies for social justice movements. *Antipode, 51*(4), 1296–1317. doi:10.1111/anti.12530

Roberts, J. M. (2014). Critical realism, dialectics, and qualitative research methods. *Journal for the Theory of Social Behaviour, 44*(1), 1–23. doi:10.1111/jtsb.12056

Rodríguez-Mojica, C., Briceño, A., & Muñoz-Muñoz, E. R. (2019). Combating linguistic hegemony: Preparing and sustaining bilingual teacher educators in the United States. *Teacher Education Quarterly, 46*(3), 57–78.

Rondeau, G., Brodeur, N., Brochu, S., & Lemire, G. (2001). Dropout and completion of treatment among spouse abusers. *Violence and Victims, 16*(2), 127–143. doi:10.1891/0886-6708.16.2.127 PMID:11345474

Rose, G. (2013). On the relation between "visual research methods" and contemporary visual culture. *The Sociological Review, 62*(1), 24–46. doi:10.1111/1467-954X.12109

Rosenbaum, A., Gearan, P. J., & Ondovic, C. (2001). Completion and recidivism among court- and self-referred batterers in a psychoeducational group treatment program: Implications for intervention and public policy. *Journal of Aggression, Maltreatment & Trauma, 5*(2), 199–220. doi:10.1300/J146v05n02_12

Ross, T., Catronova, E., & Wagner, G. (2012). Empirical research methods in virtual worlds. In Methods in Urban and Planning Studies: Design and Outcomes (pp. 299-311). Information Science Reference. doi:10.4018/978-1-4666-0074-4.ch018

Rothman, E. F., & Coutinho, P. (2007). Batterer intervention program enrollment and completion Massachusetts. *Violence Against Women, 13*(5), 527–543. doi:10.1177/1077801207300720 PMID:17478677

Roulston, K. (2014). Analysing Interviews. In U. Flick (Ed.), *The SAGE handbook of Qualitative Data Analysis - Analysing Interviews* (pp. 297–313). SAGE Publications. doi:10.4135/9781446282243

Rubin, H. J., & Rubin, I. S. (2011). Qualitative interviewing: The art of hearing data. *Sage (Atlanta, Ga.).*

Rudolph, J. W., Morrison, J. B., & Carrol, J. S. (2009). The dynamics of action-oriented problem solving: Linking interpretation and choice. *Academy of Management Review, 34*(4), 733–356.

Russell, C. K., & Gregory, D. M. (2003). Evaluation of qualitative research studies. *Evidence-Based Nursing, 6*(2), 36–40. doi:10.1136/ebn.6.2.36 PMID:12710415

Sabelli, N., & Dede, C. (2013). Empowering design-based implementation research: The need for infrastructure. *Yearbook of the National Society for the Study of Education, 112*, 464–480.

Sage. (2019). Learn to use an exploratory sequential mixed method design for instrument development. *Research Methods Datasets*. Retrieved from https://methods.sagepub.com/base/download/DatasetStudentGuide/exploratory-sequential-mixed-method-instrument-development-united-states

Sah, P. K. (2019). Academic Discourse Socialization, Scaler Politics of English, and Racialization in Study Abroad: A Critical Autoethnography. *Qualitative Report, 24*(1), 174–192.

Saldaña, J. (2009). The coding manual for qualitative researchers. *Sage (Atlanta, Ga.).*

Saldaña, J. (Ed.). (2005). *Ethnodrama: An anthology of reality theater*. AltaMira Press.

Salmons, J. (2017). *Methodspace*. Sage Publishing. https://www.methodspace.com/seeing-hearing-problem-using-video-qualitative-research/

Saltman, K. (2016). Corporate schooling meets corporate media: Standards, testing, and technophilia. *Review of Education, Pedagogy & Cultural Studies*, *38*(2), 105–123. doi:10.1080/10714413.2016.1155953

Samaras, A., & Freese, A. (2006). *Self-study of teaching practices*. Pete Lang Publishing, Inc.

Samaras, A., & Freese, A. (2009). Looking back and looking forward: An historical overview of the self-study school. In C. A. Lassonde, S. Galman, & C. Kosnik (Eds.), *Self-study research methodologies for teacher educators* (pp. 3–19). Sense Publishers.

Samaras, A., & Roberts, L. (2011). Flying solo. Teachers take charge of their learning through self-study research. *Journal of Staff Development*, *32*(5), 42–45.

Sam, C. (2012). Activity theory and qualitative research in digital domains. *Theory into Practice*, *51*(2), 83–90. doi:10.1080/00405841.2012.662856

Samuels, J. (2004). Breaking the Ethnographer's frames: Reflections on the use of photo Elicitation in understanding Sri Lankan monastic culture. *The American Behavioral Scientist*, *47*(12), 1528–1550. doi:10.1177/0002764204266238

Sassen, K. (1991). The polarization lidar technique for cloud research: A review and current assessment. *Bulletin of the American Meteorological Society*, *72*(12), 1848–1866. doi:10.1175/1520-0477(1991)072<1848:TPLTFC>2.0.CO;2

Saunders, S. G., Barrington, D. J., Sridharan, S., Meo, S., Hadwen, W., Shields, K. F., Souter, R., & Bartram, J. K. (2016). Addressing WaSH challenges in Pacific Island Countries: A participatory marketing systems mapping approach to empower informal settlement community action. *Habitat International*, *55*, 159–166. doi:10.1016/j.habitatint.2016.03.010

Savage, M., & Burrows, R. (2007). The coming crisis of empirical sociology. *Sociology*, *41*(5), 885–899. doi:10.1177/0038038507080443

Savan, B., Flicker, S., Kolenda, B., & Mildenberger, M. (2009). How to facilitate (or discourage) community-based research: Recommendations based on a Canadian survey. *Local Environment*, *14*(8), 783–796. doi:10.1080/13549830903102177

Saxena, D. (2019). The Search for Mechanisms in Business Research: Reflections on Retroductive Analysis in a Multilevel Critical Realist Case Study. *Electronic Journal of Business Research Methods*, *17*(1), 17–27.

Sayem, A. M., & Nury, A. T. M. S. (2013). An assessment of attitude towards equitable gender norms among Muslim women in Bangladesh. *Women's Studies International Forum*, *40*, 102–110. doi:10.1016/j.wsif.2013.04.007

Sayer, A. (2010). *Method in social science* (2nd ed.). Routledge. doi:10.4324/9780203850374

Schoonenboom, J., & Johnson, R. B. (2017). How to Construct a Mixed Methods Research Design. *Kölner Zeitschrift für Soziologie und Sozialpsychologie*, *69*(S2, Suppl 2), 107–131. doi:10.100711577-017-0454-1 PMID:28989188

Schubert, W. H. (2004). Curriculum and pedagogy for reconstruction and reconceptualization. *Journal of Curriculum and Pedagogy, 1*(1), 19–21. doi:10.1080/15505170.2004.10411471

Schultze, U., & Avital, M. (2011). Designing interviews to generate rich data for information systems research. *Information and Organization, 21*(1), 1–16. doi:10.1016/j.infoandorg.2010.11.001

Schwandt, T. A. (2001). *Dictionary of qualitative inquiry*. Sage Publications.

Schwandt, T. A. (2014). *The SAGE Dictionary of Qualitative Inquiry*. SAGE Publications.

Scott, K. W., & Howell, D. (2008). Clarifying analysis and interpretation in grounded theory: Using a conditional relationship guide and reflective coding matrix. *International Journal of Qualitative Methods, 7*(2), 1–15. doi:10.1177/160940690800700201

Seale, C. (1999). *The quality of qualitative research. Introducing qualitative methods series*. Sage.

Segal, E. A. (2011). Social empathy: A model built on empathy, contextual understanding, and social responsibility that promotes social justice. *Journal of Social Service Research, 37*(3), 266–277. doi:10.1080/01488376.2011.564040

Shaw, D. (2013). A new look at an old research method: Photo-elicitation. *TESOL Journal, 4*(4), 785–799. doi:10.1002/tesj.108

Shcroder, K. C. (2012, November). Methodological pluralism as a vehicle of qualitative generalization. *Participations, 9*(2), 798–825.

Sherman, F., & Torbert, W. (Eds.). (2000). *Transforming social inquiry, transforming social action*. Kluwer Academic Publishers.

Shirazi, F. (2014). *Interrogating Iran's restricted public cloud: An Actor Network Theory perspective*. Academic Press.

Shotter, J. (2014). Agential realism, social constructionism, and our living relations to our surroundings: Sensing similarities rather than seeing patterns. *Theory & Psychology, 24*(3), 305–325. doi:10.1177/0959354313514144

Silver-Pacuilla, H. V. (2004). The meanings of literacy: A participatory action research project involving women with disabilities. *Women's Studies Quarterly, 32*(1/2), 43–58.

Simon, R. I., & Dippo, D. (1986). On critical ethnographic work. *Anthropology & Education Quarterly, 17*(4), 195–202. doi:10.1525/aeq.1986.17.4.04x0613o

Sinclair, N. (2003). Aesthetic Values in Mathematics: A Value-Oriented Epistemology. *International Group for the Psychology of Mathematics Education, 4*, 199–205.

Sinner, A., Leggo, C., Irwin, R., Gouzouasis, P., & Grauer, K. (2006). Arts-based education research dissertations: Reviewing the practices of new scholars. *Canadian Journal of Education, 29*(4), 1223–1270. doi:10.2307/20054216

Small, M. L. (2011). How to conduct a mixed methods study: Recent trends in a rapidly growing literature. *Sociology*, *37*(1), 57–86. doi:10.1146/annurev.soc.012809.102657

Solórzano, D. G., & Yosso, T. J. (2001). From racial stereotyping toward a critical race theory in teacher education. *Multicultural Education*, *9*, 2–8.

Solórzano, D. G., & Yosso, T. J. (2002). Critical race methodology: Counter-storytelling as an analytical framework for education research. *Qualitative Inquiry*, *8*(1), 23–44. doi:10.1177/107780040200800103

Sontag, S. (1977). *On photography*. Penguin.

Spasser, M. A. (2002). Realist activity theory for digital library evaluation: Conceptual framework and case study. *Computer Supported Cooperative Work*, *11*(1-2), 81–110. doi:10.1023/A:1015288305397

Spillane, J. (2012). Data in practice: Conceptualizing the data-based decision-making phenomena. *American Journal of Education*, *118*(2), 113–141. doi:10.1086/663283

Spivak, G. C. (1988). Can the subaltern speak? In C. Nelson & L. Grossberg (Eds.), *Marxism and the interpretation of culture* (pp. 271–315). University of Illinois Press. doi:10.1007/978-1-349-19059-1_20

Spring, J. (2003). *American education*. McGraw Hill.

Stage, E. K., Asturias, H., Cheuk, T., Daro, P. A., & Hampton, S. B. (2013). Opportunities and challenges in next generation standards. *Science*, *340*(6130), 276–277. doi:10.1126cience.1234011 PMID:23599466

Stake, R. E. (1978). The case study method in social inquiry. *Educational Researcher*, *7*(2), 5–8. doi:10.3102/0013189X007002005

Stake, R. E. (1995). The art of case study research. *Sage (Atlanta, Ga.)*.

Stake, R. E. (2005). Qualitative case studies. In N. K. Denzin & Y. S. Lincoln (Eds.), *Handbook of qualitative research* (pp. 443–466). Sage.

Stanley, N., Miller, P., Foster, H. R., & Thomson, G. (2010). *Children and families experiencing domestic violence: Police and children's social services' responses*. Issue January.

Stanton, T. K., Giles, D. E. Jr, & Cruz, N. I. (1999). *Service-Learning: A Movement's Pioneers Reflect on Its Origins, Practice, and Future*. Jossey-Bass.

Steenbruggen, J., Nijkamp, P., & van der Vlist, M. (2014). Urban traffic incident management in a digital society: An actor-network approach in information technology use in urban Europe. *Technological Forecasting and Social Change*, *89*, 245–261. doi:10.1016/j.techfore.2013.08.038

Stinson, D. W. (2008). Negotiating sociocultural discourses: The counter-storytelling of academically (and mathematically) successful African American male students. *American Educational Research Journal*, *45*(4), 975–1010. doi:10.3102/0002831208319723

Stokoe, E. (2010). "I'm not gonna hit a lady": Conversation analysis, membership categorization and men's denials of violence towards women. *Discourse & Society*, *21*(1), 59–82. doi:10.1177/0957926509345072

Stovall, D. (2006). Forging community in race and class: Critical race theory and the quest for social justice in education. *Race, Ethnicity and Education*, *9*(3), 243–259. doi:10.1080/13613320600807550

Strauss, A., & Corbin, J. (1990). Basics of qualitative research: Grounded theory procedure and techniques. *Sage (Atlanta, Ga.)*.

Strauss, A., & Corbin, J. (1990). Open coding. Basics of qualitative research. *Grounded Theory Procedures and Techniques*, *2*, 101–121.

Strauss, A., & Corbin, J. M. (1998). *Basics of qualitative research: techniques and procedures for developing grounded theory*. Sage Publications.

Strier, R. (2014). Fields of paradox: University-community partnerships. *Higher Education*, *68*(2), 155–165. doi:10.100710734-013-9698-5

Supovitz, J. A. (2008). Implementation as iterative refraction. In J. A. Supovitz & E. H. Weinbaum (Eds.), *The implementation gap: Understanding reform in high schools* (pp. 151–172). Teachers College Press.

Susman, G. I., & Evered, R. D. (1978). An assessment of the scientific merits of action research. *Administrative Science Quarterly*, 582–603.

Swap, W., Leonard, D., & Mimi Shields, L. A. (2001). Using mentoring and storytelling to transfer knowledge in the workplace. *Journal of Management Information Systems*, *18*(1), 95–114. doi: 10.1080/07421222.2001.11045668

Sweat, K., Yedgarian, V., Rigler, K., Maddox, S., & Throne, R. (2021). Doctoral practitioner researcher agency and the practice-based research agenda. In R. Throne (Ed.), *Practice-based and practice-led research for dissertation development*. IGI Global.

Sweet, E. L., & Ortiz Escalante, S. (2017). Engaging territorio cuerpo-tierra through body and community mapping: A methodology for making communities safer. *Gender, Place and Culture*, *24*(4), 594–606. doi:10.1080/0966369X.2016.1219325

Syed, M., Goza, B. K., Chemers, M. M., & Zurbriggen, E. L. (2012). Individual Differences in Preferences for Matched-Ethnic Mentors Among High Achieving Ethnically Diverse Adolescents in STEM. *Child Development*, *83*(3), 896–910. doi:10.1111/j.1467-8624.2012.01744.x PMID:22506789

Tarver, D. D. (2006). *An explanatory sequential mixed method study of well-being, religious coping, and service utilization patterns of African American caregivers of ADRD elders* [Unpublished doctoral dissertation]. The University of Nebraska-Lincoln.

Tate, W. (2001). Science education as a civil right: Urban schools and opportunity-to-learn considerations. *Journal of Research in Science Teaching*, *38*(9), 1015–1028. doi:10.1002/tea.1045

Tate, W. F. (1997). Race-ethnicity, SES, gender, and language proficiency trends in mathematics achievement: An update. *Journal for Research in Mathematics Education*, *28*(6), 652–679. doi:10.2307/749636

Tatum, B. (2003). *Why Are All the Black Kids Sitting Together in the Cafeteria?* (Rev. ed.). Basic Books.

Teräs, H., & Kartoglu, U. (2017). A grounded theory of professional learning in an authentic online professional development program. *International Review of Research in Open and Distributed Learning*, *18*(7), 191–212. doi:10.19173/irrodl.v18i7.2923

Teuton, C. B. (2012). *Cherokee stories of the Turtle Island Liars' Club*. University of North Carolina Press.

Thomas, G. (2011). A typology for the case study in social science following a review of definition, discourse, and structure. *Qualitative Inquiry*, *17*(6), 511–521. doi:10.1177/1077800411409884

Thomas, J. (1993). *Doing critical ethnography*. Sage Publications. doi:10.4135/9781412983945

Thornberg, R., & Charmaz, K. (2014). Grounded theory and theoretical coding. In *The SAGE Handbook of Qualitative Data Analysis*. SAGE Publications. doi:10.4135/9781446282243.n11

Throne, R. (2020). Dispossession of land cultures: Women and property tenure among Lowcountry heirs in the Gullah Geechee Corridor. In U. T. Onyebadi (Ed.), *Multidisciplinary issues surrounding African Diasporas* (pp. 152–174). IGI Global. doi:10.4018/978-1-5225-5079-2.ch007

Tierney, W. (1993). *Building communities of difference: Higher education in the twenty-first century*. Bergin and Garvey.

Tilley, D. S., & Brackley, M. (2005). Men who batter intimate partners: A grounded theory study of the development of male violence in intimate partner relationships. *Issues in Mental Health Nursing*, *26*(3), 281–297. doi:10.1080/01612840590915676 PMID:16020047

Tilley-Lubbs, G. A. (2016). Critical autoethnography and the vulnerable self as researcher. In *Re-telling our stories* (pp. 1–15). Brill Sense. doi:10.1007/978-94-6300-567-8_1

Torbert, W. R., & ... (2004). *Action inquiry: The secret of timely and transforming leadership*. Berrett-Koehler Publishers.

Torbert, W., Livne-Tarandach, R., McCallum, D., Nicolaides, A., & Herman-Baker, E. (2010). Developmental action inquiry: A distinct integral theory that actually integrates developmental theory practice and research. In S. Esbjörn-Hargens (Ed.), *Integral research in action* (pp. 413–430). SUNY Academic Press.

Torre, M. E. (2009). Participatory action research and critical race theory: Fueling spaces for nos-otras to research. *The Urban Review*, *41*(1), 106–120. doi:10.100711256-008-0097-7

Torre, M. E., & Ayala, J. (2009). Envisioning participatory action research entremundos. *Feminism & Psychology*, *19*(3), 387–393. doi:10.1177/0959353509105630

Tosey, P., & Marshall, J. (2017). The demise of inquiry-based HRD programmes in the UK: Implications for the field. *Human Resource Development International*, 1–10.

Toulmin, S. (1996). Is action research really 'research'? *Concepts and Transformation*, *1*(1), 51–62. doi:10.1075/cat.1.1.05tou

Towson, L., & Pulla, V. (2015). Preparing social workers for empowerment: The place of positionality and its continual intricacies. In V. Pulla & B. B. Mamid (Eds.), *Some aspects of community empowerment and resilience* (pp. 31–49). Allied Publishers.

Trinder, L. (2000). A critical appraisal of evidence-based practice. In L. Trinder & S. Reynolds (Eds.), Evidence-based practice, a critical appraisal (pp. 212-241). Blackwell Science. doi:10.1002/9780470699003.ch10

Tsang, E. W. K. (2014). Case studies and generalization in information systems research: A critical realist perspective. *The Journal of Strategic Information Systems*, *23*(2), 174–186. doi:10.1016/j.jsis.2013.09.002

Tsui, L. (2007). Effective strategies to increase diversity in STEM fields: A review of the research literature. *The Journal of Negro Education*, 555–581.

Tulloch, S., Kusugak, A., Chenier, C., Pilakapsi, Q., Uluqsi, G., & Walton, F. (2017). Transformational bilingual learning: Re-engaging marginalized learners through language, culture, community, and identity. *Canadian Modern Language Review*, *73*(4), 438–462. doi:10.3138/cmlr.4052

Turan, F., & Bayram, O. (2013, February). Information access and digital library use in University students' education: The case of Ankara University. *Procedia: Social and Behavioral Sciences*, *73*, 736–743. doi:10.1016/j.sbspro.2013.02.112

Turney, L., & Pocknee, C. (2004). *Virtual focus groups: New technologies, new opportunities, new learning environments*. www.ascilite.org.au/conferences/perth04/procs/pdf/turney.pdf

Tyack, D. B. (1974). *The one best system: A history of American urban education*. Harvard University Press. doi:10.2307/j.ctv136c61j

Tygret, J. A. (2017). The influence of student teachers on student achievement: A case study of teacher perspectives. *Teaching and Teacher Education*, *66*(1), 117–126. doi:10.1016/j.tate.2017.04.005

Uppal, V., & Chindwani, G. (2013, July). An Empirical Study of Application of Data Mining Techniques in Library System. *International Journal of Computers and Applications*, *74*(11), 42–46. doi:10.5120/12933-0008

Valentine, K. B. (2006). Unlocking the doors for incarcerated women through performance and creative writing. In D. S. Madison & J. Hamera (Eds.), *The Sage handbook of performance studies* (pp. 309–324). Sage. doi:10.4135/9781412976145.n17

Van de Ven, A. H. (1992). Suggestions for studying strategy process: A research note. *Strategic Management Journal, 13*(5), 169–188. doi:10.1002mj.4250131013

Van de Ven, A. H., & Johnson, P. E. (2006). Knowledge for theory and practice. *Academy of Management Review, 31*(4), 802–821.

Van de Ven, A. H., & Poole, M. S. (1995). Explaining development and change inorganizations. *Academy of Management Review, 20*(3), 510–540. doi:10.5465/amr.1995.9508080329

Van de Ven, A. H., & Sun, K. (2011). Breakdowns in implementing models of organization change. *The Academy of Management Perspectives, 25*(3), 58–74.

Vidich, A. J., & Lyman, S. M. (1994). Qualitative methods: their history in Sociology and Anthropology. In Handbook of Qualitative Research (pp. 23-59). Sage Publications.

Vio Grossi, F. (1982). Peasant participation, adult education, and agrarian reform in Chile. In B. Hall, A. Gillette, & R. Tandon (Eds.), Creating knowledge: A monopoly (pp. 153–174). Society for Participatory Research in Asia.

Vio Grossi, F. (1981). Socio-political implications of participatory research. *Convergence (Toronto), 14*, 43–51.

Volkoff, O., & Strong, D. M. (2013). Critical realism and affordances: Theorizing IT-associated organizational change processes. *Management Information Systems Quarterly, 37*(3), 819–834. doi:10.25300/MISQ/2013/37.3.07

Volkoff, O., & Strong, D. M. (2018). Affordance theory and how to use it in IS research. In R. D. Galliers & M. K. Stein (Eds.), *The Routledge Companion to Management Information System*. Routledge.

Walcott, H. F. (2009). *Writing up qualitative research* (3rd ed.). Sage. doi:10.4135/9781452234878

Walker, D. D., Neighbors, C., Mbilinyi, L. F., O'Rourke, A., Zegree, J., Roffman, R. A., & Edleson, J. L. (2010). Evaluating the impact of intimate partner violence on the perpetrator: The perceived consequences of domestic violence questionnaire. *Journal of Interpersonal Violence, 25*(9), 1684–1698. doi:10.1177/0886260509354592 PMID:20139346

Waller, B. (2016). Aggression and violent behavior broken fixes : A systematic analysis of the effectiveness of modern and postmodern interventions utilized to decrease IPV perpetration among Black males remanded to treatment. *Aggression and Violent Behavior, 27*, 42–49. doi:10.1016/j.avb.2016.02.003

Walters, D. (2013). Grounded practice: Putting the 'self' back into self-evaluation. *Educational Action Research, 22*(1), 72–92. doi:10.1080/09650792.2013.851612

Walton, G., & Niblett, B. (2013). Investigating the problem of bullying through photo elicitation. *Journal of Youth Studies, 16*(5), 646–662. doi:10.1080/13676261.2012.733810

Wang, C. (1999). Photovoice: A participatory action research strategy applied to women's health. *Journal of Women's Health, 8*(2), 185–192. doi:10.1089/jwh.1999.8.185

Wang, C., & Burris, M. A. (1994). Empowerment through photo Novella: Portraits of participation. *Health Education & Behavior, 21*(2), 171–186. doi:10.1177/109019819402100204 PMID:8021146

Wang, C., & Burris, M. A. (1997). Photovoice: Concept, methodology, and use for participatory needs assessment. *Health Education & Behavior, 24*(3), 369–387. doi:10.1177/109019819702400309 PMID:9158980

Warfield, K. (2019). Becoming method (ologist): A feminist posthuman autoethnography of the becoming of a posthuman methodology. *Reconceptualizing Educational Research Methodology, 10*(2-3), 147–172. doi:10.7577/rerm.3674

Warren, M. E. (2017). A problem-based approach to democratic theory. *The American Political Science Review, 111*(1), 39–53. doi:10.1017/S0003055416000605

Watkins, K. E. (1991). *Validity in action research. Annual Meeting of the American Educational Research Association.*

Watkins, K. E., & Brooks, A. (1994). A framework for using action technologies. *New Directions for Adult and Continuing Education*, (63): 99–111.

Watkins, K., & Wilson, J. (2001). Chris Argyris: The reluctant adult educator. In P. Jarvis (Ed.), *Twentieth century thinkers in adult and continuing education* (pp. 139–205). Routledge.

Webel, C., Sheffel, C., & Conner, K. A. (2018). Flipping instruction in a fifth grade class: A caseof an elementary mathematics specialist. *Teaching and Teacher Education, 71*, 271–282. doi:10.1016/j.tate.2018.01.007

Weiss, J. (2012). Data for improvement, data for accountability. *Teachers College Record, 114*(11), 110307.

Wells, J. M. (2014). Investigating adult literacy programs through community engagement research: A case study. *Community Literacy Journal, 8*(2), 49–67. doi:10.1353/clj.2014.0002

Wenger, E. (2006). *Communities of practice a brief introduction.* http://www.ewenger.com/theory/

Wenger, E. (1998). *Communities of practice: learning, meaning, and identity.* Cambridge Press. doi:10.1017/CBO9780511803932

Wesselink, A., Paavola, J., Fritsch, O., & Renn, O. (2011). Rationales for public participation in environmental policy and governance: Practitioners' perspectives. *Environment & Planning, 43*(11), 2688–2704. doi:10.1068/a44161

West, C. L. (2011). *Teaching Middle School Jazz: An Exploratory Sequential Mixed Methods Study* [Unpublished doctoral dissertation]. The University of Michigan.

Wheeler, L. B., Whitworth, B. A., & Gonczi, A. L. (2014). Engineering Design Challenge. *Science Teacher (Normal, Ill.)*, *81*(9), 30–36. doi:10.2505/4/tst14_081_09_30

Whetten, D. A. (1989). What constitutes a theoretical contribution? *Academy of Management Review*, *14*(4), 490–495. doi:10.5465/amr.1989.4308371

Whiting, R., Symon, G., Roby, H., & Chamakiotis, P. (2016). Who's Behind the Lens? A Reflexive Analysis of Roles in Participatory Video Research. *Organizational Research Methods*, *21*(2), 316–340. doi:10.1177/1094428116669818

Whorf, B. (1956). Language, mind, and reality. In B. Carrol (Ed.), *Language, thought, and reality: Selected writings of Benjamin Lee Whorf* (pp. 246–270). M.I.T. Press.

Whyte, W. (1994). *Participant observer: An autobiography*. ILR Press.

Williams, O. J. (1994). Group work with African American men who batter: Toward more ethnically sensitive practice. *Journal of Comparative Family Studies*, *25*(1), 91–104. doi:10.3138/jcfs.25.1.91

Wilson, S. (2008). *Research is ceremony: Indigenous research methods*. Fernwood.

Wong Fillmore, L. (1991). When learning a second language means losing the first. *Early Childhood Research Quarterly*, *6*(3), 323–346. doi:10.1016/S0885-2006(05)80059-6

Woodby, L. L., Williams, B. R., Wittich, A. R., & Burgio, K. L. (2011). Expanding the notion of researcher distress: The cumulative effects of coding. *Qualitative Health Research*, *21*(6), 830–838. doi:10.1177/1049732311402095 PMID:21393618

Wood, J. T. (2004). Monsters and victims: Male felons' accounts of intimate partner violence. *Journal of Social and Personal Relationships*, *21*(5), 555–576. doi:10.1177/0265407504045887

Wood, L., & McAteer, M. (2017). Levelling the playing fields in PAR: The intricacies of power, privilege, and participation in a university–community–school partnership. *Adult Education Quarterly*, *67*(4), 251–265. doi:10.1177/0741713617706541

World Bank. (1994a). *The World Bank and participation*. Operations Policy Department.

World Federation of the Deaf. (2016, October 24). *Deaf culture - WFD | world federation of the deaf*. https://wfdeaf.org/our-work/focus-areas/deaf-culture

Wuest, J. (2000). Negotiating with helping systems: An example of grounded theory evolving through emergent fit. *Qualitative Health Research*, *10*(1), 51–70. doi:10.1177/104973200129118246 PMID:10724752

Wuest, J., & Merritt-Gray, M. (1999). Not going back: Sustaining the separation in the process of leaving abusive relationship. *Violence Against Women*, *5*(2), 110–133. doi:10.1177/1077801299005002002

Wynn, D. E. Jr, & Williams, C. K. (2020). Recent advances and opportunities for improving critical realism-based case study research in IS. *Journal of the Association for Information Systems*, *21*(1), 50–89. doi:10.17705/1jais.00592

Wynn, D. Jr, & Williams, C. K. (2012). Principles for conducting critical realist case study research in information systems. *Management Information Systems Quarterly*, *36*(3), 787–810. doi:10.2307/41703481

Xerri, D. (2017). Poetry education research as an anchorage of thought: Using poetry as interview stimulus material. *International Journal of Research & Method in Education*, *40*(4), 379–388. doi:10.1080/1743727X.2015.1119110

Yazan, B. (2015). Three approaches to case study methods in education: Yin, Merriam, and Stake. *Qualitative Report*, *20*(2), 134–152.

Yin, R. K. (2012b). Case study methods. In H. M. Cooper & P. M. Camic (Eds.), APA handbook of research methods in psychology, Vol 2: Research designs: Quantitative, qualitative, neuropsychological, and biological (pp. 141–155). Washington, DC: American Psychological Association. doi:10.1037/13620-009

Yin, R. (2013). Validity and generalization in future case study evaluations. *Evaluation*, *19*(3), 321–332. doi:10.1177/1356389013497081

Yin, R. K. (1981). The case study crisis: Some answers. *Administrative Science Quarterly*, *26*(1), 58–65. doi:10.2307/2392599

Yin, R. K. (2009a). *Case study research: Design and methods* (4th ed.). SAGE Publications.

Yin, R. K. (2009b). How to do better case studies. In L. Bickman & D. J. Rog (Eds.), *The Sage handbook of applied social research methods* (pp. 254–282). SAGE Publications. doi:10.4135/9781483348858.n8

Yin, R. K. (2012a). *Applications of case study research* (3rd ed.). Sage.

Yin, R. K. (2017). Case study research and applications: Design and methods. *Sage (Atlanta, Ga.)*.

Yin, R. K. (2018). *Case study research and applications: Design and methods* (6th ed.). Sage.

Yoshihama, M. (2000). Reinterpreting strength and safety in a socio-cultural context: Dynamics of domestic violence and experiences of women of Japanese descent. *Children and Youth Services Review*, *22*(3), 207–229. doi:10.1016/S0190-7409(00)00076-1

Yosso, T. J. (2002). Toward a critical race curriculum. *Equity & Excellence in Education*, *35*(2), 93–107. doi:10.1080/713845283

Yosso, T. J. (2005). Whose culture has capital? A critical race theory discussion of community cultural wealth. *Race, Ethnicity and Education*, *8*(1), 69–91. doi:10.1080/1361332052000341006

Yosso, T. J., Parker, L., Solorzano, D. G., & Lynn, M. (2004). From Jim Crow to affirmative action and back again: A critical race discussion of racialized rationales and access to higher education. *Review of Research in Education*, *28*(1), 1–25. doi:10.3102/0091732X028001001

Zachariadis, M., Scott, S., & Barrett, M. (2013). Methodological implications of critical realism for mixed-methods research. *Management Information Systems Quarterly*, *37*(3), 855–879. doi:10.25300/MISQ/2013/37.3.09

Zaniewski, A. M., & Reinholz, D. (2016). Increasing STEM success: A near-peer mentoring program in the physical sciences. *International Journal of STEM Education*, *3*(1), 1–12. doi:10.118640594-016-0043-2

Zartler, U., & Richter, R. (2012). My family through the lens. Photo interviews with children and sensitive aspects of family life. *Children & Society*, *28*(1), 42–54. doi:10.1111/j.1099-0860.2012.00447.x

Zeichner, K. (2009). *Teacher education and the struggle for social justice*. Routledge. doi:10.4324/9780203878767

Zeichner, K. (2016). Advancing social justice and democracy in teacher education: Teacher preparation 1.0, 2.0, and 3.0. *Kappa Delta Pi Record*, *52*(4), 150–155. doi:10.1080/00228958.2016.1223986

Zeichner, K. (2020). Preparing teachers as democratic professionals. *Action in Teacher Education*, *42*(1), 38–48. doi:10.1080/01626620.2019.1700847

Zeichner, K., Bowman, M., Guillen, L., & Napolitan, K. (2016). Engaging and working in solidarity with local communities in preparing the teachers of their children. *Journal of Teacher Education*, *67*(4), 277–290. doi:10.1177/0022487116660623

Zilonka, R., Cai, S., Medina, N. E. C., & Chung, G. Y. (2019). "Where are we from?": A critical community autoethnography of place, space, and belonging by Ph. D. international female student-scholar-activists in the US. *The International Journal of Critical Pedagogy*, *10*(1).

Zygmunt, E., & Cipollone, K. (2018). A pedagogy of promise: Critical service-learning as praxis in community-engaged, culturally responsive teacher preparation. In T. D. Meidl & M. M. S. Dowell (Eds.), *Handbook of research on service-learning initiatives in teacher education programs*. IGI Global. doi:10.4018/978-1-5225-4041-0.ch018

Zygmunt, E., & Clark, P. (2016). *Transforming teacher education for social justice*. Teachers College Press.

About the Contributors

Viktor Wang is a full-time faculty member in the Education Division, The University of Montana Western where he teaches foundation courses and oversees accreditation. Dr. Wang has brought leadership to the study of education. He has solidified scholars' understanding of how to conduct research into the complexities of the learning process. Alongside the production of 240+ refereed publications, Dr. Wang has provided many opportunities for his peers and students to develop their scholarly capabilities and stimulated the research agendas of numerous colleagues. His reputation as an empirical and interpretive researcher has resulted in his receiving the 2016 national Presidential Award for Exceptional and Innovative Leadership in addition to multiple institutional awards both at home and abroad.

* * *

Emmanuel Adugu, Ph.D., is a research methodologist and lecturer at the Department of Government, Sociology, Social Work and Psychology at the University of West Indies, Cave Hill Campus. He is an Ohio State University-trained Development Sociologist and a certified Survey Research Methodologist. His research and publication profile focus on: climate change in the context of sustainable development; automation/future of work and social trust; civic engagement; digital activism; and political consumption. He teaches courses such as: social statistics; survey design & analysis; qualitative research methods; and industrial & development sociology.

Jacquelynne Anne Boivin, Ph.D., is an Assistant Professor of Elementary and Early Childhood Education at Bridgewater State University in Bridgewater, MA. She is a former elementary school teacher who uses her past experience in the field to contextualize her instruction in teacher-preparation. Dr. Boivin's research specialization is in multicultural education and anti-racist pedagogy. She is the author of the book, "Exploring the Role of the School Principal in Predominantly White Middle Schools: School Leadership to Promote Multicultural Understanding" which highlights an explanatory sequential mixed methods study she conducted in order to earn her Ph.D. in Educational Leadership from Lesley University in Cambridge, MA. Her book delineates the study through a Critical Race Theory framework with contemporary social justice considerations to emphasize necessary, future changes to truly dismantle racial segregation of schools in the US.

Michael L. Boucher, Jr. is Assistant Professor of Curriculum and Instruction at Texas A & M University, San Antonio. He received his Ph.D. from Indiana University and has published on Photo elicitation and White teachers who build relationships of solidarity with their African American students. He teaches instructional methods, research methods, developmental theory, and diversity classes to graduate students in central Texas. Before becoming a teacher educator, he taught 9th and 10th graders to change the world at South High School in Minneapolis, Minnesota.

Samuel R. Brower is a Clinical Associate Professor at the University of Houston in the Department of Curriculum and Instruction. He coordinates the undergraduate social studies education program and serves as the program director for the M.Ed. in Curriculum and Instruction. Dr. Brower's work is situated at the intersection of critical pedagogy, social studies education, and teaching for justice.

Meagan Call-Cummings (PhD, Indiana University Bloomington) is an Assistant Professor of Research Methods at George Mason University's College of Education and Human Development. Dr. Call-Cummings' research most often takes critical, feminist, and participatory forms. Her most recent project, Courageous Conversations, is a long-term youth participatory action research project that engages arts-based inquiry methods to encourage youth to have their voices heard on political and social issues that are important to them. Dr. Call-Cummings most often writes on ontological, epistemological, methodological, and ethical implications of participatory inquiry.

Shea Culpepper, PhD. a Clinical Full Professor of Math Education and the Director of Teacher Education at the University of University of Houston. She is interested in the clinical preparation of teachers.

Sharmistha Dey, PhD, worked in the IT Industry for several years before pursuing a career in academia. A lecturer in the School of Information and Communication Technology at Griffith University, she has extensive experience with Enterprise Systems and implementation of large business software, having worked and taught in this area for over 15 years. Dr Sharmistha Dey is interested in real-world IT problems and their solution. She has published in several peer reviewed conference papers and journals. Her research interests are: Enterprise Systems, User Expertise, Gamification and Business Processes.

Teresa Edgar is Associate Dean of Undergraduate Studies at University of Houston. Dr. Edgar's research interests include professional development of preservice and in-service teachers in addition student success.

Chrissy Eubank is currently a doctoral student in the Curriculum and Instruction and Teacher Education program at Texas Tech University. She has served in the public education realm for 20 years as an elementary teacher, instructional coach, and most recently, as an administrator in both elementary and secondary public schools. Her research interests include mentoring practices with novice school leaders and building teacher efficacy through mentoring/coaching roles. Chrissy completed her undergraduate work in teacher education at the University of North Texas in 2000 and earned her masters degrees from Angelo State University as a Reading Specialist and in School Administration in 2008 and 2011 respectively.

Bahareh Ghodoosi, Griffith University, Nathan campus, Q Australia, has an academic and career background in business and information systems. Bahareh Ghodoosi's work centres on the data literacy imperative for organisations and is exploring how to more effectively build data literacy talent in business organisations.

Melissa Hauber-Özer is a doctoral candidate in International Education in George Mason University's School of Education. Her research focuses on language and literacy education in migration contexts and employs critical participatory methodology to examine issues of equity and access for linguistically and culturally diverse learners.

Deena Khalil is an assistant professor of Curriculum and Instruction in the School of Education at Howard University. She received her joint PhD in 2012 from Rutgers University, New Jersey Institute of Technology, and the University of Medicine and Dentistry of New Jersey in urban systems, with an emphasis on urban education policy and mathematics education. With a scholar-practitioner's focus on policy and practice, Dr. Khalil's transdisciplinary research centers on issues related to accessing and constructing diverse and equitable learning environments, both nationally and internationally, and how such access is linked to the micro socio-cultural and macro socio-political dynamics of P-20 education. Due to her experience as a fellow in the National Science Foundation MetroMath Center (a partnership between Rutgers University, The Graduate Center of the City University of New York, the University of Pennsylvania), Dr. Khalil enthusiastically pursued and has been awarded several highly competitive intramural and extramural grants totaling over three million dollars. These grants allow Dr. Khalil and her co-investigators to work with pre- and in-service teachers to iteratively disrupt and critically transform the design of STEM ecologies for nondominant, minoritized and trans-indigenous learners. Most recently, she is serving on two National Science Foundation grants as principal investigator of *E-Communities: Investigating How a Collaborative between Engineers and Teachers Influences Underserved Youth's Participation in Engineering Design*, and co-principal investigator of *Building Connections and Learning Communities at HBCUs*, where she focuses on co-constructing blended communities of practice that affectively engage educators while simultaneously empowering their agential literacy. In addition to critically designing research and implementation, Dr. Khalil has presented in over 35 national and international peer-reviewed conferences such as the North American Chapter of the International Group for the Psychology of Mathematics Education, the annual meeting of American Education Research Association, and the International Congress on Mathematics Education. Additionally, Dr. Khalil's work has also appeared in *Urban Learning, Teaching, and Research, International Journal of Qualitative Studies in Education, Peabody Journal of Education, Studies in Educational Leadership, Psychological Trauma: Theory, Research, Practice, and Policy, Leadership and Policy in Schools,* and *International Journal of Adult Vocational Education and Technology.*

Meredith Kier is an assistant professor of Science Education at the College of William and Mary, preparing secondary and elementary preservice teachers to teach science. Her research and teaching is closely connected through a commitment to identifying, implementing, and assessing equitable instructional methods in science to reach all learners. She has worked closely with K-12 students to understand how they are engaged in STEM content and practices, in-service teachers to co-construct strategies that connect all students to STEM, preservice teachers to explore the effects of preparation on their efficacy to engage future students in culturally relevant STEM teaching, and community stakeholders to connect their knowledge to K-12 STEM education. Dr. Kier develops and studies culturally responsive interventions in middle school science classrooms, specifically in classrooms with minoritized learners. Her research focuses on investigating how communities of practice can be formed between teachers, university educators, and industry professionals to devise equitable STEM learning opportunities for disenfranchised schools. Dr. Kier has diverse experiences with teacher professional development, working with teachers in grades K-12, supporting them to incorporate scientific investigations, engineering, and project based learning in their classrooms. She co-created *E-Communities*, an NSF funded Innovative Technology Strategies for Students and Teachers (ITEST) program to bring together teachers and underrepresented engineers in STEM to co-plan and implement culturally-specific engineering design challenges to influence urban, middle school students' awareness of, interest in, and intent to pursue STEM careers. She also investigates how math and science teachers develop their self-efficacy to teach engineering in regards to strategies used in professional development. This work is an extenuation of her dissertation study, where she examined how students in rural, identified with STEM careers and STEM professionals. Dr. Kier also serves as a Co-PI on an NSF Noyce grant, where she trains preservice teachers to teach science in materially underresourced schools. Dr. Kier has published in research and practitioner-based interdisciplinary STEM education journals and book chapters to disseminate work on teaching and learning in STEM. Journals include *International Journal of Adult Vocational Education and Technology*, *Research in Science Education*, *Teaching and Teacher Education*, and publications for the National Science Teachers Association. She has presented her work at over 40 local, national, and international conferences including the American Education Research Association, the National Association for Research in Science Teaching, the Association for Science Teacher Education, the and the National Science Teachers Association.

Victoria J. Marsick is Professor of Adult Learning & Leadership, Department of Organization & Leadership, Teacher's College, Columbia University. She holds a Ph.D. in Adult Education from the University of California, Berkeley, and an M.P.A. in International Public Administration from Syracuse University. Victoria co-directs (with Martha A. Gephart) the J.M. Huber Institute for Learning in Organizations and is a faculty member and coach in the Cahn Fellows Program. Victoria's scholarship—often in collaboration with Karen E. Watkins—examines naturally occurring, informal learning at work and in schools—in individual learners, and in, and on behalf of, groups, communities, and organizations.

Jennifer Morrison, Assistant Professor of English at Texas Tech University, grew up on the Cherokee Nation in Spavinaw, Oklahoma and lived there until she moved to go to college in Springfield. She has taught at various institutions while living here in Springfield, including Missouri State University and Springfield Public Schools. Her husband of 28 years and she have 2 adult sons who are pursuing their own careers in the fields of writing and linguistics. She is passionate about language education, equity in education and society in general, and about indigenous language revitalization. In her free time, she enjoys growing and canning vegetables, spending time in nature, and reading.

Chau H. P. Nguyen is currently a doctoral student in Curriculum Studies and Teacher Education in the College of Education at Texas Tech University. She is interested in the phenomenon of learning to teach, the methodology of phenomenology, and the philosophy of existentialism.

Deepak Saxena is an Assistant Professor at Birla Institute of Technology & Science Pilani, India. Prior to this assignment, he worked with Trinity College Dublin as Research Fellow, and with Dublin Institute of Technology as Assistant Lecturer. He is an active member of the Critical Realism Network, a community of scholars and educators who approach the social sciences & humanities from a realism perspective. He has published with the journals such as Electronic Journal of Business Research Methods, Irish Journal of Management, Journal of Information Science, Australasian Journal of Information Systems, and Electronic Journal of IS Evaluation.

Stacey Sneed is an instructor for Teacher Education in the College of Education at Texas Tech University. Prior to working at Texas Tech, she worked for two years as an elementary teacher and thirteen years as a middle school science teacher. With fifteen years of experience teaching and a passion to help improve curriculum for diverse classrooms, she is now actively supporting new and experienced educators both in the field and at Texas Tech University. Currently, her research focus is on teacher candidates' use of multicultural education and science education.

Amber Thompson, is a Clinical Associate Professor and Associate Chair for Teacher Education at the University of Houston in the Department of Curriculum and Instruction. She coordinates the teacher education program and teaches science methods. Dr. Thompson's work is focuses on school-university partnerships and incorporating practice based teacher education pedagogies into programming and coursework.

Robin Throne's research agenda continues to consider doctoral researcher positionality and agency as well as voice and land dispossession from various social justice research approaches. She is the author of Autoethnography and Heuristic Inquiry for Doctoral-Level Researchers: Emerging Research and Opportunities (IGI Global Research Insight Series, 2019) and editor for Practice-Based and Practice-Led Research for Dissertation Development (IGI Global, 2021) and Indigenous Research of Land, Self, and Spirit (Advances in Religious and Cultural Studies Series, IGI Global, 2021).

Geraldine Torrisi-Steele, PhD, is a senior lecturer in School of Information and Communication Technology at Griffith University and teaches in human computer interaction and information systems. Her interest in human computer interaction, usability, technology in learning contexts, learning theory, and teaching practices distil from her pre-academia experience in teaching and the design, development, and project management of digital interactive media in higher education. Cutting across her interests is a belief in 'design with the human at the centre'.

Zeynep Turhan, PhD, is currently an assistant professor in the Social Work at Bartin University in Turkey. Her studies focus on research and theory of violent and abusive behaviour in family settings, specifically on improving strategies and approaches in domestic violence perpetrator interventions. Her research interests include qualitative inquiry, intersectionality theory, black and minority ethnic men's experiences in domestic violence perpetrator interventions and behavioural change processes among perpetrators of domestic violence.

Karen E. Watkins is Professor of Learning, Leadership and Organizational Development, College of Education, the University of Georgia. She holds a Ph.D. from the University of Texas at Austin and an M.A. from the University of Wisconsin, Madison. Karen's scholarly interests include organizational learning assessment, informal and incidental learning, and action research. Watkins & Marsick developed the *Dimensions of the Learning Organization Questionnaire* (1997), used in over ninety published studies. She was inducted into the International Adult & Continuing Education Hall of Fame in 2003 and the AHRD Scholar Hall of Fame in 2014.

Jahnette Wilson is a Clinical Professor in the College of Education at the University of Houston. Her research interests include clinical practice in Teacher Education, as well as professional develop of pre-service and inservice teachers and leaders in the area of early childhood mathematics.

Aaron Zimmerman is an Assistant Professor of Curriculum and Instruction in the College of Education at Texas Tech University.

Index

A

C

D

E

G

H

Recommended Reference Books

IGI Global's reference books can now be purchased from three unique pricing formats: Print Only, E-Book Only, or Print + E-Book.
Shipping fees may apply.

www.igi-global.com

ISBN: 978-1-5225-7065-3
EISBN: 978-1-5225-7066-0
© 2019; 278 pp.
List Price: US$ 185

ISBN: 978-1-5225-2512-7
EISBN: 978-1-5225-2513-4
© 2017; 365 pp.
List Price: US$ 205

ISBN: 978-1-5225-0830-4
EISBN: 978-1-5225-0831-1
© 2017; 343 pp.
List Price: US$ 185

ISBN: 978-1-5225-9707-0
EISBN: 978-1-5225-9709-4
© 2020; 246 pp.
List Price: US$ 165

ISBN: 978-1-7998-1131-2
EISBN: 978-1-7998-1133-6
© 2020; 319 pp.
List Price: US$ 180

ISBN: 978-1-7998-1025-4
EISBN: 978-1-7998-1027-8
© 2020; 315 pp.
List Price: US$ 185

Do you want to stay current on the latest research trends, product announcements, news, and special offers?
Join IGI Global's mailing list to receive customized recommendations, exclusive discounts, and more.
Sign up at: **www.igi-global.com/newsletters.**

Publisher of Peer-Reviewed, Timely, and Innovative Academic Research

IGI Global
PUBLISHER of TIMELY KNOWLEDGE

www.igi-global.com Sign up at www.igi-global.com/newsletters facebook.com/igiglobal twitter.com/igiglobal